Chorological phenomena
in plant communities

*Proceedings of the 26th International Symposium of the International
Association for Vegetation Science, held at Prague, 5–8 April 1982*

Edited by

R. NEUHÄUSL, H. DIERSCHKE
and J.J. BARKMAN (General Editor)

Reprinted from Vegetatio, volume 59

1985 **DR W. JUNK PUBLISHERS**
a member of the KLUWER ACADEMIC PUBLISHERS GROUP
DORDRECHT / BOSTON / LANCASTER

Distributors

for the United States and Canada: Kluwer Academic Publishers, 190 Old Derby Street, Hingham, MA 02043, USA
for the UK and Ireland: Kluwer Academic Publishers, MTP Press Limited, Falcon House, Queen Square, Lancaster LA1 1RN, UK
for all other countries: Kluwer Academic Publishers Group, Distribution Center, P.O. Box 322, 3300 AH Dordrecht, The Netherlands

Library of Congress Cataloging in Publication Data

```
International Association for Vegetation Science.
   International Symposium (26th : 1982 : Prague,
   Czechoslovakia)
   Chorological phenomena in plant communities.

   (Advances in vegetation science ; v. 5)
   Reprinted from Vegetation, v. 59.
   1. Plant communities--Congresses.  2. Phytogeography
--Congresses.  3. Plants--Migration--Congresses.
I. Neuhäusl, Robert.  II. Dierschke, Hartmut.
III. Barkman, J.J.  IV. Title.  V. Series.
QK911.I47  1982      581.5'247      85-9882
ISBN 90-6193-515-6
```

ISBN 90-6193-515-6 (this volume)
ISBN 90-6193-893-7 (series)

Reprinted from *Vegetatio*, Vol. 59, 1985

PRINTED IN THE NETHERLANDS

CHOROLOGICAL PHENOMENA IN PLANT COMMUNITIES

Advances in vegetation science   5

Edited by
EDDY VAN DER MAAREL

1985 **DR W. JUNK PUBLISHERS**
a member of the KLUWER ACADEMIC PUBLISHERS GROUP
DORDRECHT / BOSTON / LANCASTER

Participants of the 26th International symposium of IAVS on Chorological Phenomena in Plant Communities. Prague, 5–8 April 1982.

Contents

H. Dierschke
Eröffnung des Symposiums   5

R. Neuhäusl
Opening Speech   7

D. Lausi & P. L. Nimis
Quantitative phytogeography of the Yukon Territory (NW Canada) on a chorological-phytosociological basis   9

G. Jahn
Chorological phenomena in spruce and beech communities   21

J. Moravec
Chorological and ecological phenomena in the differentiation and distribution of the *Fagion* associations in Bohemia and Moravia   39

J. C. Rameau
L'intérêt chorologique de quelques groupements forestiérs du Morvan, France   47

J. J. Barkman
Geographical variation in associations of juniper scrub in the central European plain   67

J.-M. Géhu & J. Franck
Données synchorologiques sur la végétation littorale européenne   73

J. M. Royer
Liens entre chorologie et différenciation de quelques associations du Mesobromion erecti d'Europe occidentale et centrale   85

W. Pietsch
Chorologische Phänomene in Wasserpflanzengesellschaften Mitteleuropas   97

E. Balátová-Tuláčková
Chorological phenomena of the *Molinietalia* communities in Czechoslovakia   111

F. Krahulec
The chorologic pattern of European *Nardus*-rich communities   119

L. Mucina & D. Brandes
Communities of *Berteroa incana* in Europe and their geographical differentiation   125

H. Passarge
Syntaxonomische Wertung chorologischer Phänomene   137

F. J. A. Daniëls
Floristic relationship between plant communities of corresponding habitats in southeast Greenland and alpine Scandinavia   145

K. Dierssen & B. Dierssen
Corresponding *Caricion bicolori-atrofuscae* communities in western Greenland, northern Europe and the central European mountains   151

L. M. Fliervoet & M. J. A. Werger
Vegetation structure and microclimate of three Dutch *Calthion palustris* communities under different climatic conditions   159

H. Dierschke
Anthropogenous areal extension of central European woody species on the British Isles and its significance for the judgement of the present potential natural vegetation   171

J. Duty
Die *Fagus*-Sippen Europas und ihre geographisch-soziologische Korrelation zur Verbreitung der Assoziationen des *Fagion* s.l.   177

N. Boscaiu & F. Täuber
Die zönologischen Verhältnisse der dazischen und dazisch-balkanischen Arten aus dem rumänischen Karpatenraum   185

4

H.-D. Krausch
Ozeanische Florenelemente in aquatischen Pflanzengesellschaften der D.D.R.   193
G. Karrer
Contributions to the sociology and chorology of contrasting plant communities in the southern part of the 'Wienerwald' (Austria)   199
J. E. Bjørndalen
Some synchorological aspects of basiphilous pine forests in Fennoscandia   211
A. Miyawaki & Y. Sasaki
Floristic changes in the *Castanopsis cuspidata* var. *sieboldii*-forest communities along the Pacific Ocean coast of the Japanese Islands   225
G. Wiegleb & W. Herr
The occurrence of communities with species of *Ranunculus* subgenus *Batrachium* in central Europe – preliminary remarks   235
S. Hejny
Expansion and retreat of aquatic macrophyte communities in south Bohemian fishponds during 35 years (1941-1976)   243
Discussions   247
List of participants   265
List of lectures   269
Author index   270

Eröffnung des Symposiums

H. Dierschke

Verehrte Gäste, liebe Kollegen und Freunde!

Zum 26. Symposium der Internationalen Vereinigung für Vegetationskunde begrüße ich Sie alle sehr herzlich. Ich darf Ihnen zunächst die Grüße und besten Wünsche unseres Präsidenten, Prof. Dr. J. Lebrun aus Brüssel, übermitteln. Sein hohes Alter erlaubt es ihm nicht, wie er mir vor kurzem schrieb, persönlich in Prag dabei zu sein.

Als wir uns vor einem Jahr am Ende des 25. Symposiums in Rinteln verabschiedeten, waren viele in einer wehmütigen Stimmung. Es war wohl allen klar, daß mit dieser Jubiläums-Tagung, die gleichzeitig dem Gedenken unserer verstorbenen Meister Braun-Blanquet, Tüxen und Whittaker gewidmet war, eine Epoche zu Ende ging. Sie bedeutete nicht nur den Abschied von maßgebenden Pflanzensoziologen der 1. Generation sondern gleichzeitig den Verlust einer vielen lieb gewordenen Tradition, sich jedes Jahr vor Ostern in der kleinen Stadt Rinteln zu treffen. Mancher mag mit dem unguten Gefühl abgefahren sein, hier wirklich ein Ende erlebt zu haben. Andererseits gab das 25. Symposium mit seiner großen Teilnehmerzahl aus 22 Ländern aller Kontinente die Hoffnung, daß es nicht nur wie bisher weitergehen würde, sondern daß unsere Vereinigung mit neuen Impulsen und Aktivitäten sich weiter entwickeln könnte, aufbauend auf dem, was ihre Mitglieder gerade in der Zeit seit Beginn der Symposien, also seit 1953 in Stolzenau erarbeitet haben.

Eine internationale Arbeitsgemeinschaft von Feldbiologen, die sich durch die vielen Symposien, Arbeitstreffen und Exkursionen auch zu einem Kreis internationaler Freundschaft entwickelt hat, sollte lebendig und kräftig genug sein, auch größere schmerzliche Einschnitte zu verkraften. So stellt gerade das heute hier beginnende 26. Symposium einen wichtigen Schritt in die Zukunft dar, indem Bewährtes mit Neuem verknüpft wird. Neu is nicht nur der Tagungsort, neu sind auch viele organisatorische Grundlagen, die hier diskutiert und festgelegt werden sollen. Neu ist auch der Teilnehmerkreis.

Ich habe die Bedenken mancher nie geteilt, daß der Einschritt auch ein Rückschritt sein könnte.

Mein Lehrer Reinhold Tüxen hat mir neben vielen anderen Dingen eine optimistische Grundhaltung vermittelt, die ich hier bestätigt sehe. Gerade der Teilnehmerkreis dieses Symposiums in Prag zeigt, daß der für dieses Jahr und für die Zukunft beschlossene örtliche Wechsel unserer Tagungen Altes mit Neuem in hoffnungsvoller Weise verbindet. Viele vertraute mischen sich mit mir neuen Gesichtern und lassen erwarten, daß sich unsere Vereinigung von Jahr zu Jahr vergrößern wird. Mit der personellen Ausweitung, besonders durch viele junge Vegetationskundler verschiedenster Arbeitsrichtungen, wird sich auch unsere wissenschaftliche Basis erweitern. Ich hoffe sehr, daß unsere Vereinigung die Kraft besitzt, die verschiedensten Richtungen der Vegetationskunde weltweit zu integrieren, was unserer Wissenschaft zu noch größerer Bedeutung in einer Zeit wachsender Umweltkrisen verhelfen sollte.

Unsere vergangenen 25 Symposien haben gezeigt, wie vielfältig die wissenschaftlichen Grundlagen sind, die uns beschäftigen und in vielen Fällen mit Nachbarwissenschaften verbinden. Trotzdem ist es nicht leicht, jedes Jahr wieder ein neues Rahmenthema zu finden, das weit genug ist, im möglichst viele zu interessieren. Natürlich kommen manche wichtige Themen immer wieder zur Sprache.

Dieses Mal ist es gelungen, ein wirklich neues Thema vorzuschlagen, das wir bisher nicht in aller Ausführlichkeit beleuchtet haben. Chorologische Fragen sind sicher so alt wie die Vegetationskunde selbst. Nicht ohne Grund sprach man ja früher von Pflanzen- und später von Vegetationsgeographie. Viele synchorologische Aspekte lassen sich aber erst erörtern, wenn eine fundierte pflanzensoziologische Kenntnis größerer Gebiete vorhanden ist. Über vikariierende Assoziationen, geographische Rassen und Ähnliches kann man erst eine abgewogene Meinung haben, wenn man das Areal und die räumliche floristische Abwandlung von Pflanzengesellschaften genügend kennt. Hier ist in der Vergangenheit viel gesündigt worden, was nicht selten mehr zur Verwirrung als zur Klärung beigetragen hat.

1964 wurde in Stolzenau erstmals über die Vorarbeiten für einen Prodromus der Pflanzengesellschaften Europas beraten. Damals schien die Zeit reif, einen umfassenden Überblick der Vegetation großer Gebiete zu erarbeiten. Inzwischen ist viel weitere Arbeit geleistet worden, wenn auch ein Prodromus immer noch aussteht. Vielleicht kann dieses Symposium hierfür neue Impulse geben, indem es einmal den synchorologischen Kenntnisstand und außerdem mögliche Wege für eine großräumige Vegetationsanalyse und -synthese aufzeigt.

Ich möchte nicht schließen, ohne schon jetzt den Dank an die örtlichen Organisatoren abzustatten. Ich weiß aus eigener Erfahrung, wie viel Arbeit die gründliche Vorbereitung eines Symposiums kostet. Mir scheint, daß hier alles sehr vorzüglich vorbereitet ist. Mit dem Dank an das gesamte Organisationskomitee eröffne ich das 26. Symposium der Internationalen Vereinigung für Vegetationskunde.

Opening Speech by the local National Organizer

R. Neuhäusl

Ladies and gentlemen, dear guests, dear colleagues and friends,

I regard it as a great privilege and pleasure to welcome this distinguished gathering of vegetation scientists from 5 continents on behalf of the National Organizing Committee. I am very happy to have the opportunity of greeting the representatives of the International Association for Vegetation Science, the Secretary Prof. Dr. H. Dierschke, and Members of the Advisory Council; it is a great pleasure to welcome such a great number of colleagues, regular guests of these spring-symposia, coming from the far East, from the western hemisphere, from Africa, from Australia, from western, southern and northern parts of Europe as well as from our neighbouring countries. With great pleasure I welcome also our guests and colleagues who are coming to the symposia of the IAVS occasionally or have come for the first time.

More than 120 participants from 17 countries have gathered at this symposium. The fact that chorological aspects of vegetation are attractive for a great number of vegetation scientists is confirmed by the presence of almost 100 foreign guests. These problems are significant, both from the point of view of a theoretical advancement of science, and that of the possibilities of applying geobotany where it concerns the protection of our environment and saving the natural wealth of our planet.

The 40 registered lectures have been arranged according to their titles, into four basic problem groups:
1. Chorological Influences in the Formation and Differentiation of Natural, Seminatural and Anthropogenic Plant Communities,
2. Syntaxonomic Evaluation of Chorological Phenomena, Vicarious and Corresponding Syntaxa, Transitions between Plant Communities,
3. Diagnostic Value and Distribution of Species and Chorological Groups in Plant Communities of Different Regions,
4. Distribution Areas of Syntaxa and their Changes.

Within the framework of these blocks of problem groups the lectures are sorted beginning with a general or historical theme, then with a special theme concerning natural, seminatural and anthropogenic communities. The synchorological theme is on the programme of the symposium for the first time. I suppose it is a good addition to the items which were presented in preceding symposia: Vegetationskartierung (in 1959), Tatsachen und Probleme der Grenzen in der Vegetation (1968), Landschaftsgliederung mit Hilfe der Vegetation (1974). A prevailing part of the lectures contains items of two or more problem groups. In some cases it was not possible to sort the lectures more exactly on the basis of titles only. The sequence of the lectures represents, therefore, a general scheme of problems solved in this year's symposium.

The task of preparing the 26th symposium of the IAVS in Prague was very difficult for us because it is the first time in 24 years that this traditional spring symposium has not been organized by and held in the native country of its founder, Prof. Tüxen. The personality, humanity and scientific abilities of Prof. Tüxen are impossible to replace. However, it is our collective duty to continue the work and aims initiated by him and develop all the positive features and aspects which have been stimulated by the activity of the IAVS and which have contributed to the development of vegetation science.

We have tried to ensure tolerable conditions for our common work here in Prague and now it depends only on the cooperation of all of us. We regret very much the absence of some friends and regular guests of the spring symposia. Some of them had to change their plans as late as the last days of March or later and their names are therefore given in the program and in the list of participants. All the colleagues who were forced to change their plans for various reasons, expressed sincere wishes of success to our scientific meeting. Allow me to satisfy the wishes of at least several colleagues and interpret the warm greetings from Prof. Westhoff, Prof. Ozenda, Prof. Suzuki (Hiroshima), Prof. Franz Fukarek (Greifswald), Dr. Beeftink, Prof. van der Maarel.

The aims of the symposia organized by Prof. Tüxen in Stolzenau, and later in Rinteln, including the symposium prepared by Prof. Dierschke, were not the readings of the lectures only, but especially the detailed discussions of problems from different aspects given by the scientific interests of the partic-

8

ipants as well as by their working territory. Prof. Tüxen stressed both the importance of a detailed discussion during the symposium and its publication in the reports. Allow me to cite his words from the preface to the Bericht über das Internationale Symposium 'Pflanzensoziologische Systematik' edited in 1968: 'Wir geben die Diskussionen ausführlich wieder, weil wir glauben, dass die oft freimütigen Äusserungen der einzelnen Redner nicht nur ein gutes Bild von der Atmosphäre unseres Symposion vermitteln, sondern auch vor allem jüngeren Pflanzensoziologen einen Eindruck von unseren Problemen sowie den Versuchen geben, ihrer Schwierigkeiten Herr zu werden.' It is our great wish to maintain this tradition in the Prague symposium.

This year's symposium is attended not only by the regular guests, but also by a great number of vegetation scientists, who so far didn't have regular contact with the activities of IAVS. We anticipate that their participation in the lectures and discussions will contribute not only to a deeper knowledge of the chorological phenomena in plant communities, but also to a mutual understanding and coming together. The unity of vegetation science, methodical combining or paralleling of approaches to the research work will not only ensure a faster development of geobotany, but will also strengthen its prestige on national levels as well as on an international scale. A cooperating and stable international society for association of vegetation scientists will guarantee further development of our 'scientia amabilis'.

Allow me, dear colleagues and friends to express my thanks to the Czechoslovak Academy of Sciences for making possible the organization of the 26th International Symposium in Prague. I am very grateful to the director of the Botanical Institute for releasing some of the fellow-workers who participated in all the preparations for this symposium and to all my fellow workers for their help. My thanks belong especially to you, dear foreign guests, who, despite the effort, expense and other difficulties, have come in such a great number to this gathering. It can be considered a real assurance of a good and fruitful cooperation and of successful work for all of us. The tradition of the spring symposia of the IAVS which has brought so many scientific results and settled friendly relations between a great number of vegetation scientists has to be kept for the future.

Quantitative phytogeography of the Yukon Territory (NW Canada) on a chorological-phytosociological basis*

D. Lausi & P. L. Nimis**

Istituto Botanico, Cas. Università, I 34100 Trieste, Italy

Keywords: Beringia, Boreal vegetation, Chorology, North America, Phytosociology, Yukon Territory

Abstract

This study is based on the analysis of the chorological spectra from 19 vegetation types obtained from a numerical classification of ca. 400 phytosociological relevés taken during a vegetation survey in the Yukon Territory (NW Canada).

All vegetation types are well characterized in terms of their chorological features. This allowed an ecological-historical interpretation of the vegetation in the study area. The distribution of the various chorological categories within the vegetation types is strongly correlated with the main environmental influences, whose action led to the present floristical and vegetational characteristics of the area, such as glaciation, fire, permafrost and water availability.

The results show how the phytosociological approach constitutes an effective methodological tool for clarifying the phytogeographical aspects in the historical-ecological interpretation of a large area.

Introduction

The northwestern part of North America included in the State of Alaska (U.S.A.) and the Yukon Territory (NW Canada) are of great biogeographical importance, for two main reasons:

1. The connection of the region, with northeastern Asia. A land bridge connecting Siberia and North America existed during the Pleistocene, permitting intermingling of the Siberian flora with the flora of Alaska-Yukon (Hultén, 1937; Hopkins, 1967; Gjaerevoll, 1980).

2. Most of the area was ice-free during the glacial period. A connection with the regions located south of the North American ice-sheet was ensured by a narrow deglaciated corridor, that separated the Cordilleran glaciers from the continental ice-sheet during xeric interglacials (Douglas, 1970).

Alaska-Yukon is therefore an important refugial area with intensive migration, chiefly through the Bering Land Bridge and through the Cordilleran corridor. This is true both for plants (Hultén, 1937), animals (Youngman, 1975; Dillon, 1956) and man (Müller-Beck, 1967; Laughlin, 1967). Important contributions to the phytogeography and ecology of the Pacific northwest include Daubenmire (1969), Kruckeberg (1969), Schofield (1969) and Wolfe (1969), and of the Yukon area in particular Porsild (1945, 1951, 1966, 1974), Porsild & Cody (1980), Jeffrey (1959), (see also Hultén, 1968, for a bibliography and for important distribution maps). Only a few local studies were performed on the basis of the phytosociological approach (La Roi, 1967; La Roi & Stringer, 1976; Hanson, 1953; Spetzman,

---

* Nomenclature follows Hultén (1968), otherwise author names are specified.
** The field work was completed in the summer of 1978. We are grateful to Dr W. Stanek, Canadian Forestry Service, for coordination of the survey and for soil data, and to Prof. L. Orlóci for organization. Partial financial support was received from the Italian C.N.R.

Vegetatio 59, 9–20 (1985).

10

1959; Douglas, 1974; Hoefs, Kowan & Krajina, 1975).

In the present study, we analyze the chorological differentiation of the vascular flora within different vegetation types, described according to the Braun-Blanquet approach, and elaborated numerically, and we relate these chorological facts both to ecological factors and to historical evidence. Numerical methods may provide a new methodological basis for phytogeographical studies. In this sense, the present paper represents also a methodological contribution towards a 'Quantitative Biogeography' (Crovello, 1981).

General features of the study area

The phytosociological relevés have been taken along the Alaska Highway, in southern Yukon, from Watson Lake to Beaver Creek, and along the Klondyke and Dempster Highways, from Whitehorse to the Richardson Mountains, in central and northern Yukon. The relevés are scattered along two transects, the first running in an east–west direction, the second in a south–north direction, both with an approximate length of ca. 900 km. A detailed phytosociological study will appear in a future monograph (Lausi & Nimis, in prep.).

The climate is continental, with long cold winters and short warm summers. The mean yearly temperature is always below zero. Different parts of the study area are further characterized by somewhat different climatic patterns. The southern part of the region is located at the leeside of the St. Elias Mountain Range, that effects a partitioning of the water-laden clouds coming from the Pacific Ocean; as a result, precipitation is low, particularly in the central portions around Kluane Lake (280 mm/yr at Haines Junction), where strong dry winds of the Foehn type frequently blow from the mountains. The southwestern portion of the Yukon Territory receives moderate marine influences from the Pacific Ocean and here precipitation is somewhat higher (412 mm/yr at Beaver Creek). A strongly continental climate occurs throughout southeastern and central Yukon, where winter temperatures are lower and summer temperatures higher than in the areas under maritime influence. The northern part of the Yukon is influenced by weather patterns from the Arctic Ocean, and climate is cold in winter and cool

in summer. Precipitation is very low (125 mm/yr at Komokuk B.). Permafrost is scattered in southern Yukon, where it is mostly confined to depressions in lowlands, widespread in central Yukon, and continuous approximately above 67° C latitude (Oswald & Senyk, 1977; Burne, 1974).

The geological pattern of the area is extremely complex (Douglas, 1970); the most peculiar feature is that the whole study area lies outside the Canadian Shield, so that limestone is much more frequent here than in other parts of North America at corresponding latitudes. In southern and central Yukon, volcanic ash has been abundantly deposited in the upper soil horizons from 1220 to 1900 yr B.P. (Hughes, Rampton & Rutter, 1972). Locally, above all in the region surrounding Kluane Lake, loess deposition is still active. Given the complex geological pattern, the pedogenesis is correspondingly complex. In general true podzols are scarcely developed, even on acid parent material, because of the low rainfall. The prevailing soil types under the boreal forest vegetation are more or less leached brown soils with an acid or subneutral reaction. Low pH values prevail under coniferous woods, where loess deposition does not occur, whereas subneutral reaction prevails under coniferous woods in zones with active loess deposition, and under *Populus tremuloides* stands. Cryosols occur where permafrost is present: they are most common in the western part of the Alaska Highway and in northern Yukon. Alkaline regosols, supporting a steppe-like grassland occur in areas characterized by strong active loess deposition, chiefly in the Kluane National Park (Orlóci & Stanek, 1979). Cryoturbational phenomena, like polygons, striped ground, and solifluction lobes, are evident at high elevations and in the northern part of the region.

As far as the vegetation is concerned, the principal formations present in the area are:
- *Picea mariana*-muskeg: a taiga-like open woodland, dominated by mosses (*Aulacomnium*, and sometimes *Sphagnum*-species), with scattered old dwarf trees (*Picea mariana, Larix laricina*), always occurring on cryosols with acid or more rarely subneutral reaction and high organic matter content in the upper soil horizons.
- *Picea glauca*-forests: including associations dominated by *Picea glauca,* in closed stands on relatively well-drained brunisols with an acid or subneutral reaction. The early successional stages

after fire leading towards a *Picea glauca* forest are dominated by *Pinus contorta* in southeastern Yukon, and by *Populus tremuloides* in the rest of the region.

- Grasslands: particularly common in the more continental parts of southern and central Yukon, above all on loess. They are dominated by *Artemisia* and *Agropyron* species and occur on subneutral to strongly alkaline regosols.
- Subalpine *Salix*-thickets: a *Salix pulchra*-dominated association prevails in the mountains of northern Yukon, near the treeline.
- Alpine tundra: localized at high elevations in southern Yukon, and widespread in the mountains of northern Yukon.
- Bogs and lakes: are very rare in the area. Tundras and bogs have not been included in the present study.

Methods

Four hundred phytosociological relevés taken along the two transects were submitted to numerical classification in order to obtain vegetation types. Typification occurred by stratifying the whole data set according to formation types, and by submitting each subset to Complete Linkage Clustering with the Correlation Coefficient as similarity measure (Anderberg, 1973; Orlóci, 1978). In order to give a synthetic view of the floristic similarity between all types, a dendrogram was constructed on the basis of the synoptic tables taken from all described vegetation types, using the Similarity Ratio for presence–absence data (Westhoff & van der Maarel, 1973). The types were also ordinated on the basis of the synoptic tables, with Principal Component Analysis (PCA), on a Correlation Coefficient matrix (Orlóci, 1978; Wildi & Orlóci, 1980). Soil data for almost all relevés include texture, organic matter content, pH, and active layer thickness (Orlóci & Stanek, 1979). This permitted a direct interpretation of the indirect gradient analysis performed with PCA on the basis of purely floristic data.

All vascular species present in the relevés were subdivided into categories according to their distribution patterns. The frequencies of each type of distributional range within the vegetation types were calculated. The relations between the chorological affinities of the associations and the ecological factors were studied taking the main trend revealed by the ordination (dotted line in Fig. 2) as an axis on which the frequencies of the various types of ranges within the vegetation types have been plotted.

The relations between vegetation types and range types were studied with Canonical Concentration Analysis (Feoli & Orlóci, 1979), performed on a contingency table based on the percentages of the types of ranges in the vegetation types.

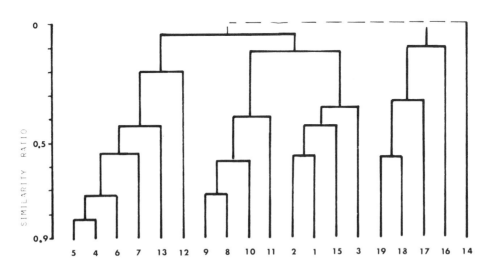

Fig. 1. Dendrogram of vegetation types. Numbers refer to associations defined by numerical analysis of the relevés.

12

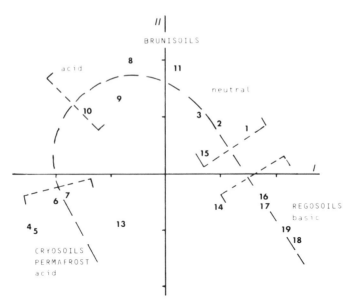

Fig. 2. Principal Component Analysis of vegetation types. Association numbers as in Figure 1.

Results

Nineteen main vegetation types have been defined. Their numerical relationships are presented in the dendrogram of Figure 1. At the similarity level of 0.12, 4 principal clusters are formed, which correspond fairly well to four principal formations:

Cluster 1: includes associations of the muskeg-type (Nos. 5, 4, 6, 7, 13, 12).

Cluster 2: includes associations of the boreal forest belt dominated by *Picea glauca* and their secundary successional stages following fire. This group can be further subdivided into two subgroups. The first includes associations dominated by *Picea glauca* and *Pinus contorta* on acid brunisols (Nos. 9, 8, 10, 11), the second includes associations dominated by *Picea glauca* and *Populus tremuloides* on subneutral brunisols (Nos. 2, 1, 15, 3).

Cluster 3: includes associations of grasslands on subneutral to alkaline regosols (Nos. 19, 18, 17, 16).

Cluster 4: this consists of one association dominated by *Salix pulchra* from the subalpine vegetation belt of the mountains in northern Yukon (No. 14).

The results of PCA performed on the synoptic tables of the associations are presented in Figure 2.

The numbers representing the associations are disposed along a horseshoe-shaped curve. The sequence of the types along the curve, from right to left, is as follows: natural grasslands on regosols, early successional stages of the boreal forest vegetation (*Populus*-stands) on subneutral brunisols, boreal forest associations (*Picea glauca* and *Pinus contorta*-stands) on acid brunisols, muskeg- associations on subacid cryosols. A clear pedological trend is evident from the right to the left of the curve in Figure 2, represented by a progressive evolution of soils. This trend is paralleled by a progressive increase of the organic matter content of the upper soil horizons and by a tendency towards impeded drainage conditions, from the excessively drained regosols to waterlogged cryosols.

A synthetic view of the relationships between phytogeographical categories, vegetation types and ecological factors is provided for each species group. In these figures, the horseshoe-shaped curve of Figure 2 has been informally stretched with unaltered positions of the vegetation types. For each areal type the percentages of the species in the various vegetation types has been calculated. The individual figures will be presented in the monograph. Here we show only one example (Fig. 3) and a summary graph (Fig. 4).

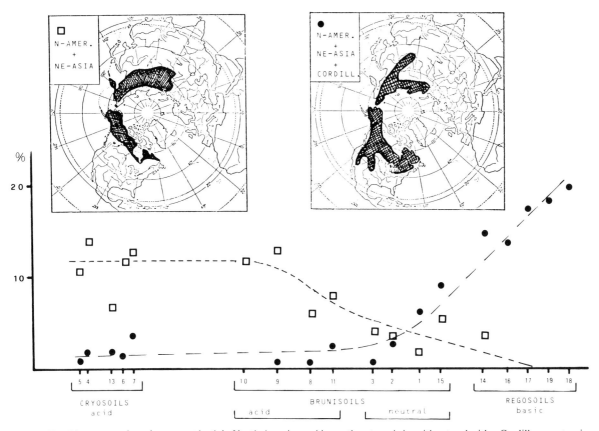

Fig. 3. Total frequency of species present both in North America and in northeastern Asia, without and with a Cordilleran extension. Frequency values indicated per vegetation type, arranged according to the gradient in Figure 2.

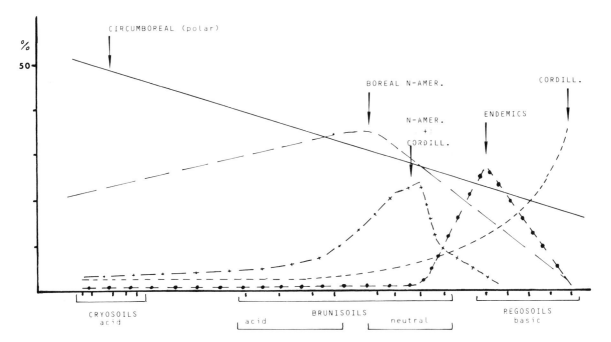

Fig. 4. Ecological behaviour of the main chorological categories (explanation in text).

Circumboreal (circumpolar) species

This category includes species with a distribution extending through North America and Eurasia, within the arctic, subarctic and boreal vegetation belts. This element has a clear maximum in the muskeg vegetation on cryosols, and decreases linearly towards the grassland vegetation on regosols (Fig. 4). Circumboreal (circumpolar) species common both to muskeg and to *Picea glauca* forests are relatively scarce.

The most frequent species of this category that occur both in the muskegs and in the boreal forest are: *Empetrum nigrum, Equisetum arvense, Salix glauca* s.l., *Vaccinium vitis-idaea* s.l. The most common species occurring chiefly in the muskeg are: *Andromeda polifolia, Carex aquatilis, Carex capillaris, Carex rostrata, Carex vaginata, Equisetum scirpoides, Eriophorum vaginatum, Oxycoccus microcarpus, Potentilla fruticosa, Rubus arcticus* s.l., *Rubus chamaemorus, Salix reticulata, Selaginella selaginoides, Tofieldia pusilla, Vaccinium uliginosum.* The most frequent species chiefly occurring in the boreal forest are: *Calypso bulbosa, Corallorhiza trifida, Epilobium angustifolium, Goodyera repens, Linnaea borealis* s.l.,*Lycopodium annotinum, Lycopodium complanatum, Moneses uniflora, Pyrola grandiflora, Pyrola secunda, Pulsatilla patens, Rosa acicularis, Rubus idaeus* s.l.

Most of the circumboreal(-polar) species occurring mainly in the muskeg are common components of the artic and alpine tundra throughout the study area. The muskeg vegetation itself can be considered as a tundra-like formation with scattered dwarf trees. According to Johnson & Packer (1967), tundra climates persisted in the amphiberingian area during the whole Quaternary, and the Bering Land Bridge was covered by a flat tundra. Evidently the Bering Strait has not been an effective barrier against tundra plant dispersal, which may explain the high frequency of circumboreal (-polar) species here.

The distribution patterns of the species chiefly occurring in the boreal forest-associations differ from the ones of most of the muskeg species, in that they extend further south into the boreal zone. In this sense, they are mostly true circum*boreal* species, whereas the circum*polar* element prevails in the muskeg-vegetation. None of these circumboreal species has its optimum in tundra-like vegetation on permafrost. Most of them have strongly disjoint stations south of the boreal vegetation belt, e.g. in central China, Japan and the Himalayan mountains. Such a distribution pattern suggests that these species had an ancient broad distribution before the ice age, which later became fragmented in the southern part of their ranges. This, together with an intolerance to permafrost, makes the hypothesis of their possible migration through the Bering Land Bridge as tundra-plants less probable. The history of this species group is probably different from the one of the muskeg species: they seem to have survived glaciations in refugia located mostly south of the North American ice-sheet (or in Beringia itself), from where they later extended in connection with the expansion of the boreal forest during the postglacial period.

Boreal North American species

In this group distributions are restricted to the boreal vegetation belt throughout North America, sometimes extending up to the subarctic and arctic vegetation belts. These species have a clear maximum in the boreal forest on brunisols, and in its secondary degradation stages following fire (Fig. 4).

The number of boreal North American species occurring in the muskeg is quite small. The most frequent species are *Betula glandulosa* and *Salix myrtillifolia;* they extend from coast to coast with a small southern extension along the Cordillera. They probably survived glaciation south of the North American ice-sheet, and penetrated into Alaska-Yukon after the opening of the Bering Strait. This could explain their complete absence from northeastern Siberia.

The most important difference between *Picea mariana* muskeg and *Picea glauca* forest associations is the higher frequency of boreal North American species in the latter type. Some of them have a coast to coast North American distribution. The most frequent species are: *Carex concinna, Geocaulon lividum, Ledum groenlandicum* (reaching Greenland), *Picea glauca, Platanthera hyperborea, Salix planifolia* and *Viburnum edule.* Other species, although present both in eastern and western North America, have a discontinuous range, mainly with gaps located south of Hudson Bay: *Achillea nigrescens, Amerorchis rotundifolia, Cypripedium passerinum, Gentiana propinqua, Juniperus horizontalis, Listera borealis.* The common absence of these species from the Hudson Bay region can be explained by the much colder climate, compared to the rest of North America at corresponding latitudes. In the area between Hudson Bay and the Great Lakes, the boreal forest occurs at much lower latitudes than in the rest of North America, forming the narrowest belt in the whole of the North American continent (between 48° and 50° latitude; Rowe, 1977). This region is therefore to be considered as an important phytogeographical transition zone, as it is also evident from the fact that most species of the *Gaultherio-Piceetalia* (see below) have their northwestern distributional limit just between the Great Lakes and Hudson Bay.

A last point concerns the principal tree species present in the study area, which can all be included in the present category. They are: *Picea mariana, Larix laricina, Picea glauca* and *Pinus contorta.* The former two species are confined to cryosols, the latter occur mainly on brunisols. All are restricted to the North American continent. In eastern Siberia *Larix laricina* is replaced by *Larix dahurica,* whereas such species as *Picea obovata, Larix sibirica* and *Pinus sibirica* occur further west, in the less continental regions of western Siberia. Not a single coniferous tree species is shared by the two continents. These facts suggest that the boreal forests of North America and eastern Asia did not form a continuous belt during the Pleistocene, despite the presence of the Bering Land Bridge. This assumption is supported by palynological and paleobotanical evidence. The separation of the forests of the two continents seems to date back to late Miocene or early Pliocene (Johnson & Packer, 1967; Wolfe & Leopold, 1967), and seems to have lasted throughout the Pleistocene (Colinvaux, 1967; Giterman & Golubeva, 1967). The high percentage of boreal North American species in the boreal forest associations on brunisols is a further indication of the different history of this formation type compared to the muskeg vegetation dominated by *Picea mariana.*

North American + Cordilleran species

These species have ranges very much resembling the ones of boreal North American species, the difference being a pronounced lobe projecting southwards along the Cordilleran Mountains (Fig. 4).

The most frequent species are: *Anemone multifida, Antennaria pulcherrima, Arabis hoelbelii, Draba aurea, Dryas drummondii, Geum perincisum, Mertensia paniculata* s.l., *Oxytropis deflexa, Petasites sagittatus, Potentilla pennsylvanica, Platanthera obtusata, Populus tremuloides, Shepherdia canadensis.* Their frequencies in the various vegetation types are generally low, with the exception of the degradation stages of the boreal forest on neutral brunisols dominated by deciduous trees and shrubs (*Populus tremuloides, Populus balsamifera, Salix* spp.). Compared to the previous group, the present one has a frequency maximum more towards the right along the pedological gradient, i.e. within more xerophytic vegetation types. The group includes *Populus tremuloides*, which is the most important early successional tree on neutral brunisols following fire. It seems that *Populus tremuloides* woods moved as a whole from their probable refugial stations along the Cordillera, a fact that is now reflected in the high number of species with similar distribution occurring in these communities.

Cordilleran species

This category includes species whose distribution is restricted to the Cordilleran ridges of western North America. They have a clear frequency maximum in the grasslands on regosols under xeric conditions as far as both soil and climate are concerned (Fig. 4).

The Cordilleran species can be subdivided into three groups, according to their present distribution in the study area.

The first group includes species whose northern distributional limits correspond with the limits of the ancient periglacial steppes of southeastern Alaska and southwestern Yukon. The most frequent are: *Arnica cordifolia, Astragalus tenellus, Carex filifolia, Draba oligosperma, Penstemon procerus, Sedum lanceolatum.* It is difficult to say whether the northern distributional limit of these species is the consequence of present climatical and pedological conditions, or an indication that they survived glaciations in the periglacial steppes that extended in the area from the glacial period onwards, from which they later expanded southwards along the deglaciated corridor. Hultén (1937) seems to be inclined towards the latter hypothesis.

The second subgroup includes Cordilleran species whose northern distributional limits are located well above southwestern Yukon and sometimes reach the Polar Circle. They are: *Antennaria rosea, Crepis elegans, Delphinium glaucum, Erigeron compositus, Erigeron grandiflorus, Linum perenne* ssp. *lewisii, Oxytropis campestris* var. *sericea, Oxytropis viscida, Polemonium pulcherrimum, Plantago canescens, Senecio cymbalarioides, Solidago decumbens* var. *oreophila, Solidago multiradiata* var. *scopulorum, Zygadenus elegans.* Most of these species are less strictly linked to typically steppic vegetation than the ones included in the previous subgroup. They probably survived glaciation at the southern border of the North American ice-sheet, and penetrated into the area during xeric intergla-

cials along the deglaciated corridor. A peculiar case is the distribution of *Solidago multiradiata* s.l.: the species is known in two varieties: the var. *scopulorum* is Cordilleran, with its northern distributional limit in central Alaska, the var. *multiradiata* extends throughout the North American continent along the arctic coasts. This fact probably reflects the separation of a previously homogeneous population into two different demes, respectively located to the south and to the north of the North American ice-sheet.

The last subgroup includes Cordilleran species with a large gap between their main southern distributional center, and a small enclave in southwestern Yukon (and sometimes southeastern Alaska). The most frequent are: *Artemisia cana, Erigeron caespitosus, Eurotia lanata, Orobanche fasciculata, Townsendia hookeri.* They were probably components of an old steppic vegetation in the study area, that has been completely separated from the southern populations by glaciations. *Erigeron gormani* could be also included in this group, since a closely related species, *Erigeron eriantherus*, occurs further south along the Cordillera (Hultén, 1968). In this case, geographic separation contributed to the morphological differentiation.

Summarizing, the Cordilleran element is mainly present in grassland vegetation on regosols, under continental, xeric conditions. The history of these species seems to be linked with the existence of the ice-free corridor. The past ecological conditions along the corridor were such as to allow a relatively easy migration of grassland-species from the periglacial steppes of Beringia southwards and from the central North American prairies northwards.

Species present both in North America and northeastern Asia

These species have distribution ranges extending from North America (mainly in the boreal zone) to northeastern Asia (sometimes with isolated outposts in Japan, China etc.), and are absent from northwestern Asia and Europe. They differ from the true amphiberingian species (see below) for having broader ranges, not restricted to the remnants of ancient Beringia.

In Figure 3 the species group is presented in its extension over the soil types as suggested in Figure 2. There are two subgroups, one with ranges extending along the Cordilleran Mountains, the other without this extension. To the first subgroup belong species such as *Calamagrostis purpurascens. Trisetum spicatum, Festuca brachyphylla, Chamaerhodos erecta, Epilobium latifolium, Artemisia frigida, Crepis nana, Festuca altaica* s.l. The second group includes such species as *Carex membranacea, Luzula rufescens, Tofieldia coccinea, Cornus canadensis, Pyrola asarifolia, Arctostaphylos rubra, Pedicularis labradorica, Carex scirpoidea.* The difference in distribution patterns is reflected in a very different behaviour along the ecological gradient. The species not extending along the Cordillera are most frequent at the left side of the gradient (i.e. on acid cryosols and brunisols with impeded to moderate drainage), whereas the species of the former group are most frequent at the right side of the gradient (i.e. on neutral brunisols and regosols with moderate to excessive drainage). The curves representing the behaviour of the two subgroups along the gradient cross each other at the point where the soil reaction turns from acid to alkaline. The gradient of

Figure 2 can also be interpreted as a gradient in soil moisture, and this probably explains the different behaviour. Only those species that were able to survive in a relatively xeric environment had the possibility to expand southwards along the deglaciated corridor during the time of ice retreat.

Endemic species

The endemic element is concentrated in the grassland vegetation on regosols, and behaves very much like the Cordilleran element (Fig. 4). This fact can be better explained if we consider all endemic taxa of the Yukon, not only those that are represented in our data set.

The endemic element is mainly concentrated in two areas. The first group (Fig. 5) includes taxa whose range is restricted to southwestern Yukon, sometimes extending to southeastern Alaska. They are: *Agropyron yukonense, Artemisia rupestris* ssp. *woodii, Aster yukonensis, Astragalus viciifolius, Castilleja yukonensis, Castilleja villosissima, Douglasia gormani, Erigeron purpuratus, Oxytropis huddelsonii, Oxytropis scammanniana, Penstemon gormani, Salix setchelliana*. These species are concentrated in the region of periglacial steppes at the leeside of the St. Elias Mountains. Most of them are grassland plants. The presence of such a high number of endemics in the grasslands of southern Yukon is an indication of the relatively ancient origin of natural grasslands in the area. There is extensive paleobotanical and palynological evidence (see Hopkins, 1967), that the southeastern part of Beringia (Fig. 7) supported a parasteppic periglacial vegetation in which *Artemisia*-species were particularly abundant. The existence of such a vegetation was mainly due to the strong continentality of the local climate and there is reason to assume that the local conditions did not substantially change from the late Pliocene until now. After the retreat of the

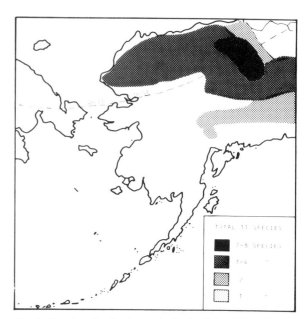

Fig. 6. Density of endemic species in the Yukon flora, whose ranges are centered in N Yukon and N Alaska.

glaciers some of the grassland species were able to migrate through the Cordilleran Corridor, others ('rigid species' sensu Hultén, 1937), did not substantially change their distributions. The latter are the endemics of the first group.

The second group of endemics (Fig. 6) includes mainly arctic-alpine taxa presently occurring in the alpine and subalpine vegetation belts of the mountains in northern Yukon. They are: *Campanula aurita, Draba ogilviensis, Erysimum angustatum,*

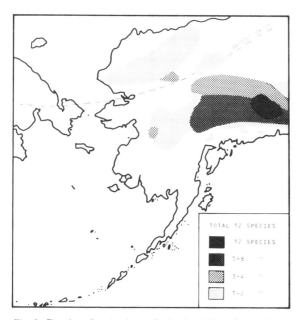

Fig. 5. Density of endemic species in the Yukon flora, whose ranges are centered in SW Yukon. (explanation in text).

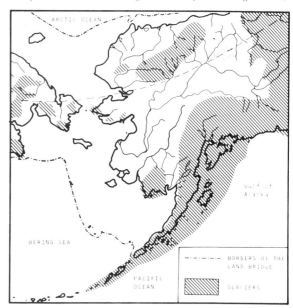

Fig. 7. Beringia during the height of the Wisconsin (Würm) glaciation. (drawn from D. M. Hopkins, 1967).

Festuca ovina ssp. *alaskensis, Lesquerella calderi, Papaver keelei* Porsild, *Papaver macconnelli, Poa porsildii* Gjaerevoll, *Senecio hyperborealis, Senecio yukonensis, Smelowskia borealis.* These taxa are concentrated in the unglaciated mountains of northern Yukon and northern Alaska (Fig. 6). They are the remnants of an old arctic-alpine flora that must have been more widespread before glaciation, and survived as a consequence of this area being ice-free during the whole glacial period (Fig. 7). Today they still occur as frequent components of the arctic-alpine tundra in the mountains of northern Yukon.

It is possible to sketch a picture of the ancient landscape of the Yukon, dominated by two principal formation types: xeric grasslands in the south and a rich alpine tundra in the mountains of the north, with scattered woodlands in-between. Grasslands and alpine tundra were the main centers for the differentiation and conservation of the endemic element in the Yukon flora.

Boreal northwestern American species

This group is restricted to boreal northwestern America. It differs from the endemic group by having broader species ranges, extending southwards into British Columbia and eastwards into the Northwest Territories. The frequency of these species is low in all vegetation types. The most important are: *Aquilegia brevistyla,*

Betula papyrifera, Carex microchaeta, Lupinus arcticus, Pinus contorta, Ribes hudsoniamum, Ribes oxyachanthoides and *Saussurea angustifolia.*

Pinus contorta is one of the dominant species in the secondary woods following fire in southeastern Yukon. There is evidence that before the glaciations this was the only *Pinus*-species growing in the west (Mirov, 1937). The species seems to have persisted in the study area in ice-free refugia during at least the last Wisconsin glaciation (Hultén, 1937). Its subsequent spreading northwards was chiefly favoured by fire and volcanic ash deposition (Hansen, 1943, 1947; Mirov, 1967). Jeffers & Black (1963) contributed to the elucidation of the infraspecific variation of *Pinus contorta* on the basis of multivariate methods. Their results confirm the subdivision of the species into two main varieties, respectively found in the inland (var. *latifolia*) and in the coastal provinces (var. *contorta*). The absence of the species from Siberia can be explained on the same grounds discussed for *Picea glauca,* i.e. the absence of forest vegetation on the Bering Land Bridge during the glacial period. The same could apply for most of the species of this group, since all of them, but two, seem to be permafrost intolerant, occurring mainly on brunisols. The two species that have also been found in muskegs on cryosols are *Carex microchaeta* and *Saussurea angustifolia.* Their ranges, if taken in a wider sense, are actually amphiberingian: a species

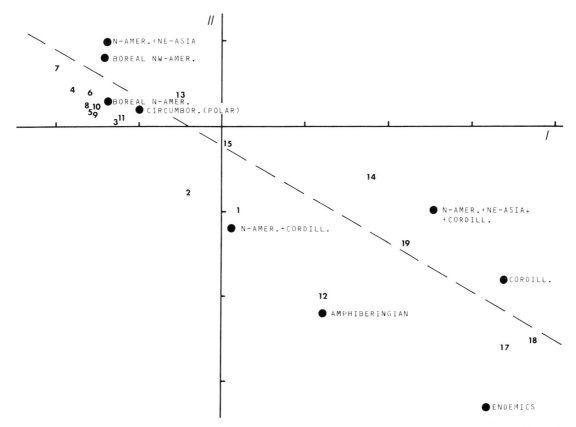

Fig. 8. Joint distribution of vegetation types (Nos. as in Fig. 1) and chorological categories according to the first two canonical variates. The method is AOC.

closely related to *Saussurea angustifolia*, i.e. *Saussurea pseudangustifolia* occurs in northeastern Asia, and a close relative of *Carex microchaeta*, viz. *Carex melanostoma*, occurs in northeastern Siberia. The possibility of their migration through the Bering Land Bridge as tundra plants cannot be excluded, in which case speciation took place after the opening of the Bering Strait.

Amphiberingian species

The ranges of these species correspond fairly well with the largely unglaciated parts of eastern Siberia and Alaska-Yukon that constituted Beringia during the glacial period.

The most common amphiberingian species present in our data set are: *Aconitum delphinifolium, Alnus crispa* ssp. *sinuata, Anemone richardsonii, Arnica frigida, Artemisia arctica, Carex consimilis, Gentiana glauca, Lagotis glauca, Petasites hyperboreus, Polemonium acutiflorum, Rhodiola integrifolia, Rumex arcticus, Salix pulchra, Valeriana capitata.*

The frequency of amphiberingian species in the various vegetation types is generally low (1–8%), with the remarkable exception of *Salix pulchra*-thickets in the subalpine belt of the mountains of northern Yukon, where they reach 30% of the total. The *Salix pulchra*-association was the only vegetation type from northern Yukon included in our data set; single relevés of other vegetation types from northern Yukon are also very rich in amphiberingian species. Most species are growing on cryosols, none is present in grasslands on regosols. It therefore seems that most of the amphiberingian species are remnants of the old arctic-alpine flora in the study area, that migrated through the Bering Land Bridge as tundra plants, or were preserved on the unglaciated mountains of Beringia.

Joint distribution of vegetation types and chorological groups

The results of the Canonical Concentration Analysis is presented in Figure 8. The first axis has been rotated informally to enhance the interpretation. A clear correlation is evident between vegetation types and phytogeographical elements. The areas of the various types of ranges tends to decrease from the left to the right on the axis, a fact that is even more evident from the results of PCA (see Fig. 4). If the gradient of Figure 2 is interpreted as a soil moisture gradient, a possible conclusion could be that the presence of water is linked with the presence of species with broad distribution patterns, and that xeric environments are more favourable for species with narrower distribution patterns. This hypothesis could perhaps be extended beyond the limits of the study area, as suggested by a preliminary study made by us in the central European and Mediterranean regions.

General discussion

The existence of an ice-free area (Beringia, Fig. 7) including large portions of Alaska-Yukon is reflected in the high number of endemic and amphiberingian species in the flora of the Yukon Territory. Two principal centers of conservation and differentiation were located in the southwestern and northern part of the region (Figs. 5, 6). They correspond with a periglacial steppic vegetation and with an arctic-alpine tundra respectively. The Bering Land Bridge was a most effective migration route for tundra-plants. The ice-free corridor connecting Beringia with the ice-free areas south of the North American ice-sheet constituted the main migration route for steppic species.

The main chorological feature of the muskeg communities is the high frequency of species with a broad circumboreal (-polar) distribution. This is probably due to the ecological affinities between muskeg and tundra, a fact that allowed an easy migration of muskeg species through the Bering Land Bridge. For this reason, the muskeg associations of the Yukon Territory can be included in the class *Oxycocco-Sphagnetea*.

The boreal forest vegetation dominated by *Picea glauca* and *Pinus contorta* is characterized by a smaller share of the circumboreal element, which is partly replaced by boreal North American species. These species are probably confined to the North American continent because during the Pleistocene the Bering Land Bridge did not support a closed forest vegetation. From the syntaxonomical point of view, the boreal forest associations of the Yukon should be included into a high unit endemic of boreal North America. The order *Gaultherio–Piceetalia,* proposed by Braun-Blanquet, Sissingh & Vlieger (1939) cannot be taken into consideration, since its characteristic species are mainly restricted to the area along the Atlantic (and a few along the Pacific) coast. Most of these species survived glaciation in refugia located along the coasts. Some of them are more or less oceanic species (Hultén, 1937), not able to expand in the more continental parts of North America.

The chorological study of the grasslands of southern Yukon reveals their importance as a center of conservation and differentiation (high percentage of endemics) and their affinities both with the xeric grasslands of the Rocky Mountains and with the prairies of central North America (high frequency of Cordilleran species). The latter affinities seem to be due to relatively recent events, chiefly to migration through the Cordilleran Corridor during xeric interglacials. A probably more ancient connection with the steppes of western Siberia can be postulated on the basis of the presence of such species as *Artemisia rupestris* s.l. and the lichen *Caloplaca tominii* Savicz (Nimis, 1981a, b). The two species have a range centered in the steppes of central Asia, with two disjunctions, one in the most continental parts of Scandinavia, the other in the study area. Recently, *Caloplaca tominii* has also been found in the *Artemisia* grasslands of Montana and Idaho (Rosentreter, in litt.), being at present the only species connecting the *Artemisia*-steppes of central Asia with the ones of central North America.

The woods dominated by deciduous trees (*Populus tremuloides, P. balsamifera, Betula papyrifera*) are mainly early successional stages after burning of the boreal forest. Their main chorological feature, i.e. the high frequency of species with boreal North American distribution extending southwards along the Cordillera, reflects their intermediate position between the boreal forest and the grasslands both from the syndynamical and ecological points of view. On the east side of the Rockies in Washington and British Columbia too, groves of *Populus tremuloides* are a conspicuous feature of landscapes on the margin of the steppe, where they are confined to locally moist places (Daubenmire, 1969).

The results underline the great phytogeographical relevance of the study area, that appears to be one of the most important regions for the study of the vegetational and floristic history of the arctic and boreal biomes in the northern hemisphere.

Conclusions

Two general methodological conclusions may be presented:
- Vegetation types may have a high predictive value with regard to chorological categories.
- Chorological categories may have a high predictive value with regard to vegetation types. Such a result is also due to a particular methodological approach: the chorological categories adopted

in the present study have not been chosen 'a priori' on the basis of a previous formal classification of range types. They have been defined in such a way as to maximize the information content with regard to the ecology of the vegetation types, on the basis of a trial-and-error procedure. Smaller or larger categories would have obscured the main trends revealed by the numerical analysis of the data.

The high correlation between chorological categories and vegetation types, represents a synthesis of floristical, vegetational, ecological and historical elements that reflect the evolution of the vegetational landscape in the study area.

References

Anderberg, M. R., 1973. Cluster Analysis for Applications. Academic Press, New York.

Braun-Blanquet, J., Sissingh, G. & Vlieger, J., 1939. Prodromus der Pflanzengesellschaften. 6. Klasse der Vaccinio-Piceetea. Comm. S.I.G.M.A. Montpellier. 123 pp.

Burne, B. M., 1974. The climate of the Mackenzie Valley. II. Environm. Canada, Athm. Climatol. Studies, 24.

Colinvaux, P. A., 1967. Quaternary vegetational history of Arctic Alaska. In: D. M. Hopkins (ed.), The Bering Land Bridge, pp. 207–231. Stanford University Press.

Crovello, J. J., 1981. Quantitative biogeography: An overview. Taxon 30: 563–575.

Daubenmire, R., 1969. Ecologic plant geography of the Pacific Northwest. Madroño 20: 11–128.

Dillon, L. S., 1956. Wisconsin climate and life zones in North America. Science 123 (3188): 167–176.

Douglas, G. W., 1974. Montane zone vegetation of the Alsek River Region, southwestern Yukon. Can. J. Bot. 52: 2505–2532.

Douglas, R. J. W. (ed.), 1970. Geology and Economic Minerals of Canada. Dept. of Energy, Mines and Resources, Ottawa. 838 pp.

Feoli, E. & Orlóci, L., 1979. Analysis of concentration and detection of underlying factors in structured tables. Vegetatio 40: 49–54.

Giterman, R. E. & Golubeva, L. V., 1967. Vegetation of eastern Siberia during the anthropogene period. In: D. M. Hopkins (ed.), The Bering Land Bridge, pp. 207–231. Stanford Univ. Press.

Gjaerevoll, O., 1980. A comparison between the alpine plant communities of Alaska and Scandinavia. Acta Phytogeogr. Suec. 68: 83–88.

Hansen, H. P., 1943. Palaeoecology of the sand dune bogs on the southern Oregon coast. Am. J. Bot. 30: 335–340.

Hansen, H. P., 1947. Postglacial vegetation of the Northern Great Basin. Am. J. Bot. 34: 164–171.

Hanson, H. C., 1953. Vegetation types in northwestern Alaska and comparison with communities in other Arctic regions. Ecology 34: 111–140.

Hoefs, M., McCowan, I. & Krajina, V. J., 1975. Phytosociological analysis and synthesis of Sheep Mountain, Southwest Yukon Territory, Canada. Syesis 8 (Suppl. 1): 125–228.

Hopkins, D. M. (ed.), 1967. The Bering Land Bridge. Stanford Univ. Press, Stanford, California.

Hughes, O. L., Rampton, V. N. & Rutter, N. W., 1972. Quaternary geology and geomorphology, southern and central Yukon (N Canada). Guidebook, Field Excursion. All. Int. Geol. Congr. 24th Session.

Hultén, E., 1937. Outline of the History of Arctic and Boreal Biota during the Quaternary Period. Stockholm.

Hultén, E., 1968. Flora of Alaska and neighboring Territories. Stanford Univ. Press, Stanford, California. 1008 pp.

Jeffers, J. N. R. & Black, T. M., 1963. An analysis of variability of Pinus contorta. Forestry 36: 199–218.

Jeffrey, W. W., 1959. Notes on plant occurrence along Lower Liard River, N. W. T. Nat. Mus. Can. Bull. 171: 32–115.

Johnson, A. W. & Packer, J. G., 1967. Distribution, ecology and cytology of the Ogotoruk Creek Flora and the history of Beringia. In: D. M. Hopkins (ed.), The Bering Land Bridge, pp. 245–265. Stanford Univ. Press.

Kruckeberg, A. R., 1969. Soil diversity and the distribution of plants, with examples from western North America. Madroño 20: 129–153.

La Roi, G. H., 1967. Ecological studies in the Boreal spruce-fir forests of the North American taiga. I. Analysis of the vascular flora. Ecol. Monogr. 37: 229–253.

La Roi, G. H. & Stringer, M. H. L., 1976. Ecological studies in the Boreal spruce-fir forests of the North American taiga. II. Analysis of the bryophyte flora. Can. J. Bot. 54: 619–643.

Laughlin, W. S., 1967. Human migration and permanent occupation in the Bering Sea area. In: D. M. Hopkins (ed.), The Bering Land Bridge, pp. 409–450. Stanford Univ. Press.

Mirov, N. T., 1967. The Genus Pinus. The Ronald Press Co., New York. 602 pp.

Müller-Beck, H., 1967. On migrations of hunters across the Bering Land Bridge in the upper Pleistocene. In: D. M. Hopkins (ed.), The Bering Land Bridge, pp. 373–408. Stanford Univ. Press.

Nimis, P. L., 1981a. Caloplaca tominii Savicz new to North America. Bryologist 84: 222–225.

Nimis, P. L., 1981b. Epigaeous lichen synusiae in the Yukon Territory. Cryptogamie, Bryol. Lichenol. 2: 127–151.

Orlóci, L., 1978. Multivariate Analysis in Vegetation Science, 2nd ed. Junk, The Hague. 451 pp.

Orlóci, L. & Stanek, W., 1979. Vegetation survey of the Alaska Highway, Yukon Territory: types and gradients. Vegetatio 41: 1–56.

Oswald, E. T. & Senyk, J. P., 1977. Ecoregions of the Yukon Territory. Can. Dept. Environ., Can. For. Serv. Victoria, B.C. 115 pp.

Porsild, A. E., 1945. The alpine flora of the east slope of Mackenzie Mountains, Northwest Territories. Nat. Mus. Can. Bull. 101: 1–35.

Porsild, A. E., 1951. Botany of the southeastern Yukon adjacent to the Canol Road. Nat. Mus. Can. Bull. 121: 1–400.

Porsild, A. E., 1966. Contributions to the flora of southwestern Yukon Territory. Nat. Mus. Can. Bull. 216: 1–86.

Porsild, A. E., 1974. Materials for a flora of central Yukon Territory. Nat. Mus. Publ. Bot. 4: 1–78.

Porsild, A. E. & Cody, W. J., 1980. Vascular plants of Continental Northwest Territories, Canada. Publ. Div. Nat. Mus. Canada. 676 pp.

Rowe, J. S., 1972. Forest Regions of Canada. Can. Dept. Environ., Can. For. Serv., Publ. 1300. Ottawa. 177 pp. and map.

Schofield, W. B., 1969. Phytogeography of northwestern North America: Bryophytes and vascular plants. Madroño 20: 155–207.

Spetzman, L. A., 1959. Vegetation of the Arctic slope of Alaska. U.S. Geol. Surv. Prof. Pap. 302: 1–58.

Westhoff, V. & van der Maarel, E., 1973. The Braun-Blanquet approach. In: R. H. Whittaker (ed.), Handbook of Vegetation Science 5: 619–726.

Wildi, O. & Orlóci, L., 1980. Management and multivariate analysis of vegetation data. Swiss. Fed. Inst. Forestry Res., Rep. No. 215. 68 pp.

Wolfe, J. A., 1969. Neogene floristic and vegetational history of the Pacific Northwest. Madroño 20: 83–109.

Wolfe, J. A. & Leopold, E. B., 1967. Neogene and early Quaternary vegetation of northwestern North America and northeastern Asia. In: D. M. Hopkins (ed.), The Bering Land Bridge, pp. 193–206. Stanford Univ. Press.

Youngman, P. M., 1975. Mammals of the Yukon Territory. Nat. Mus. of Canada, Publ. in Zool. 10: 1–192.

Accepted 25.2.1984.

Chorological phenomena in spruce and beech communities*

Gisela Jahn

Institute of Silviculture, University of Göttingen, Büsgenweg 1, 3400 Göttingen-Weende, F.R.G.

Keywords: Beech communities, Classification, Europe, *Fagus sylvatica, Picea abies,* Spruce communities, Synecology

Abstract

The problems and tasks arising from chorological phenomena in plant communities may be summarized as (a) the recognition and description of syngeographical changes and their causes with the aid of vegetation tables comparing communities over large regions and of a study of the environmental conditions; (b) the identification of analogous or vicariant communities and the detection of synonyms; (c) the classification of these communities within the phytosociological system.

These phenomena will be demonstrated with spruce and beech communities as examples. In Table 1 the floristic similarities and differences of the European spruce communities are depicted. Table 2 compares the most important spruce communities that have been described up to now and shows the large number of synonyms for communities which are closely related from the sociological and ecological points of view. Their systematic position is briefly discussed. In Tables 3–5, chorological phenomena of the suballiance Eu–Fagenion are described, viz. the association groups of beech woods on limestone and of beech communities rich in species on rich siliceous substrate in the montane and submontane zones. Tables 7 and 8 contain a survey of how chorological phenomena have been treated in phytosociological systematics by various authors. Usually regional classification is carried out at the level of the alliance or suballiance, or, more frequently, at the level of the association.

Introduction

The knowledge of two complexes is necessary in order to recognize and judge chorological phenomena in plant communities:
- environmental conditions, and
- the complete combination of plant species as occurring from vegetation tables.

Besides the 'floristic combination' in vegetation tables, further statements in synecology, 'life form', 'periodicity', 'aspect', and 'succession' are desirable (Rübel, 1932). The 'changes caused by environmental differences' must be thoroughly studied, as well as changes caused by geographical differences. By 'collecting all these facts', 'a good picture of the plant community with all its different parts' will be the result. 'A large problem in classifying the different communities is constituted by the fact that changes occur in two directions: namely regional and ecological'.

'One direction is easy to investigate. We know that the flora changes from one country to the other, from one region to the other. We leave the area of one species and reach that of another. For this reason we find a number of regional floristic facies in Europe. Are they to be named associations or only facies' (i.e. subcommunities of the associations)? 'It is the old question: How many species may change without forcing us to speak of a new association – or how many must change in order to

* Nomenclature follows Oberdorfer (1979).

Vegetatio 59, 21–37 (1985).

speak of a new association?'

We shall find a whole series 'more or less, of well-defined associations in each country. Then the difficult task of comparing and combining them will follow because many of them will prove to be synonyms'.

Within these lines, the frame of our symposium is drawn. One may agree with me, but one would be wrong in thinking that these thoughts are new. They were expressed exactly 50 years ago by Rübel (1932) in his introduction to the 'Buchenwälder Europas', and they were supported by Tansley, Szafer and their contemporaries. In spite of their relatively high age these themes are still actual and the problems mentioned have not really been solved in spite of the large amount of scientific material which has become available. These problems are to be discussed with spruce and beech communities as examples. Some attention will be given to the problematic use of old vegetation tables and tables from others. These difficulties were discussed by Jahn (1977).

There are several possible scales to approach chorological problems, e.g. worldwide, northern hemisphere, Europe, and smaller geographical regions. Here the European scale will be used for spruce communities and for beech communities the central European one.

Chorological phenomena in spruce communities

Table 1 (from Jahn, 1977) documents spruce communities from the E and W Alps, the Balkan, the E and W Carpathians, the hercynic mountains, western (W Norway + Sweden) and eastern Fennoscandia, including Russia as far as the area of *Picea abies* is concerned. In this table mostly typical spruce communities rich in *Vaccinium* are compared. Species not occurring in the community are marked by a dash, species not occurring in the region by a zero. The table is arranged in a chorological – that is climatical – order. This arrangement was not clear from the beginning. Passarge (1971) subdivided first and foremost on the trophic level and then subordinated on the chorological aspect. A classification with the aid of computer programs (Schmid & Kuhn, 1970), however, resulted in a clear prevalence of chorological phenomena as division criteria. The table contains only

88 out of the 184 species collected in the basic table. It shows what *Picea* communities in Europe have in common and where they are different. All have in common the dominance of *Picea abies* and the occurrence of *Sorbus aucuparia* in the tree layer. In northern Europe *Pinus sylvestris* increases where continentality increases.

Shrubs are restricted to communities in certain regions. The following species are never absent in the field layer:

Vaccinium myrtillus	CC
Vaccinium vitis-idaea	CC
Lycopodium annotinum	OC
Melampyrum sylvaticum ssp.	OC
Avenella flexuosa	Co
Oxalis acetosella	Co
Hylocomium splendens	
Pleurozium schreberi	
Polytrichum attenuatum	
Dicranum scoparium	

(CC = class character-species;
OC = order character-species;
Co = companion.)

There is no one of these species which is restricted to *Picea* associations or even *Picea* alliances. In such an 'expansive' view there are no true character-species of association or alliance, a fact, as Braun-Blanquet (1938) already pointed out.

The following species occur relatively often as well (but not in the Harz and Thuringia mountains) *Homogyne alpina* and *Luzula sylvatica*, especially ssp. *sieberi*.

Furthermore *Calamagrostis villosa* is typical of central European *Picea* communities in mountains with a natural spruce zone, hence not of the Swiss Jura and the Black Forest or the Balkan mountains.

The N European communities are delimited very clearly from the central European ones, positively, through numerous arctic, northern and northern-continental geographic differential species and by *Linnaea borealis*, and negatively, by the absence of numerous central European geographic differential species and the absence of *Homogyne alpina* and *Calamagrostis villosa*. The latter can be seen as a character-species of certain vicariants of central European communities.

Clear sociological differences also occur within central Europe between the spruce communities of the higher (Alps) and the lower mountains, the western and the eastern Alps, the western and the eastern Carpathians, and smaller geographic units. The ecological subdivision is many-sided and it has led to a vast nomenclature. Yet it is possible to build 4 groups, with a relatively uniform sociology and ecology, differing mainly in their chorology. They are pointed out in Table 2 (Jahn, 1977), where variations due to local differences, especially soil and macroclimate, are compared.

Chorological phenomena in beech communities

The examples are limited to beech communities of the *Fagion*, suballiance *Eu–Fagenion* = *Galio odorati–Fagenion* Ob. 57, in the central European highlands. Table 3 contains mesophile beech forests on limestone with a uniform foundation of species. Without considering the differences caused by altitude (*Abies alba* in montane areas), three subunits become apparent:

- the *Dentario heptaphyllidi–Fagetum* in Switzerland, characterized positively through the geographic differential species
  Dentaria heptaphyllos,
  Dentaria pentaphyllos,
  Adenostyles glabra,
  Lonicera alpigena,
  Lonicera nigra;
- the 'normal' *Lathyro–Fagetum* without geographic differential species;

- and the subatlantic *Elymo–Fagetum*, characterized negatively through the absence of
  Lathyrus vernus,
  Hepatica nobilis,
  Aquilegia vulgaris.

Table 4 demonstrates expansive differences through geographic causes in the association groups of beech communities rich in species on rich siliceous substrate in the montane zone.

The *Dentario glandulosae–(Abieti–)Fagetum* of the W Carpathians (Beskiden) and the *Dentario enneaphyllidi–(Abieti–)Fagetum* of the Sudetic mountains are rich in species, containing, except beech *(Fagus sylvatica)*, the admixed species *Abies alba;* in certain areas *Larix decidua* and *Picea abies. Dentaria glandulosa* is the geographic character-species and differential species which demonstrates the difference between the above-mentioned forest community and the three other beach forest types rich in species in the montane zone.

The species-poor *Abieti–Fagetum* in the Black Forest lacks *Dentaria enneaphyllos* and *Dentaria bulbifera*. Still poorer in species, the *Dentario–Fagetum* of the northwestern highlands lacks all species of conifers.

A survey of the '*Melica uniflora*' beech forest of central Europe (Jahn, 1980) (Table 5) demonstrates fewer significant geographic differences. These differences are more strongly expressed through the complete species combination and their abundance

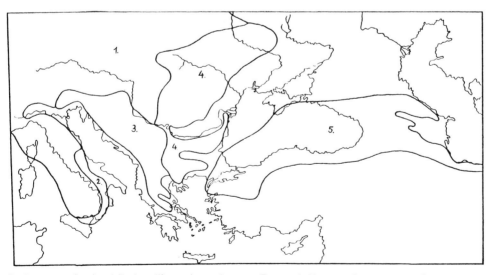

Fig. 1. Distribution areas of regional *Fagion* alliances in southeastern Europe. 1. *Fagio medio-europaeum,* 2. *F. austro-italicum,* 3. *F. illyricum,* 4. *F. dacicum,* 5. *F.* (resp. *Fagetalia*) *orientalis* (draft by A. Borhidi & R. Soó).

and dominance.

As to the syngeographical vicariants of beech communities, Soó (1964) discovered in southeastern Europe, as shown in Figure 1, four vicariant communities which he considered parallel to the atlantic *Scillo–Fagion* and the central European *Fagion medio-europaeum*.

In addition to the expansive chorological phenomena shown here, differences caused by climate in the species combinations of various communities are recognized in smaller areas (Jahn, 1972; Oberdorfer, 1957). Without the indication of authentic geographical differential species, the differences are expressed through the predominance, occurrence, or absence of individual species or groups of species and through variable quantity proportions. Such chorological units are often defined as 'regional units' or 'races'.

Considering chorological phenomena in phytosociological systematics

Although much attention has been given to this topic, there are still no generally accepted answers to the questions of how and when a new vegetation unit should be acknowledged and how it should be ranked. Only a few proposals are known. The following scheme shows a proposal from Knapp (1942).

Classification scheme according to Knapp (1942)	
Class Order Union Main association	marked by character-species
Association Subassociation Variant Subvariant	marked by differential species

Hartmann & Jahn (1967, p. 4/5) have thoroughly discussed geographical divisions. They summarize vicariant associations as association groups, as Oberdorfer (1957) sometimes does (see also Table 3). A. & W. Matuszkiewicz (1973) use the territorially defined association, the so-called 'regional association', as a basic unit of the phytosociological system in the meaning used by Tüxen (1960). This 'regional association' is identical to the vicariant association of Hartmann & Jahn (1967). The association can be subdivided according to a three-dimensional system:

– subassociations and their subunits corresponding to site differentiation;
– altitudinal forms corresponding to altitudinal differentiation;
– geographic races and local variants (Hartmann & Jahn: regional variants), taking into account further regional differences.

The Tables 6 and 7 show further examples of how different authors have tried to solve the problem.

Chorological phenomena in plant communities have been known for a long time. These phenomena become more apparent, after more material has been collected from various geographical regions. The extraordinary variety of names of sociologically and ecologically related communities and the disunitary consideration in the sociological systematic show that the problems recognized fifty years ago by Rübel (1932) have yet to be solved. I hope the International Association for Vegetation Science will provide further platforms for a discussion of these problems!

References

Borhidi, A., 1966. Die pflanzenökologische Stellung der illyrischen Buchenwälder. Angewandte Pflanzensoziologie XVIII/XIX: 19–22. Springer-Verlag.

Braun-Blanquet, J., 1928, 1951, 1964. Pflanzensoziologie. Springer-Verlag, Wien (1964). 865 pp.

Braun-Blanquet, J. & Moor, M., 1938. Prodromus der Pflanzengesellschaften. 5, Verband des Bromion erecti. 64 pp.

Braun-Blanquet, J., Sissingh, G. & Vlieger, J., 1939. Prodromus der Pflanzengesellschaften. 6, Klasse der Vaccinio-Piceetea. Montpellier. 123 pp.

Doing-Kraft, H. & Westhoff, V., 1958. De plaats van de beuk (Fagus sylvatica) in het midden- en west-europese bos. Jaarb. Ned. Dendrol. Ver. 21: 226–254.

Hartmann, F. K. & Jahn, G., 1967. Waldgesellschaften des mitteleuropäischen Gebirgsraumes nördlich der Alpen. Gustav Fischer Verlag, Stuttgart. 636 pp.

Jahn, G., 1972. Forstliche Wuchsraumgliederung und waldbauliche Rahmenplanung in der Nordeifel. Dissertationes botanicae 16. Verlag Cramer, Lehre. 294 pp.

Jahn, G., 1974. Vegetationskundliche Übersicht der Fichtenwälder Europas. Vortr. d. Tagung Arbeitsgem. forstl. Vegetationskde. 4: 19–41. Göttingen.

Jahn, G., 1977. Die Fichtenwaldgesellschaften in Europa. In: H. Schmidt-Vogt (ed.), Die Fichte, 1, pp. 468–560. Verlag Paul Parey, Hamburg und Berlin.

Jahn, G., 1980. Das Melico-Fagetum in seinen Beziehungen zur Umwelt. Berichte der Internationalen Vereinigung für Vegetationskunde, Epharmonie (Rinteln 1979): 209–233. Cramer, Vaduz.

Kielland-Lund, J., 1981. Die Waldgesellschaften Südost-Norwegens. Phytocoenologia 9: 53–250.

Knapp, R., 1942. Zur Systematik der Wälder, Zwergstrauchheiden und Trockenrasen des eurosibirischen Vegetationskreises. Arb. Zentralst. Vegetationskartierung d. Reiches, Beil. z. 12. Rundbrief (als Manuskript gedruckt).

Kramer, H., 1963. Der Einfluß von Großklima und Standort auf die Entwicklung von Waldbeständen am Beispiel langfristig beobachteter Versuchsflächen von Douglasie, Fichte, Buche und Eiche. Schriftenreihe der Forstl. Fakultät der Universität Göttingen, 31/32. Sauerländer's Verlag, Frankfurt/M. 140 pp.

Matuszkiewicz, A. & W., 1973. Przeglad fitosocjologiczny zbiorowisk leśnych Polski. CZ I. Lasy bukowe (Pflanzensoziologische Übersicht der Waldgesellschaften von Polen. 1. Die Buchenwälder.). Phytocoenosis 2: 143–202. Warszawa-Bialowieza.

Mitscherlich, G., 1950. Die Bedeutung der Wuchsgebiete für das Bestandeswachstum von Buche, Eiche, Erle, Birke. Forstwiss. Centralbl. 4: 10–89. Berlin.

Moor, M., 1960. Zur Systematik der Querco-Fagetea. Mitt. Flor.-soz. Arbeitsgem. NF 8: 263–293. Stolzenau/Weser.

Müller, Th. & Görs, S. 1958. Zur Kenntnis einiger Auenwaldgesellschaften im württembergischen Oberland. Beitr. Naturkundl. Forsch. SW Deutschland. 17: 88–165. Karlsruhe.

Müller, Th., 1964. Niederschrift über die Besprechung (über geographische Differenzierung von Waldgesellschaften) G. Buck-Feucht, S. Görs, F. K. Hartmann, G. Jahn, Th. Müller in Hann. Münden, Manuskript.

Neushäusl, R., 1977. Comparative ecological study of European Oak–Hornbeam forests. Nat. Can. 104: 109–117.

Oberdorfer, E., 1957. Süddeutsche Pflanzengesellschaften. Pflanzensoziologie 10. Fischer, Jena. 564 pp.

Oberdorfer, E., 1979. Pflanzensoziologische Exkursionsflora, 4th ed. Ulmer, Stuttgart. 997 pp.

Passarge, H. & Hofmann, G., 1968. Pflanzengesellschaften des nordostdeutschen Flachlandes II. Pflanzensoziologie 16. Fischer, Jena. 298 pp.

Passarge, H., 1971. Zur soziologischen Gliederung mitteleuropäischer Fichtenwälder. Feddes Rep. 81: 577–604. Berlin.

Rübel, E. (ed.), 1932. Die Buchenwälder Europas. Veröff. Geobot. Inst. Rübel. Bern. 507 pp.

Schmid, P. & Kuhn, N., 1970. Automatische Ordination von Vegetationsaufnahmen in pflanzensoziologischen Tabellen. Naturwiss. 57, 9: 462.

Soó, R., 1964. Die regionalen Fagion-Verbände und Gesellschaften Südosteuropas. Stud. Biol. Acad. Sci. Hung. 1. Akadémiai Kiadó, Budapest. 104 pp.

Traczyk, T., 1962. Próba podsumowania badań (1) nad ekologicznym zróżnicowaniem gradów w Polsce. (Essai d'une synthèse des élaborations sur la différentation du Querco-Carpinetum en Pologne.) Acta Soc. Bot. Pol. V 31: 610–635. Warschau.

Tüxen, R., 1960. Zur Systematik der west- und mitteleuropäischen Buchenwälder. Bull. Inst. Agron. Stat. Rech. Gembloux, H.S. I: 45–56. Gembloux.

Walter, H., 1954. Grundlagen der Pflanzenverbreitung II Arealkunde. Einführung in die Phytologie III. Verlag Eugen Ulmer, Stuttgart. 245 pp.

Accepted 4.4.1984.

Appendix

Table 1. Picea communities in Europe.

	Alps West	Alps East (subalpine)	Alps (montane)	Balkan	Carpathians East	Carpathians West	Hercynian Mountains Bohemian Sudetes Mts.	Hercynian Mountains Thuringian Mts. Harz	Swiss Jura Black Forest	Fennoscandia West	Fennoscandia East
Number of relevés	52	123	137	163	53	63	61	44	30	691	209
Cons. Numbers	1	2	3	4	5	6	7	8	9	10	11
Trees											
Picea abies	(4) 5	V 5	V 5	V 5	(4) 5	V 5	V 5	(3) 5	(2) 5	V 5	V 5
Sorbus aucuparia	(4) 4	V 4	IV 3	V 4	(4) 4	V 5	IV 3	(3) 3	(2) 4	V 3	V 4
Pinus sylvestris	(1) 4	–	III 1	I 2	–	–	I 3	–	(1) 1	III 1	V 4
Betula pubescens	–	–	–	I 1	–	–	–	–	(2) 3	III 1	IV 4
Fagus sylvatica	(1) 3	–	III 1	IV 3	(1) 1	I 4	I 2	(1) 1	(2) 2	I 1	0
Abies alba	(3) 2	III 1	I 2	IV 4	(1) 1	II 1	II 2	(1) 2	(2) 5	0	0
Larix decidua	(3) 3	V 3	IV 4	0	0	0	0	0	0	0	0
Pinus cembra	(1) 2	IV 3	–	–	0	I 1	0	0	0	0	0
Alnus viridis	(2) 2	II 2	–	0	0	0	0	0	0	0	–
Populus tremula	–	–	–	–	–	–	–	–	–	I 2	III 2
Shrubs											
Lonicera nigra	(4) 4	V 3	V 2	IV 3	(3) 3	IV 1	I 1	0	(1) 1	0	0
Rosa pendulina	(2) 4	II 2	III 2	III 3	(2) 1	III 1	0	0	0	0	0
Lonicera caerulea	–	IV 2	III 1	–	–	–	–	–	–	–	I 1
Rhododendron ferrugineum	(2) 1	IV 2	–	0	0	0	0	0	0	0	0
Frangula alnus	–	–	–	I 1	–	–	–	–	(1) 1	III 1	III 3
Herb layer											
Character-species of *Vaccinio–Piceetea* class, *Vaccinio–Piceetalia* order and *Vaccinio–Piceion* alliance											
Vaccinium myrtillus	(4) 5	V 5	V 5	V 5	(4) 3	V 5	V 5	(3) 5	(2) 5	V 5	V 5
Vaccinium vitis-idaea	(4) 3	V 5	V 4	III 3	(4) 2	III 2	III 2	(2) 2	(2) 5	V 4	V 5
Lycopodium annotinum	(2) 2	V 4	V 4	III 4	(4) 4	III 3	IV 3	(2) 1	(2) 3	IV 3	V 2
Melampyrum sylvaticum	(4) 5	V 4	V 4	III 4	(2) 3	I 1	III 1	–	(2) 2	V 3	III 3
Pyrola secunda	(3) 5	V 3	V 3	III 2	(1) 1	II 2	–	–	(1) 1	V 3	V 4
Listera cordata	–	II 4	III 1	III 2	(3) 3	V 2	II 2	0	(2) 4	III 1	II 1
Moneses uniflora	–	IV 3	V 2	III 2	(3) 2	IV 4	I 1	0	(1) 1	III 1	III 1
Homogyne alpina	(4) 3	V 5	V 3	II 2	(4) 5	V 5	V 5	0	(2) 3	0	0
Blechnum spicant	–	II 5	–	II 2	(3) 2	I 3	III 3	(1) 3	(2) 3	0	0
Calamagrostis villosa	(2) 1	V 5	V 5	III 5	(4) 4	V 3	V 5	(3) 5	0	0	0
Luzula luzulina	(4) 3	V 3	V 3	III 1	–	V 4	0	0	(1) 2	0	0
Corallorhiza trifida	–	–	–	–	(3) 1	IV 3	–	0	(1) 1	I 1	III 3
Huperzia selago	–	–	III 1	II 1	(3) 4	I 2	II 1	(2) 3	(1) 1	III 1	II 1
Trientalis europaea	–	–	–	II 2	–	–	V 3	(3) 4	(1) 1	III 4	V 4
Melampyrum pratense ssp.	–	–	–	–	–	–	I 4	(1) 1	(1) 2	III 1	V 3
Linnaea borealis	–	II 3	I 1	0	0	0	0	0	0	V 4	III 2

Table 1. Continued.

	Alps			Balkan	Carpathians		Hercynian Mountains		Swiss Jura Black Forest	Fennoscandia	
	West	East			East	West	Bohemian Sudetes Mts.	Thuringian Mts. Harz		West	East
		subalpine	montane								
Number of releves	52	123	137	163	53	63	61	44	30	691	209
Cons. Numbers	1	2	3	4	5	6	7	8	9	10	11
Geographic differential species and species occurring mainly in certain regions											
Campanula rhomboidalis	(4) 3	0	0	0	0	0	0	0	0	0	0
Luzula nivea	(4) 3	III 2	IV 3	0	0	0	0	0	0	0	0
Carex alba	–	II 2	III 4	0	0	0	0	0	0	0	0
Valeriana tripteris	(3) 2	III 2	III 2	I 5	(3) 1	III 2	0	0	0	I 1	0
Viola biflora	(2) 3	V 2	V 3	I 2	(2) 3	III 4	–	–	–	–	–
Clematis alpina	(2) 2	IV 2	IV 3	I 4	(2) 1	III 1	–	–	–	–	–
Veronica urticifolia	(3) 4	V 2	V 4	I 3	(2) 3	–	0	0	0	0	0
Cirsium erisithales	–	II 1	III 2	II 2	–	I 1	–	–	–	–	–
Prenanthes purpurea	(4) 4	V 3	V 3	II 3	(2) 1	V 3	III 2	–	0	0	0
Gentiana asclepiadea	–	III 2	I 1	V 4	(3) 2	V 5	III 2	–	–	–	–
Aremonia agrimonoides	0	0	0	IV 3	–	0	0	0	0	0	0
Spiraea chamaedryfolia	0	0	0	I 1	x	0	0	0	0	0	0
Pulmonaria filarizkyana	0	0	0	–	(2) 2	0	0	0	0	0	0
Dentaria glandulosa	0	0	0	0	(3) 2	IV 3	0	0	0	0	0
Soldanella carpatica	0	0	0	0	(4) 3	II 3	0	0	0	0	0
Galium harcynicum	–	–	–	–	–	–	I 1	(3) 3	(1) 1	I 2	III 4
Convallaria majalis	–	–	–	–	–	–	–	–	–	I 2	–
Calamagrostis phragmitoides	–	–	–	–	–	–	–	–	–	II 2	–
Differential species on better sites											
Gymnocarpium dryopteris	(3) 3	V 3	IV 3	I 2	(4) 4	IV 4	III 1	–	(1) 1	III 1	III 3
Athyrium filix-femina	–	IV 3	III 3	II 3	(2) 4	IV 3	III 2	x	(1) 2	III 2	III 2
Polygonatum verticillatum	–	III 1	II 2	II 3	(4) 2	V 3	IV 3	(1) 1	(1) 1	I 1	–
Luzula albida	–	V 3	IV 4	III 3	(1) 2	V 1	III 1	–	(1) 1	–	–
Calamagrostis arundinacea	–	–	I 2	II 4	(2) 1	IV 2	III 2	(1) 1	–	III 3	IV 4
Carex digitata	(3) 3	II 2	III 3	I 1	–	I 2	–	–	–	IV 4	III 4
Hepatica nobilis	(2) 4	I 2	IV 3	I 2	–	–	–	–	–	II 3	II 3
Melica nutans	–	II 2	IV 3	–	–	–	–	–	–	III 3	IV 2
Rubus saxatilis	(3) 1	IV 2	V 3	II 2	–	–	–	–	–	III 3	IV 4

Table 1. Continued

	Alps	Alps	Alps	Balkan	Carpathians	Carpathians	Hercynian Mountains	Hercynian Mountains	Swiss Jura	Fennoscandia	Fennoscandia
	West	East subalpine	East montane		East	West	Bohemian Sudetes Mts.	Thuringian Mts. Harz	Black Forest	West	East
Number of relevés	52	123	137	163	53	63	61	44	30	691	209
Cons. Numbers	1	2	3	4	5	6	7	8	9	10	11
Companions											
Avenella (= Deschampsia) flexuosa	(3) 5	V 5	V 4	IV 3	(2) 3	IV 4	V 3	(3) 5	(1) 5	V 4	III 3
Oxalis acetosella	(4) 4	V 5	V 4	V 4	(4) 5	V 5	V 5	(2) 2	(2) 3	IV 4	IV 5
Dryopteris dilatata	–	V 4	IV 3	II 4	(4) 5	III 5	V 5	(3) 4	(2) 4	II 2	III 3
Majanthemum bifolium	(4) 2	V 2	V 4	III 4	(1) 1	IV 2	V 4	–	(2) 3	V 5	IV 4
Luzula sylvatica	(4) 4	V 2	IV 2	III 3	(4) 4	V 5	V 4	(2) 4	(2) 2	0	0
Fragaria vesca	(3) 4	V 3	IV 4	IV 3	(2) 3	III 1	–	–	–	II 4	IV 4
Hieracium sylvaticum	(4) 5	V 5	V 5	IV 4	–	V 3	–	–	(1) 1	III 3	–
Solidago virgaurea	–	V 3	V 4	I 2	–	I 3	III 1	–	–	III 3	V 4
Pteridium aquilinum	–	I 2	I 3	II 4	–	I 3	–	–	(1) 1	I 3	III 4
Equisetum sylvaticum	–	I 3	–	–	–	I 1	–	–	(2) 2	III 1	III 2
Moss layer											
Character-species (see above)											
Ptilium crista-castrensis	–	V 3	III 1	–	(3) 4	III 1	II 2	–	(2) 3	IV 4	IV 3
Mnium spinosum	–	V 2	IV 2	I 2	(2) 5	V 4	–	–	(1) 1	III 3	IV 3
Barbilophozia lycopodioides	–	V 4	III 2	–	(4) 5	IV 2	III 4	(3) 2	(2) 1	III 4	I 5
Bazzania trilobata	–	II 2	–	I 2	(3) 1	I 1	III 3	(2) 2	(2) 3	I 1	–
Sphagnum girgensohnii	–	II 5	–	II 2	(3) 2	I 3	III 3	(1) 3	(2) 3	–	–
Plagiothecium undulatum	–	II 2	–	–	(4) 4	I 4	V 3	(3) 3	(2) 2	I 1	–
Hylocomium umbratum	–	IV 3	–	–	(2) 3	IV 4	I 1	–	(1) 1	III 1	–
Barbilophozia floerkei	–	–	–	–	–	I 2	III 3	(2) 1	(1) 1	III 1	–
Other mosses and lichens											
Hylocomium splendens	(4) 5	V 5	V 5	II 5	(4) 4	V 4	III 3	(1) 1	(2) 5	V 5	IV 4
Pleurozium schreberi	(2) 3	V 5	V 4	III 3	(4) 4	V 4	V 4	(3) 3	(2) 3	V 5	V 5
Polytrichum attenuatum	(1) 2	V 4	IV 3	II 4	(4) 5	V 4	V 5	(3) 5	(3) 3	III 1	II 1
Dicranum scoparium	(4) 3	V 5	V 4	V 4	(4) 5	V 4	V 4	(3) 5	(2) 5	IV 4	III 3
Rhytidiadelphus loreus	(1) 2	II 5	–	II 5	(2) 2	I 5	III 3	(3) 2	(2) 5	I 1	–
Plagiochila asplenioides	(1) 1	V 3	V 3	II 3	–	III 1	III 2	–	(2) 5	I 2	–
Lichenes div. spec.	–	V 4	III 2	II 3	(1) 4	IV 1	II 2	(1) 1	(1) 3	II 2	II 3
Sphagna div. spec.	–	IV 3	I 1	II 2	(2) 3	I 1	V 3	(3) 4	(2) 4	I 1	I 3
Polytrichum commune	–	II 3	–	I 4	(4) 2	III 1	IV 2	(2) 2	(2) 4	III 1	I 5
Hypnum cupressiforme	–	I 1	IV 3	III 4	–	I 1	I 1	(2) 4	(1) 1	I 1	II 2
Dicranum undulatum	–	–	–	–	(1) 2	I 1	–	(1) 2	(1) 1	III 1	IV 3
Dicranum majus	–	I 1	–	–	(1) 4	–	II 1	–	–	IV 3	III 3

Table 2. Vicariant associations of *Picea* communities in Europe.

Picea communities rich in *Vaccinium myrtillus*

Name	Distribution	Geographic differential species
Pessière à Myrtille	Western Alps	*Campanula rhomboidalis*
Piceetum subalpinum sphagnetosum		
and *myrtilletosum = Sphagno Piceetum*		*Larix decidua*
calamagrostietosum villosae		
Homogyno–Piceetum	Eastern Alps	
Piceetum excelsae	Balkan	*Aremonia agrimonoides*
with geographical supplement		
Hieracio transsylvanico–Piceetum	Eastern Carpathians	*Hieracium transsylvanicum*
Piceetum exc. myrtilletosum	Western Carpathians	*Soldanella carpatica,*
Sorbeto–Piceetum		*Dentaria glandulosa*
		(all Carpathians)
Calamagrostio villosae–Piceetum,	Herzynian-sudetic	*Trientalis europaea*
Lophozio–Piceetum	mountains	*Galium harcynicum*
Eu–Piceetum	Norway	
Dwarfshrub and fern-dwarfshrub type	Sweden	
Myrtillus and *Oxalis–Myrtillus* type	Finland	*Linnaea borealis*
Branch moss–spruce woodlands	Russia	
Fresh bilberry–woodlands	Estonia	

Picea communities (on limestone) rich in herbs

Name	Distribution
Adenostylo glabrae–Piceetum	Eastern Alps
Piceetum croaticum	Balkan
Piceetum normale	Western Carpathians
Melico–Piceetum	Norway
Herb-type	Sweden
Spruce woodlands rich in herbs	Russia
Spruce groves	Estonia

Picea communities rich in tall herbs

Name	Distribution	Geographic differential species
Pessière à Adènostyle	Western Alps	Central Europe
Piceo–Adenostyletum	Swiss Alps	*Adenostyles alliariae*
Adenostylo alliariae–Piceetum	Eastern Alps	
Chrysanthemo rotundifolio–Piceetum	Eastern Carpathians	
Adenostyleto–Piceetum, Piceetum	Western Carpathians	
tatricum subnormale, partly		
Aceri–Piceetum		
Athyrio distentifolio–Piceetum	Hercynian-sudetic	
	mountains	Northeastern Europe
Melico–Piceetum with *Aconitum*	Norway	*Aconitum septentrionale*
septentrionale		
Fresh herb-dwarfshrub type	Sweden	
Aconitum type	Finland	

Picea communities rich in Sphagnum

Name	Distribution	Geographic differential species
Picea communities rich in *Sphagnum*	Central Europe,	
(different sociological units)	mainly northern Alps	
	and uplands	

Table 2. Continued.

Name	Distribution	Geographic differential species
Chamaemoro–Piceetum	Norway	
Wet dwarfshrub type	Sweden	
Vaccinium bog woodlands	Finland	*Rubus chamaemorus*
Longmoss–spruce woodlands	Russia	
Sphagnum–spruce woodlands		
Boglike woodlands rich in bilberries	Estonia	

Table 3. Vicariant associations of *Fagus* communities on limestone (extract).

Order	Fagetalia				
Alliance	*Fagion*				
Association group	Montane *Fagus* woods on limestone		Mesophilous *Fagus* woods on limestone		
	Swiss Jura	Suebian Alb	Swiss Jura	South and central Germany	Rhenish Schistmountains
Association	*Dentaria heptaphyllae– Abieti–Fagetum*	*Lathyro verni– Abieti–Fagetum*	*Dentario heptaphyllae– Fagetum*	*Lathyro verni–Fagetum*	*Elymo–Fagetum*
Trees					
Fagus sylvatica					
Abies alba					
Acer pseudoplatanus					
Tilia platyphyllos					
Ulmus scabra					
Carpinus betulus					
Prunus avium					
Tilia cordata					
Acer campestre					
Quercus robur					
Quercus petraea					
Pinus sylvestris					
Quercetalia pubescenti-petr.					
Sorbus torminalis					
Sorbus aria					
Shrubs					
Lonicera xylosteum					
Corylus avellana					
Crataegus spec.					
Viburnum lantana					
Cornus sanguinea					
Prunus spinosa					
Ligustrum vulgare					
Berberis vulgaris					

Table 3. Continued.

Order	Fagetalia				
Alliance	*Fagion*				
Association group	Montane *Fagus* woods on limestone		Mesophile *Fagus* woods on limestone		
	Swiss Jura	Suebian Alb	Swiss Jura	South and central Germany	Rhenish Schistmountains
Herb layer					
D Swiss Jura					
Dentaria heptaphyllos					
Dentaria pentaphyllos					
Adenostyles glabra					
Adenostyles alliariae					
Lonicera alpigena					
Lonicera nigra					
D Lathyro–(Abieti–)Fagetum					
Geranium robertianum					
Primula elatior					
Stachys sylvatica					
Athyrium filix-femina					
Fagion-alliance					
Prenanthes purpurea					
Senecio fuchsii					
Festuca altissima					
Polygonatum verticillatum					
Galium odoratum					
Neottia nidus-avis					
Actaea spicata					
Mercurialis perennis					
Lathyrus vernus					
Anemone hepatica					
Lilium martagon					
Aquilegia vulgaris					
Carex digitata					

Table 4. Chorological differences in the association group of the montane *Fagus* woods rich in species in central Europe.

	Beskids	Sudetic chain to Bohemian Mountains	Southwestern Germany with Black Forest	Northwestern Germany, highlands
	Dentario glandulosae–(Abieti–)Fagetum	*Dentario enneaphillidis–(Abieti–)Fagetum* Hartm. 53	*Abieti–Fagetum* (rhenanum) Oberd. 38	*Dentario bulbiferae–Fagetum* (Büker 41) Hartm. 53
Fagus sylvatica	x	x	x	x
Abies alba	x	x	x	–
Picea abies	x	x	–	–
Larix decidua	x	Sudetes only	–	–
Dentaria enneaphyllos	x	x	–	–
Dentaria bulbifera	x	x	–	x
Dentaria glandulosa	x	–	–	–
Species richness (relative)	high	high	middle	small

Table 5. The *Melico–Fagetum* in central Europe.

Author and year	F. Celiński, 1962	F. K. Hartmann, 1933	Scamoni, 1960	Passarge, 1960	Scamoni, 1967	Passarge & Hofmann, 1968	Passarge & Hofmann, 1968	Markgraf, 1932	Rühl, 1957	Tüxen, 1937	Tüxen, 1937	Tüxen, 1931
Number of relevés	9	30	289	46	52	108	29	6	9	17	15	5
No.	1	2	3	4	5	6	7	8	9	10	11	12
Trees												
Fagion character-species												
Fagus sylvatica	V	V	V	V	V	V	V	V	V	V	V	V
Acer pseudoplatanus	I	–	III	I	–	–	–	–	I	IV	II	–
Abies alba	–	–	–	–	–	–	–	–	–	–	–	–
Carpinion character-species												
Carpinus betulus	I	–	IV	II	I	I	I	–	–	I	I	II
Prunus avium	–	I	III	I	–	–	–	–	–	–	–	–
Tilia cordata	–	–	III									
Fagetalia and Querco–Fagetea character-species												
Fraxinus excelsior	IV	–	III	–	–	–	–	–	–	III	III	III
Quercus petraea	I	I	IV	–	I	II	III	–	–	–	–	–
Quercus robur	I	–	IV	III	I	–	I	–	III	–	II	I
Acer platanoides	–	–	II	–	–	–	–	–	–	III	I	–
Acer campestre	–	I	–	–	–	–	–	–	–	–	–	–
Companions												
Picea abies	–	–	–	–	–	–	–	–	–	–	–	–
Sorbus aucuparia	II	–	IV	II	–	I	II	–	–	II	–	–
Shrubs												
Crataegus spec.	–	–	–	–	–	–	–	–	–	–	–	IV
Lonicera xylosteum	–	–	–	–	–	–	–	–	–	–	–	I
Herbs and mosses												
Asperulo–Fagenion Tx 55 and Fagion character-species												
Galium odoratum	V	V	V	V	V	V	V	V	V	V	V	IV
Elymus europaeus	–	II	III	–	–	–	–	–	–	–	–	V
Dentaria bulbifera	I	I	II	–	–	I	–	–	II	III	–	I
Festuca altissima	IV	II	IV	–	I	–	–	–	III	II	I	I
Fagetalia character-species (Carpinion character-species included = C)												
Viola reichenbachiana	V	V	V	V	IV[1]	V	V	V	IV	III	IV	IV
Lamium galeobdolon	V	IV	V	IV	I	IV	II	I	III	IV	V	V
Melica uniflora	V	IV	V	V	I	IV	–	V	V	IV	III	IV
Scrophularia nodosa	III	II	IV	I	II	II	III	–	II	III	IV	I
Dryopteris filix-mas	V	III	IV	I	II	I	I	I	III	III	V	I
Epilobium montanum	–	II	V	I	I	II	–	–	–	IV	III	III
Carex sylvatica	IV	IV	V	III	I	X	X	–	I	III	V	III
Catharinaea undulata	V	III	V	V	II	IV	III	I	–	III	V	–
Polygonatum multiflorum	–	II	IV	III	–	I	–	–	II	I	I	III
Daphne mezereum	O	O	O	O	O	O	O	O	O	II	–	IV
Arum maculatum	–	I	I	–	–	–	–	–	–	I	I	–
Senecio fuchsii	–	–	–	–	–	–	–	–	–	II	–	–
Sanicula europaea	I	II	V	I	–	X	–	–	I	–	–	–
Neottia nidus-avis	I	–	I	–	–	–	–	–	–	–	I	–
Campanula trachelium	–	–	–	–	–	–	–	–	II	I	–	III
Pulmonaria officinalis	–	II	IV	X	–	–	X	–	–	–	–	–
Dactylis polygama C	I	I	V	I	II	I	I	–	I	–	–	III
Stellaria holostea C	I	III	IV	IV	–	–	–	I	III	I	III	II
Galium sylvaticum C	–	–	–	–	–	–	–	–	–	II	I	IV

Legend: 1– 9 Baltic region – without *Abies alba*.
   10–27 Northern hills of central Europe – without *Abies alba*.
   28–32 Southern hills of central Europe – with *Abies alba*. (See Jahn, 1980).

Rödel, 1970	Eichner, 1976	Bornkamm & Eber, 1967	Blosat & Schmidt, 1975	Tüxen, 1954	Bauer, 1978	Bauer, 1978	Grüneberg & Schlüter, 1957	Seibert, 1954	Trautmann, 1957	Lohmeyer, 1967	Jahn, 1972	F. K. Hartmann, 1953	Déthioux, 1969	Zeidler, 1953	Knapp & Ackermann, 1952	Oberdorfer, 1957	Müller, 1977	Mayer, 1964
12	222	6	16	5	39	7	11	13	36	13	17	9	9	10	16	14	13	11
14	15	16	17	18	19	20	21	22	23	24	25	26	27	28	29	30	31	32
V	V	V	V	V	V	V	V	V	V	V	V	V	V	V	V	V	V	V
III	II	–	–	II	I	–	III	I	III	–	I	III	I	II	I	II	–	–
–	–	–	–	–	–	–	I	–	–	–	–	–	–	V	–	IV	IV	II
II	II	–	I	–	I	I	–	–	I	I	I	II	I	–	IV	III	I	–
–	–	–	–	–	–	–	–	–	I	–	–	–	I	I	–	III	II	I
V	II	–	–	IV	II	–	I	–	V	III	–	I	I	–	IV	II	I	II
–	–	IV	I	–	–	I	–	–	I	–	III	–	II	I	V	V	I	–
–	–	–	I	–	–	I	–	–	I	–	–	–	II	–	–	–	–	I
III	II	–	–	II	–	–	–	–	I	–	–	–	I	I	–	II	I	–
I	I	–	–	–	–	–	–	–	I	I	–	–	I	–	V	–	I	–
–	–	–	–	–	–	–	II	–	–	–	–	I	–	V	–	–	–	V
I	–	–	–	I	–	I	–	–	I	II	I	–	II	III	–	–	–	–
IV	–	–	–	–	–	–	–	I	I	V	–	–	II	–	–	–	I	I
II	–	–	–	–	–	–	–	–	–	I	–	–	–	–	I	–	–	II
V	V	V	II	I	–	–	V	–	V	V	V	V	IV	V	III	V	V	V
IV	III	II	–	–	III	II	II	V	V	–	I	II	–	II	–	–	–	–
–	I	–	–	–	–	–	III	I	–	–	–	II	–	II	–	I	–	–
–	–	–	–	–	–	–	I	–	–	–	–	IV	–	V	–	IV	I	–
II	III	I	I[2)	II	III	II	II	II	V	V	IV	IV	III[2)	V	V	V	V	IV
III	IV	–	IV	IV	III	I	III	II	IV	III	I	IV	V	V	III	IV	IV	II
V	III	V	IV	V	V	V	V	V	III	II	III	V	III	IV	V	V	IV	–
I	II	III	I	I	I	–	II	II	II	III	IV	II	II	III	III	III	–	I
V	III	II	I	II	II	–	V	–	I	III	III	IV	IV	V	–	V	IV	V
I	–	I	–	I	I	I	I	I	II	II	II	IV	II	II	–	III	I	I
–	III	–	III	–	II	IV	–	II	V	V	I	IV	III	I	–	II	V	V
–	–	III	I	–	I	II	II	II	II	–	I	II	IV	II	–	II	II	IV
III	–	–	III	–	–	–	–	I	II	III	–	II	II	I	–	II	II	–
I	I	–	–	–	I	I	–	I	I	–	I	II	I	I	III	–	IV	III
I	III	–	·	IV	II	–	–	–	V	V	–	II	–	IV	–	III	I	–
–	–	–	–	I	II	III	II	–	–	–	IV	III	II	–	–	II	–	I
–	–	–	–	I	–	–	–	II	I	I	–	II	–	–	–	II	I	I
–	–	–	–	–	I	–	–	–	–	II	–	II	I	I	III	–	–	–
I	–	–	–	–	–	–	I	–	–	–	–	I	I	–	–	–	–	–
II	I	V	II	–	II	–	II	II	I	–	–	II	–	–	II	–	–	–
II	–	II	II	–	II	–	II	–	–	–	–	–	III	II	–	IV	–	–
–	–	I	–	–	I	–	I	–	–	–	I	II	–	–	IV	–	–	–

Table 5. Continued.

Author and year	F. Celiński, 1962	F. K. Hartmann, 1933	Scamoni, 1960	Passarge, 1960	Scamoni, 1967	Passarge & Hofmann, 1968	Passarge & Hofmann, 1968	Markgraf, 1932	Rühl, 1957	Tüxen, 1937	Tüxen, 1937	Tüxen, 1931
Number of relevés	9	30	289	46	52	108	29	6	9	17	15	5
No.	1	2	3	4	5	6	7	8	9	10	11	12
Querco–Fagetea character-species												
Milium effusum	V	V	V	V	III	V	V	IV	V	I	IV	II
Anemone nemorosa	III	II	V	V	II	V	III	V	II	III	II	IV
Poa nemoralis	–	III	V	V	V	IV	V	–	–	IV	II	IV
Mycelis muralis	II	III	V	III	V	IV	IV	–	II	IV	II	I
Moehringia trinervia	I	I	II	I	II	III	IV	–	–	III	II	–
Phyteuma spicatum	–	I	III	II	–	–	–	–	–	I	–	I
Brachypodium sylvaticum	II	II	III	–	III	X	X	–	–	I	II	–
Carex digitata	I	I	IV	–	III	I	III	–	–	–	–	–
Convallaria majalis	–	–	II	–	I	–	–	–	I	–	–	I
Differential species												
1. of units on very nutritious soils, mostly on limestone												
Mercurialis perennis	II	I	IV	I	–	X	–	–	III	III	–	IV
Lathyrus vernus	–	–	III	–	–	–	–	–	–	I	–	IV
Hepatica nobilis	–	I	IV	–	–	–	–	–	–	–	–	II
2. of units on moist sites												
Athyrium filix-femina	III	IV	V	II	II	X	X	II	III	V	V	III
Stachys sylvatica	IV	II	V	–	I	X	X	–	II	III	IV	–
Geranium robertianum	III	IV	V	I	I	X	X	–	–	III	II	I
Urtica dioica	IV	IV	V	II	II	X	X	–	I	IV	III	–
Circaea lutetiana	IV	II	V	II	–	X	X	II	II	I	IV	–
Deschampsia caespitosa	II	II	V	V	I	X	X	–	I	–	–	III
Carex remota	II	II	III	I	I	X	X	–	–	II	IV	–
Festuca gigantea	II	II	IV	–	II	X	X	I	I	I	II	I
Impatiens noli-tangere	III	II	III	–	–	–	–	–	I	III	II	–
Primula elatior	–	I	III	–	–	–	–	–	–	–	I	II
Gymnocarpium dryopteris	–	I	I	–	I	–	–	–	–	V	V	–
Veronica montana	III	II	III	–	–	–	–	–	–	I	II	–
3. of the suballiance Luzulo–Fagenion Lohm. et Tx 54												
Luzula albida	O	O	O	O	O	O	O	O	O	IV	II	–
Polytrichum attenuatum	III	I	III	I	I	X	X	–	–	–	II	–
Deschampsia flexuosa	–	–	III	–	II	X	X	–	–	–	–	I
Calamagrostis arundinacea	–	–	II	–	I	–	–	–	–	II	–	–
4. of different climatic regions (geographic, altitudinal levels)												
Hedera helix (atl.)	–	II	IV	I	–	–	–	–	–	–	II	IV
Ilex aquifolium (atl.)	–	–	–	–	–	–	–	–	II	–	–	–
Polygonatum verticillatum (mont.)	–	–	–	–	–	–	–	–	–	II	–	I
Prenanthes purpurea (praealpin)	O	O	O	O	O	O	O	O	O	–	–	–
Companions												
Oxalis acetosella	V	V	V	V	V	V	V	V	V	V	V	II
Vicia sepium	–	II	IV	I	I	II	–	–	II	–	–	III
Dryopteris carthusiana	II	–	III	II	II	I	I	–	I	II	IV	I
Luzula pilosa	I	II	IV	II	II	IV	IV	–	–	–	II	III
Majanthemum bifolium	II	I	V	I	I	II	II	II	II	–	–	–

Rödel, 1970	Eichner, 1976	Bornkamm & Eber, 1967	Blosat & Schmidt, 1975	Tüxen, 1954	Bauer, 1978	Bauer, 1978	Grüneberg & Schlüter, 1957	Seibert, 1954	Trautmann, 1957	Lohmeyer, 1967	Jahn, 1972	F. K. Hartmann, 1953	Déthioux, 1969	Zeidler, 1953	Knapp & Ackermann, 1952	Oberdorfer, 1957	Müller, 1977	Mayer, 1964
12	222	6	16	5	39	7	11	13	36	13	17	9	9	10	16	14	13	11
14	15	16	17	18	19	20	21	22	23	24	25	26	27	28	29	30	31	32
III	III	II	II	II	II	III	V	III	I	III	V	V	V	III	–	V	III	II
–	–	II	V	V	II	–	–	IV	II	V	I	II	IV	I	III	II	V	I
I	–	IV	I	II	II	III	II	V	I	III	V	III	III	II	II	II	–	I
–	–	V	I	III	I	–	I	–	I	–	II	IV	–	V	–	IV	–	I
–	–	V	II	–	–	I	II	–	I	–	I	II	II	–	–	III	I	–
–	–	–	I	I	II	–	I	–	I	–	–	–	–	–	–	I	IV	–
–	–	–	–	–	–	–	I	II	–	IV	–	II	I	III	V	II	–	V
–	–	–	II	–	–	I	–	II	–	–	–	I	I	–	V	–	–	II
V	III	–	–	IV	III	I	–	II	II	I	I	IV	–	IV	–	I	I	I
I	–	–	–	–	–	–	I	II	–	–	–	–	–	I	V	–	–	–
I	–	–	–	–	–	–	I	–	I	–	–	–	–	–	–	–	–	–
–	–	III	V	I	I	I	III	–	III	IV	I	III	IV	III	–	I	II	V
III	II	–	I	II	III	–	II	–	II	IV	–	I	II	I	–	–	I	III
I	–	–	–	III	II	–	II	–	II	III	I	III	III	I	–	II	I	III
–	–	–	I	IV	II	–	II	II	I	–	I	II	–	I	–	–	–	I
–	II	–	I	I	III	–	–	–	II	V	I	–	II	I	–	–	II	IV
–	–	–	II	–	I	III	–	–	I	II	–	I	IV	–	–	–	–	III
–	–	III	I	–	–	–	–	I	I	I	–	I	II	–	–	–	–	II
III	–	–	–	I	–	–	–	–	I	–	I	II	–	II	–	–	–	III
I	I	–	–	–	–	–	–	–	I	III	–	–	–	–	–	–	II	–
–	–	–	III	–	–	–	III	–	–	–	–	I	–	–	–	–	–	–
–	–	–	–	I	–	–	–	–	I	I	–	–	–	–	–	–	–	–
–	–	V	III	–	I	IV	–	IV	I	–	V	II	III	I	–	III	–	II
–	–	–	–	–	–	V	II	–	I	–	I	–	II	–	–	–	–	III
–	–	–	–	–	–	II	–	II	–	–	I	I	I	III	–	–	–	–
–	–	–	–	–	–	II	II	–	–	–	I	–	–	I	–	–	–	–
I	I	I	I	–	I	I	–	–	I	V	–	I	I	–	V	V	V	IV
–	–	–	–	–	–	–	–	–	–	I	–	–	–	–	–	IV	V	–
–	–	–	–	–	–	–	–	–	–	–	III	I	I	III	–	–	–	–
–	–	–	–	–	–	–	–	–	–	–	–	–	–	V	V	V	I	–
I	–	–	V	IV	III	III	V	II	IV	IV	V	III	III	V	–	II	III	V
I	–	II	–	III	–	–	I	V	I	II	I	II	II	–	III	I	–	III
–	–	I	III	–	–	I	II	–	I	–	I	–	II	III	–	II	–	IV
–	–	–	II	–	I	II	–	–	–	I	–	I	–	III	–	I	II	III
–	–	–	II	–	–	–	–	–	–	I	–	–	–	II	–	–	–	I

Table 6. Examples of how chorological phenomena are considered in the phytosociological system, Class: *Vaccinio–Piceetea* Br.-Bl. 39. (The main criteria for the classification are added to the vegetation units. For further explanations see text.)

Authors	Order	Vicariant order	Alliance	Suballiance	Association group	Association	Level of chorological classification
Braun-Blanquet, Sissingh & Vlieger 39	*Vaccinio–Piceetalia* Br.-Bl. 39	*Vaccinio–Piceetalia (Gaultheri–Piceetalia)* regional	*Vaccinio–Piceion* Br.-Bl. 38	*Abieti–Piceion* *Rhododoreto–Vaccinion* *Piceion septentrionale* *Phyllodoco–Vaccinion* synecological, regional		several	suballiance
Oberdorfer 57	*Vaccinio–Piceetalia Br.-Bl. 39*	*Vaccinio–Piceetalia Br.-Bl. 39*	*Vaccinio–Piceion* Br.-Bl. 38	*Eu–Vaccinio–Piceion* Ob. 57 *Rhododendro–Vaccinion* Br.-Bl. 26 synecological	Spruce forests in the Hercynian-Mts. and northern Alps / Peat bogs with pines / synecological	*Bazzanio–Piceetum* Br.-Bl. 39 Black Forest / *Piceetum boreo–alpinum* Oberd. 50 Northern Alps / *Soldanello–Piceetum* Volk 39 Bohemian Forest / *Piceetum hercynicum* TX. 39 Harz / regional	association
Oberdorfer 79	*Vaccinio–Piceetalia Br.-Bl. 39*	*Vaccinio–Piceetalia Br.-Bl. 39*	*Dicrano–Pinion* Libb. 33 *Vaccinio–Piceion* Br.-Bl. 38 *Piceion septentrionale* Br.-Bl. 39 (= *Linnaeae–Piceion*) regional pr. p. synecol.	*Eu–Vaccinio–Piceion*	synecological	*Homogyno–Piceetum* H. May 74 / *Calamagrostio villosae–Piceetum* Hartm. 53 / *Bazzanio–Piceetum* Br.-Bl. 39 / *Asplenio–Piceetum* Koch 54 / regional	alliance pr. p. association

the vegetation units. For further explanations see (ext.)

Authors	Order	Regional order Suborder	Alliance	Regional alliance	Suballiance	Association group	Associations vicariant associations	Regional associations, regional variants, races	Level of chorological classification
Passarge & Hofmann 68	*Aegopodio Fagetalia* synecological	*Mercuriali- Fagetalia Bromo Carp. bet.* altitudinal	*Fraxino-Fagion Mercuriali-Fag. Sorbo-Fagion* synecological	*Eu-Frax.-Fagion Impatiento-Fr.- Fagion* altitudinal			*Fraxino-Fagetum Impatienti-Fagetum* synecological	several 3 'regionals' regional	each, if practicable order, alliance (and others)
Doing Kraft & Westhoff 58	*Fagetalia sylvaticae* Pawl. 28	*Aceri-Fagetalia Querco-Fagetalia* altitudinal	*Querco-Fagion Carpinion* altitudinal						
Soó 64	*Fagetalia sylvaticae* Pawl. 28	*(Fagetalia silv.) (pr. p. Fagetalia orientalis)*	*Fagion sylvaticae* Pawl. 28	*Fagion medio-europaeum austro-italicum illyricum dacicum orientalis* regional	*Carpino-Fagion Lonicero-Fag. Ostryo-Fagion* synecological	Montane woodlands of beech and fir and 'gorge' woodlands	*Lamio orvolae- Fagetum Calamintha grandi- florae-Abieti-Fag.* synecological	*dinaricum hosniacum montenegrinum albanicum* regional	(order) alliance association
Knapp 42	*Fagetalia sylvaticae* Pawl. 28		*Fagion sylvaticae* Pawl. 28		main association *Melico-Fagetum* synecological	groups of regional associations regional	regional association	possible	regional association
Hartmann Jahn 67	*Fagetalia sylvaticae* Pawl. 28		*Fagion sylvaticae* Pawl. 28		*Eu-Fagion* synecological	Montane (fir) beech woodlands, rich in species Submontane *Melica* beechwoods synecological	*Dentario ennea- phyllidis-Fagetum Abieti-Fagetum Dentario bulbiferae- Fagetum* regional	Different regional variants: Harz Rhenish Schist Mountains Hessian Hills regional	vicariant association
Oberdorfer 57	*Fagetalia sylvaticae* Pawl. 28		*Fagion sylvaticae* Pawl. 28		*Eu Fagion* Ob. 57 synecological	Submontane lowland beech woodlands Submontane beech woodlands on limestone Montane fir beech woodlands rich in species on silicate soils	*Melico-Fagetum Cephalanthero- Fagetum Abieti-Fagetum* synecological	Black Forest-Vosges race Colline Main-Neckar race Frankonia Forest race 2 races regional	race of the association
Oberdorfer 57	*Fagetalia sylvaticae* Pawl. 28		*Carpinion* Issl. 31 emend. Ob. 53	*Pulmonario- Carpinion Galio-Carpinion Tilio-Carpinion* regional		Subatlantic oak hornbeam woodlands Subcontinental oak hornbeam woodlands regional	*Stellario-Carpinetum Galio-Carpinetum* synecological regional	Upperrhenish race Main-Neckar race Eastern Bavaria race Swiss foreland race regional	suballiance, race of the association
Neuhäusl 77	*Fagetalia sylvaticae* Pawl. 28		*Carpinion* Issl. 31 emend. Ob. 53				*Endymio-Carpinetum Rusco-Carpinetum Stellario-Carpinetum Galio-Carpinetum Melampyro- Carpinetum Primulo veris-Carp. Carici pilosae-Carp. Tilio-Carpinetum Physospermo-Carp. Querco petraeae- Carpinetum illyricum* regional		vicariant association
Traczyk 62	*Fagetalia sylvaticae* Pawl. 28		*Carpinion* Issl. 31 emend. Ob. 53				*Galio-Carpinetum* Ob. 57 *Tilio-Carpinetum* Tracz. 62 regional	1 race 3 races regional	vicariant association

Chorological and ecological phenomena in the differentiation and distribution of the *Fagion* associations in Bohemia and Moravia (Czechoslovakia)*

J. Moravec
Botanical Institute, Czechoslovak Academy of Sciences, 252 43 Průhonice, Czechoslovakia

Keywords: Bohemia, Chorological and ecological causes, Distribution, *Fagion* associations, Floristic differentiation, Moravia, Syntaxonomy, Vicarious associations

Abstract

The 10 *Fagion* associations bound to the western part of Czechoslovakia are conditioned both ecologically and chorologically in their species differentiation as well as geographical distribution. The division of the alliance into suballiances follows primarily the ecological phenomena. Within the *Eu-Fagenion* the associations form 3 groups conditioned by ecological factors.

The *Tilio platyphylli–Fagetum, Tilio cordatae–Fagetum, Melico–Fagetum* and *Carici pilosae–Fagetum* represent associations of the submontane belt. The *Tilio platyphylli–Fagetum* is characterized by overlapping of *Fagion* and *Carpinion* species and by the absence of any *Dentaria* species due to chorological causes. The *Tilio cordatae–Fagetum* does not show any chorological phenomena in the species composition, however, it is limited to central, southern and western Bohemia only. The *Melico–Fagetum* is characterized by dominant *Melica uniflora* which is absent in the western and southern part of Bohemia. The *Melico–Fagetum* has even a more limited distribution occurring in northern and eastern Bohemia and in northern and central Moravia only. The *Carici pilosae–Fagetum* is characterized by *Carex pilosa* (dominant), *Cephalanthera longifolia* and *Euphorbia amygdaloides;* it is confined to the Carpathian province.

The *Dentario enneaphylli–Fagetum, Dentario glandulosae–Fagetum* and *Violo reichenbachianae–Fagetum* represent a group of vicarious associations of the montane belt. The *Dentario enneaphylli–Fagetum* occurs mainly in the geographical province Česká vysočina. Its eastern limit lies in the westernmost part of the Carpathian province where it forms a special subassociation and contacts the *Dentario glandulosae–Fagetum*. The latter association is characterized primarily by *Dentaria glandulosa*, a Carpathian endemic species. The *Violo reichenbachianae–Fagetum* is conditioned chorologically by the absence of any *Dentaria* species; it occurs in the mountains Krušné hory and Doupovské hory only.

The *Festuco–Fagetum* is the single representant of the third group conditioned mainly by ecological factors.

Introduction

The species composition of plant communities which expresses their individuality is considered to originate by the action of one or two selection mechanisms (cf. Burrichter, 1964): (1) the first selection is based on ecological factors of the habitat and initiates species assemblages; (2) the second (and final) selection is based on interactions between populations of species and produces real cenoses.

Apart from these general selection mechanisms changes in the flora itself can be mentioned: (1) evolution of species – a slow process passing through geological eras; (2) migration of species by

* Nomenclature of species follows Rothmaler *et al.* (1970).

Vegetatio 59, 39–45 (1985).

active spreading of propagules and ecesis – a process lasting hundreds of years for indigeneous flora, and decennia for neophytes.

It would be incorrect to believe that the flora of a region develops by evolution and/or migration of species as a first step followed, as a second step, by ranging of species into plant communities by means of the above selection mechanisms acting like one or two sieves. In fact, evolution of species takes place within plant communities and migration of species passes through them or with them (synmigration – cf. Moravec, 1969: p. 145).

In individual plant communities the species composition reflects both the ecological and chorological causes of its origin, the first being traced in detailed regional studies, the second in large syntaxonomic syntheses. The syntaxonomic evaluation of the first or second phenomenon with regard to the floristic composition of plant communities is not yet stabilized, and two approaches can be traced in the literature: (1) distinguishing syntaxa of different status – those conditioned by ecological factors and those conditioned by chorological phenomena – resulting in a two- or even poly-dimensional system; (2) keeping the linear hierarchical system of syntaxa and evaluating merely the 'amount' of floristic difference irrespective of its ecological or chorological causes.

The advantages and disadvantages of these two approaches were discussed earlier (Moravec, 1975, 1981). The following contribution shows that if we follow the second approach, the ecological and chorological phenomena in the floristic composition of herb-rich beech forest communities of the *Fagion* Luquet 26 in the Czech Socialist Republic (i.e. Bohemia and Moravia) alternate at different levels of syntaxonomic importance and they can be interpreted within a linear hierarchical system of syntaxa.

Common features of the *Fagion* associations

The herb-rich beech forests of the Czech Socialist Republic have been classified into 10 associations (cf. Moravec *et al.*, 1982). The differences in their species composition which support this classification are conditioned both ecologically and chorologically. The floristic differences caused by ecological factors lead to delimitation of syntaxa occupying different habitats. In a limited territory this phenomenon is well visible in the ecological pattern of plant communities showing usually clear boundaries or narrow ecotones. On the contrary, differences of syntaxa caused by species of similar ecological behaviour but of different distribution result in distinguishing vicarious syntaxa the limits of which are not so clearly visible unless they are separated by geographical gaps.

The 10 *Fagion* associations under consideration have a stock of species which are common to all of them, with a fairly high constancy. These species are mainly character- and differential species of the *Fagetalia*.

Floristic differences of the suballiances

The *Fagion* associations of Bohemia and Moravia are divided into three suballiances (cf. Moravec *et al.*, 1982) differing in ecological conditions.

The *Eu-Fagenion* Oberdorfer 57 em. Tüxen in Tüxen et Oberdorfer 58 has no diagnostic species of its own. It groups 8 associations occupying the silicate soils of the Central European 'Braunerde' in the submontane and montane belt. The representation of the diagnostic species of the *Fagion* reaches its maximum values in the montane associations (*Dentario enneaphylli–Fagetum, Dentario glandulosae–Fagetum* and *Violo reichenbachianae–Fagetum*).

The *Acerenion* Oberdorfer 57 em. Husová 82 is characterized by differential species having their origin in the subalpine tall forb communities of the *Mulgedio–Aconitetea* Hadač et Klika in Klika et Hadač 44; these are *Cicerbita alpina, Ranunculus platanifolius, Circaea alpina, Athyrium distentifolium, Rumex arifolius, Streptopus amplexifolius* and *Adenostyles alliariae*. Therefore this suballiance occurs and shows the best development in the mountains with a fairly large subalpine belt. Thus, though the particular species composition of this suballiance is conditioned primarily by ecological factors (higher humidity, lower temperatures, lower evaporation, deep snow cover of prolonged duration) a geographically limited distribution of this suballiance can be observed:

The *Cephalanthero–Fagenion* Tüxen in Tüxen et Oberdorfer 58 is conditioned primarily by edaphic factors. It is bound to carbonate rocks where it occupies the more developed soils of the rendsina series (from the moder rendsina to the terra fusca). Therefore the differential species of this suballiance comprise calciphilous and/or neutrophilous species and also some subthermophilous ones: *Corallorhiza trifida, Orthilia secunda, Campanula rapunculoides, Epipactis helleborine, Neottia nidusavis, Cornus sanguinea* and *Daphne mezereum.*

Floristic differentiation of the associations and its ecological and chorological interpretation

The associations of the *Eu-Fagenion* can be grouped in three groups conditioned primarily by ecological factors (cf. Moravec, 1983) but without a great syntaxonomic importance:

a. Associations of the submontane belt, concentrated in the altitudinal range of 400–600 m – *Tilio platyphylli-Fagetum, Tilio cordatae-Fagetum, Melico-Fagetum* and *Carici pilosae-Fagetum.*

b. Associations of the montane belt with local occurrence in the submontane belt concentrated in the altitudinal range of 500–1000 m – *Dentario enneaphylli-Fagetum, Dentario glandulosae-Fagetum* and *Violo reichenbachianae-Fagetum.*

c. Association of poorer or depauperate soils with a deep moder layer; it is scattered in small stands within the associations of the other two groups, but only in the upper part of the submontane and in the montane belt – *Festuco-Fagetum.*

The floristic differentiation of the associations of the first two groups as well as their geographical distribution is associated with chorological phenomena.

The *Tilio platyphylli-Fagetum* Klika 39 is characterized positively by overlapping of the diagnostic species group of the *Fagion* with a species group characterizing the *Carpinion* associations. In addition there is a negative character viz. the absence of any *Dentaria* species. This has a chorological cause. The absence concerns not only the rather small distribution area of the *Tilio platyphylli-Fagetum* in the western part of the České středohoří Mts. where this association occurs (Fig. 1) but the whole northwestern quadrant of Bohemia (cf. also the *Violo reichenbachianae-Fagetum*).

The *Tilio cordatae-Fagetum* Mráz 60 em. Moravec 77 has an area limited to central, southern and western Bohemia (Fig. 1). However, it does not show any chorological phenomena in the species composition. The diagnostic species – *Bromus benekenii, Vicia sylvatica, Campanula trachelium* and *Cardamine impatiens* are distributed widely outside the area of this association.

The *Melico-Fagetum* Seibert 54 is characterized by dominant *Melica uniflora* and a relatively low average number of species per stand. *Melica uniflora* shows a gap in the distribution in the western and southern part of Bohemia (cf. Moravec, 1971). The distribution of the *Melico-Fagetum* in Bohemia and Moravia is (and also has been in the past) even more limited than the distribution of *Melica uniflora*. Unfortunately, it is not possible to estimate the entire original distribution because most of the stands were cut and replaced by wood plantations of other trees. Originally, the *Melico-Fagetum* occupied mainly the mature, deep, meso- to oligotrophic Central European 'Braunerde' on flat relief and gentle slopes. The *Melico-Fagetum* is found in northern Bohemia, an isolated occurrence is known from eastern Bohemia. In Moravia this association originally formed the climax vegetation on the plateaus of the highlands Drahanská vrchovina, Nízký Jeseník and Oderské vrchy (Fig. 1).

The *Carici pilosae-Fagetum* Oberdorfer 57 is characterized by *Carex pilosa* (dominant), *Cephalanthera longifolia* and *Euphorbia (Tithymalus) amygdaloides* which form its diagnostic species. Within Czechoslovakia, *Carex pilosa* is distributed mainly in the Carpathians. Through northern and central Moravia it penetrates into eastern Bohemia, however, as a companion of the *Carpinion* communities. An isolated occurrence is also known from central Bohemia in the *Carpinion*, too. *Euphorbia amygdaloides* has a similar distribution within Czechoslovakia. Both species are also present in the *Carpinion* communities constituting the *Carici pilosae-Carpinetum* Neuhäusl R. et Z. 64. In fact, neither *Carex pilosa* nor *Euphorbia amygdaloides* can be regarded as 'Carpathian' species as they occur in some countries lying west of Czechoslovakia. However, their gaps in Bohemia should not be underestimated. Within the study area, the *Carici pilosae-Fagetum* is limited to the Carpathian province except the southeastern part of the highland Drahanská vrchovina where it penetrates

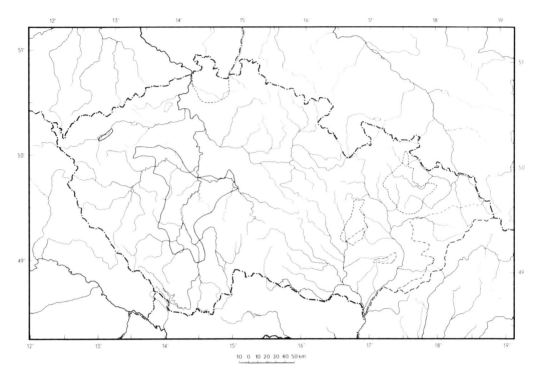

Fig. 1. Limits of distribution of the following herb-rich beech forest associations in the Czech Socialist Republic: .... *Tilio platyphyl-li–Fagetum*, —— *Tilio cordatae–Fagetum*, ---- *Melico–Fagetum*, and –.–.–.– *Carici pilosae–Fagetum*.

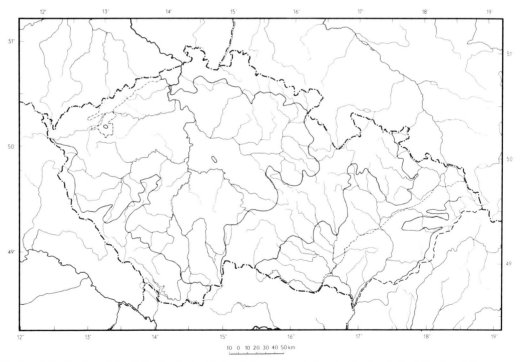

Fig. 2. Limits of distribution of the following herb-rich beech forest associations in the Czech Socialist Republic: —— *Dentario ennephylli–Fagetum*, ... *Dentario glandulosae–Fagetum*, and –.–.–.– *Violo reichenbachianae–Fagetum*, – – – – western limit of the Carpathian province.

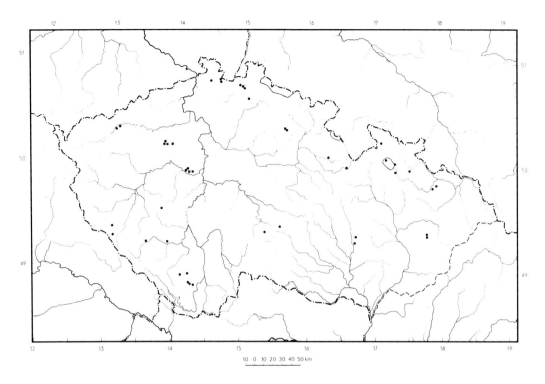

Fig. 3. Distribution of the following herb-rich beech forest associations in the Czech Socialist Republic: —— *Aceri–Fagetum,* ● *Festuco–Fagetum,* ■ *Cephalanthero–Fagetum.*

into the geographical province Česká vysočina (Czech highlands – Fig. 1).

The *Dentario enneaphylli–Fagetum* Oberdorfer ex W. et A. Matuszkiewicz 60 is characterized primarily by *Dentaria enneaphyllos* which can be regarded as its regional differential species. This species occurs in other associations in the Alps and Illyric mountains. However, in the territory lying north of the Alps it is concentrated in the *Dentario enneaphylli–Fagetum* from where it penetrates with a lower constancy into the *Tilio cordatae-Fagetum, Melico–Fagetum* and into some associations of the *Tilio–Acerion* (cf. Moravec *et al.*, 1982: Table 1). *Dentaria enneaphyllos* occurs with a rather high constancy in the Westcarpathian form of the *Dentario glandulosae–Fagetum* and, in fact, it cannot serve as a diagnostic species for delimitation of these two vicarious associations in the field. The *Dentario enneaphylli–Fagetum* is distributed primarily in the province Česká vysočina (Fig. 2) including the adjacent parts of the border mountains in the Federal Republic of Germany (Bayrischer Wald and Böhmerwald), Austria (Böhmerwald and

Freiwald), the German Democratic Republic (Lausitzer Gebirge) and Poland (Krkonoše and other mountains of the Sudeten system). The herb-rich montane beech forests of the westernmost part of the Carpathians have been classified within the *Dentario enneaphylli–Fagetum* by the author as a particular subassociation – *Dentario enneaphylli–Fagetum salvietosum glutinosae* Moravec 74.

Lower syntaxa – subassociations, variants and subvariants – have been distinguished within the *Dentario enneaphylli–Fagetum* (Moravec, 1974). Ecological and chorological phenomena alternate as causes of the floristic differentiation of these syntaxa.

The differentiation of the subassociations *Dentario enneaphylli–Fagetum typicum* and *impatientetosum,* occurring in the geographical province Česká vysočina, is conditioned by ecological factors. The *Dentario enneaphylli–Fagetum impatientetosum* is restricted to higher altitudes (700–1000 m); the *Dentario enneaphylli–Fagetum typicum* occurs mainly at altitudes of 400–800 m.

The *Dentario enneaphylli–Fagetum salvietosum glutinosae* is characterized by the differential species *Salvia glutinosa, Euphorbia (Tithymalus) amygdaloides, Isopyrum thalictroides* and *Glechoma hederaceum* ssp. *hirsutum*, none of which, however can be considered a Carpathian species though they represent a Carpathian migration stream into Bohemia and Moravia. The subassociation occurs in the Moravian Carpathians, in the mountains Moravskoslezské Beskydy, Hostýnské vrchy, Vsetínské vrchy, Javorníky and Bílé Karpaty (cf. Moravec, 1974: p. 124, Fig. 1). In the Moravskoslezské Beskydy and Javorníky Mts. it comes into contact with the *Dentario glandulosae–Fagetum* which is vicarious in the Carpathian mountains lying eastwards. The differential species of the *Dentario enneaphylli–Fagetum salvietosum* occur also in the *Dentario glandulosae–Fagetum* and some of them can be considered differential species of the latter as well. Thus, a serious objection can be raised against the above classification. The ranging of these beech forests into the *Dentario glandulosae–Fagetum* could seem more logical because in that case the limit of distribution of this association would coincide with the geographical limit of the Carpathian province. However, the absence of *Dentaria glandulosa* as well as the presence of *Festuca altissima* (though with reduced constancy) and of *Hordelymus europaeus* (penetrating into the western part of the area of the *Dentario glandulosae–Fagetum* but lacking in the typical stands in the eastern Carpathians – cf. W. et A. Matuszkiewicz, 1973) have been considered decisive for the above classification. This opinion is also supported by the fact that the differential species of the *Dentario enneaphylli–Fagetum salvietosum* are not true Carpathian species.

The *Dentario glandulosae–Fagetum* Matuszkiewicz 64 sec. auct. is characterized primarily by *Dentaria glandulosa*, an endemic Carpathian species, further by *Symphytum cordatum* (which does not reach Bohemia and Moravia) and *Salvia glutinosa*. The western limit of the distribution area of this association has been estimated according to the presence of *Dentaria glandulosa*. It runs through the mountains Moravskoslezké Beskydy and Javorníky, where this association penetrates to the catchment area of the Olše and of the upper Odra and to the spring area of the Vsetínská Bečva, respectively (Fig. 2). Recently isolated stands have been found in the Vsetínské vrchy Mts. by M. Sed-

láčková (in litt.). *Dentaria glandulosa*, however, occurs also in the Moravian Carpathians west of the limit of the *Dentario glandulosae–Fagetum*, but here it is confined to alluvial forests and does not enter the beech forests.

The *Dentario glandulosae–Fagetum* represents a vicarious Carpathian association replacing the *Dentario eneaphylli–Fagetum* in the montane belt of the Carpathians except the westernmost part.

The *Violo reichenbachianae–Fagetum* Moravec 79 is the third association of the montane herb-rich beech forests replacing the *Dentario enneaphylli–Fagetum* in the mountains Krušné hory and Doupovské hory (except the summit area of the latter). This association differs from the related associations mainly in a negative sense and the differentiation is conditioned by chorological phenomena, primarily by the absence of any *Dentaria* species that characterize the *Dentario enneaphylli–Fagetum* and the *Tilio cordatae–Fagetum*. *Melica uniflora* is lacking too, and *Festuca altissima* has only a low presence degree. An other chorological phenomenon is the substitution of *Lamium galeobdolon* ssp. *montanum* (= *Galeobdolon montanum*) occurring with a high presence degree in the *Dentario enneaphylli–Fagetum* by *Lamium galeobdolon* ssp. *galeobdolon* (= *Galeobdolon luteum*) which is typical of the submontane associations *Tilio platyphyllae–Fagetum* and *Tilio cordatae–Fagetum*, within the *Dentario enneaphylli–Fagetum typicum* it characterizes a form (Moravec, 1974: p. 120) occurring in the territories adjacent to the Krušné hory Mts. (Lužické hory and České středohoří Mts.) and in the summit area of the Doupovské hory Mts. In the *Violo–Fagetum, Hordelymus europaeus* reaches the highest constancy among the *Fagion* associations of Bohemia and Moravia (cf. Moravec *et al.*, 1982: Table 1), and becomes dominant in mature stands on deep soils. This phenomenon indicates a local optimum of this species which could be evaluated as a character species of weak fidelity and of local importance (Schwerpunktcharakterart).

The *Festuco–Fagetum* Schlüter in Grüneberg et Schlüter 57 is characterized by the dominance of *Festuca altissima* and a lower species number per stand. It occurs in relatively small stands in special sites with poor or impoverished soils and a deeper moder layer. These sites are found on summits or slope edges with shallow soil. The *Festuco–Fagetum* forms islands within the *Tilio cordatae–Fagetum*, the *Melico–Fagetum*, the *Dentario enneaphylli–Fagetum* and the *Violo–Fagetum*. The area of this association is not well known so far. In Czechoslovakia it seems to be confined to the geographical province Česká vysočina (Czech highlands), but it penetrates into the westermost part of the Carpathians (Fig. 3) as far as the mountains are built of hard rock (flysh conglomerates, coarse-grained sandstones and mesozoic sandstones) following the *Dentario enneaphylli–Fagetum*. The *Festuco–Fagetum* has been recorded from the Hostýnské vrchy Mts. and from the Moravskoslezské Beskydy Mts. So far this association is not known to the author from the Slovak Carpathians. An association similar in physiognomy but dominated by *Festuca drymeja* seems to be vicarious to the *Festuco–Fagetum* in the eastern part of Slovakia.

The *Aceri–Fagetum* J. et M. Bartsch 40 is the only representative of the *Acerenion* in the study area. The floristic differentiation as well as the occurrence is conditioned primarily by ecological factors as stated above. Here the diagnostic species of this association are identical with those of the suballiance. The *Aceri–Fagetum* is a local forest community of the supramontane belt descending sometimes into the montane belt. In Bohemia and Moravia it is rather rare and limited to the high mountains of the Sudeten system; it occurs in the mountains Krkonoše, Hrubý Jeseník, Králický Sněžník and Rychlebské hory (Fig. 3).

The *Cephalanthero–Fagetum* Oberdorfer 57 is bound to carbonate substrata which determine both its species composition and its distribution area. The distribution of this association in the study area (Fig. 3) is determined by the occurrence of carbonate substrata. It occurs in the promontory of the Šumava Mts. (Šumavské podhůří – rarely), in the hilly lands Karlštejnská pahorkatina and Džbán, in the Ještědský hřbet Mts., in isolated stands on the chalk plateau Česká křidová tabule from eastern Bohemia to western Moravia and (rarely) in the carst Moravský kras.

References

Burrichter, E., 1964. Wesen und Grundlagen der Pflanzengesellschaften. Abhandl. Landesmus. Naturkunde Münster Westfalen 26(3): 1–16.

Matuszkiewicz, W. & Matuszkiewicz, A., 1973. Przegl. fitosocjologiczny zbiorowisk leśnych Polski. Cz. 1. Lasy bukowe. (Phytosociological review of the forest communities of Poland. Part 1. Beech forests.) Phytocoenosis 2: 143–202.

Moravec, J., 1969. Succession of plant communities and soil development. Folia Geobot. Phytotax. 4: 133–164.

Moravec, J., 1971. Poznámky k výskytu strdivky jednokvěté – Melica uniflora Retz. – v dolním Posázaví. (Anmerkungen zum Vorkommen von Melica uniflora Retz. im Gebiet des Unterlaufes des Sázava-Flusses.) Zpr. Čs. Bot. Společ. 6: 185–187.

Moravec, J., 1974. Zusammensetzung und Verbreitung des Dentario enneaphylli–Fagetum in der Tschechoslowakei. Folia Geobot. Phytotax. 9: 113–152.

Moravec, J., 1975. Die Untereinheiten des Assoziation. Beitr. Naturk. Forsch. Südw. Deutschl. 34 (Oberdorfer-Festschr.): 225–232.

Moravec, J., 1981. Die Logik des pflanzensoziologischen Klassifikationssystems. Ber. Int. Sympos. IVV – Syntaxonomie, pp. 43–61.

Moravec, J., 1983. The ecological indication of the herb-rich beech forest associations in the Czech Socialist Republic (Czechoslovakia). In: W. Schmidt (ed.), Festschrift für Heinz Ellenberg. Verh. Ges. Ökol. Göttingen 11: 291–304.

Moravec, J., Husová, M., Neuhäusl, R. & Neuhäuslová-Novotná, Z., 1982. Die Assoziationen mesophiler und hygrophiler Laubwälder in der Tschechischen Sozialistischen Republik. Vegetace ČSSR A 12, Praha, 292 pp.

Rothmaler, W., 1970. Exkursionsflora von Deutschland. Kritischer Ergänzungsband. Gefässpflanzen. Volk und Wissen Volkseigener Verlag, Berlin. 622 pp.

Accepted 19.1.1984.

L'intérêt chorologique de quelques groupements forestiérs du Morvan, France*

J. C. Rameau

Laboratoire de Taxonomie Expérimentale et Phytosociologie, Faculté des Sciences, La Bouloie, 25030 Besançon Cedex, France, et Ecole Nationale du Génie Rural, des Eaux et des Forêts, Centre de Nancy, 14 Rue Girardet, 54042 Nancy Cedex, France

Keywords: *Carpinion betuli,* Chorological differentiation, Ecological differentiation, Forest, France, Morvan, *Quercetalia robori-petraeae*

Abstract

Because of its geographical position the Morvan region has certain peculiar climatological conditions which, despite the distance from the sea, permit the development of a characteristic Atlantic vegetation. Forest communities of *Quercion robori-petraeae* and *Carpinion betuli* are described and compared with the associations already distinguished for France. A chorological synthesis of the *Quercion robori-petraeae* and *Carpinion betuli* communities is made for a considerable part of France and the phytogeographical importance of the Morvan region is revealed.

Introduction

Le Morvan est un petit massif montagneux qui culmine à 900 m; il prolonge et termine au nord-est le Massif Central, au contact du Bassin Parisien. Il s'agit d'un ensemble cristallin d'âge hercynien formé de granites, rhyolites, gneiss et schistes, entouré de terrains sédimentaires: dépressions périphériques marneuses du Lias, calcaires du Jurassique parfois recouverts de chailles siliceuses, sables mio-pliocènes du Pays de Fours (Fig. 1a).

Le Morvan, malgré son altitude modeste constitue un relief assez marqué par rapport aux régions voisines; il répresente de ce fait le premier obstacle véritable aux vents d'ouest, situation à l'origine de précipitations élevées: 900 à 1 600 mm pour la façade ouest ou Haut-Morvan dont l'altitude varie entre 400 et 900 m; 800 à 900 mm pour le Bas-Morvan d'altitude comprise entre 200 et 400 m (au nord et au sud) atteignant plus rarement 600 m (à l'est, en contrebas du Haut-Morvan).

Les paysages régionaux sont dominés par les prairies au sein desquelles subsistent un bocage encore dense et par la forêt qui couvre une superficie importante. Les milieux forestiers se répartissent en trois groupes. Les forêts acidiphiles (Fig. 1b) largement répandues sur sol brun acide, sol lessivé acide, sol ocre podzolique, andosol... (*Quercion robori petraeae* et '*Luzulo–Fagenion*': Bugnon & Rameau, 1975; Rameau, 1981).

Les forêts mésoacidiphiles à neutrophiles (Fig. 1c) occupent des sols plus riches, situés en bas de versant ou sur des placages de limon dissimulant le substrat cristallin (*Carpinion betuli* et *Asperulo–Fagenion*). Enfin les fonds de vallon hébergent des forêts mésohygrophiles linéaires (*Stellario–Alnetum*: Estrade & Rameau, 1980) ou hygrophiles (*Alnion glutinosae*).

Les forêts acidiphiles

Elles constituent donc l'essentiel de la couverture forestière du Morvan. Leur composition floristique fournie par le tableau synthétique 1 est marquée par deux caractères: une certaine pauvreté et une rela-

* Nomenclature des espèces d'après Flora Europaea, par Tutin *et al.* (1964–1980); nomenclature syntaxonomique d'après Rameau (1981) et Oberdorfer (1978).

48

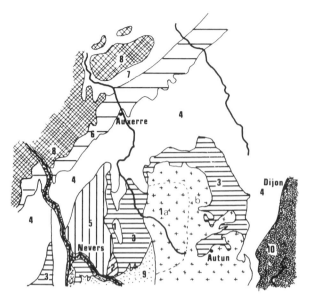

Fig. 1a. Données géologiques.
10 Alluvions anciennes et récentes; 9 Sables mio-pliocènes; 8 Argile à silex sur Crétacé; 7 Craie; 6 Terrains argilo-sableux (Crétacé inférieur); 5 Argile à chailles sur calcaires jurassiques; 4 Calcaires jurassiques; 3 Dépressions marneuses liasiques; 2 Grès triasiques; 1 Massif ancien (Morvan): a Haut-Morvan, b Bas-Morvan.

Fig. 1c. Les forêts du *Carpinion betuli.*
1 *Endymio–Carpinetum;* 2 *Poo chaixii–Carpinetum:* a race morvandelle, b race du Nivernais, c race de l'Auxois, d race de la plaine de Saône; 3 *Rusco–Carpinetum;* 4 *Stellario–Carpinetum;* 5 *Scillo–Carpinetum;* 6 *Lithospermo–Carpinetum.*

tive homogénéité de la flore (constance de *Deschampsia flexuosa, Pteridium aquilinum, Melampyrum pratense, Carex pilulifera, Holcus mollis, Vaccinium myrtillus, Teucrium scorodonia* et divers *Bryophytes*). Des variations floristiques peuvent être décelées cependant dans le tableau nous permettant d'individualiser quatre groupements végétaux principaux, dont la répartition est donnée par la Figure 1b.

Le premier caractérise le pays de Fours qui jouxte le bord ouest du Morvan (Fig. 1a: 9); l'altitude de ce terroir oscille entre 250 et 300 m; les précipitations y sont inférieures à 800 mm par an; le substrat est constitué par des sables miopliocènes. *Quercus petraea* domine la strate arborescente où *Fagus sylvatica* est moyennement représenté. Ce groupement forestier est différencié par la présence exclusive de deux taxa: *Sorbus torminalis* relativement fréquent et *Pyrus cordata* beaucoup plus rare (Tableau 1, col. 1).

Dans le Haut-Morvan (Fig. 1a:1a) deux groupements se succèdent en fonction de l'altitude. Entre 300 et 750 m une hêtraie-chênaie (Tableau 1, col. 2) couvre les sols bruns acides, les sols lessivés ou les sols ocre-podzoliques; sa physionomie est profondément marquée par l'abondance-dominance de *Ilex aquifolium*, et la présence de *Rhytidiadelphus loreus* et *Blechnum spicant.*

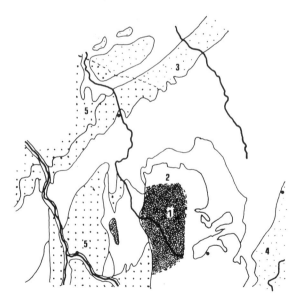

Fig. 1b. Les forêts du *Quercion robori-petraeae.*
1 *Fago–Quercetum* race atlantique morvandelle; 2 *Fago–Quercetum* race subatlantique morvandelle; 3 *Fago–Quercetum* race subatlantique de Champagne; 4 *Fago–Quercetum* race de la plaine de Saône; 5 *Peucedano–Quercetum.*

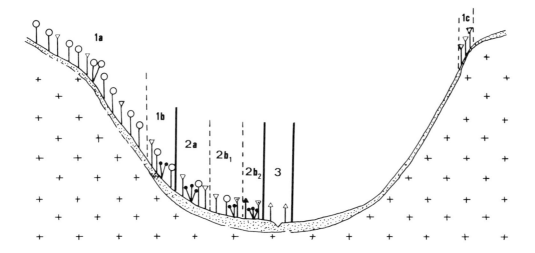

Fig. 2a. Transect théorique de la végétation forestière du Haut-Morvan.

1 *Fago-Quercetum* race atlantique: a sous-association typique, b sous-association *carpinetosum,* c chênaie sessiliflore à *Calluna vulgaris;* 2 *Endymio-Carpinetum:* a mésoacidiphile (*holcetosum*), b mésoneutrophile, 2b 1 variante mésophile, 2b 2 variante mésohygrophile;

3 *Stellario-Alnetum.*

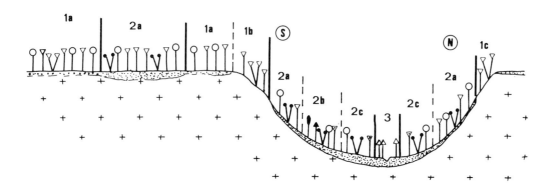

Fig. 2b. Transect théorique de la végétation forestière du Bas-Morvan.

1   *Fago-Quercetum* race subatlantique:

    1a sous-association typique,

    1b chênaie sessiliflore à *Polypodium vulgare,*

    1c chênaie sessiliflore à *Silene nutans;*

2   *Poo chaixii-Carpinetum:*

    2a sous-association mésoacidiphile,

    2b sous-association à *Polystichum setiferum,*

    2c sous-association mésoneutrophile variante mésohygrophile;

3   *Stellario-Alnetum.*

Fagus sylvatica ; Quercus petraea ; Quercus robur ; Carpinus betulus V ; Fraxinus excelsior ; Tilia cordata ; Alnus glutinosa .

Tableau 1. Tableau synthétique des forêts acidiphiles du Morvan.

Association	1					2	3	4					
Sous-association		a	b	c	d				a	b	c	d	e
Nombre de relevés	8	29	7	19	7	62	28	60	15	11	21	8	5
Altitudes	250-300 m		300 à 750 m			750 - 900 m	750 - 900 m				250 à 600 m	600 m	60
Arbres	%	%	%	%	%	%	%	%	%	%	%	%	%
Quercus petraea (Mat.) Liebl.	100	100	86	89	100	95	32 (+ à r)	55	100	100	90	62	100
Fagus sylvatica L.	100	100	100	89	—	85	100	52	100	45	48	12	—
Sorbus aucuparia L.	37	17	29	21	14	18	36	19	40	27	14	12	—
Carpinus betulus L.	—	—	43	95	14	30	—	34	—	9	90	12	—
Sorbus aria (L.) Crantz	62	24	43	16	—	18	32	11	7	27	5	12	20
Betula pendula Roth.	62	7	43	—	—	13	4 (r)	27	40	45	14	12	20
Abies alba Mill. (planté)	—	14	—	11	—	—	32	—	—	—	—	—	—
Picea abies (L.) Karsten (planté)	12	3	43	—	—	11	28	—	—	—	—	—	—
Castanea sativa Mill.	12	14	29	11	—	13	—	6	13	27	24	25	—
Quercus robur L.	87	14	—	—	—	—	11 (+)	11	—	9	—	—	—
Sorbus torminalis (L.) Crantz	87	—	—	—	—	—	—	—	—	—	—	—	—
Arbustes													
Ilex aquifolium L.	62	86	71	89	29	79 (+ à 4)	61 (+ à 3)	22	33 (+ à 2)	9	24	25	—
Cytisus scoparius (L.) Link.	62	3	43	11	57	16	4	32	7	27	24	100	60
Frangula alnus Mill.	62	31	43	—	29	19	4	18	20	45	52	12	20
Corylus avellana L.	—	—	—	11	14	11	—	27	13	9	—	—	—
Malus sylvestris Mill.	—	7	—	16	45	10	—	10	—	36	9	37	20
Juniperus communis L.	—	—	—	5	—	6	—	8	—	—	—	—	—
Mespilus germanica L.	12	3	—	5	—	r	—	r	—	18	—	—	—
Sorbus torminalis (L.) Crantz	87	—	71	89	—	—	—	—	—	—	—	—	—
Pyrus cordata DesV.	12	38	14	26	29	27	72	22	47	—	24	—	—
Ilex aquifolium L.	62	3	—	11	—	5	32	18	—	9	19	25	—
Rhytidiadelphus loreus Warnst.	—	—	—	—	—	—	11	—	—	—	—	—	—
Blechnum spicant (L.) Roth.	—	—	—	—	—	—	7	—	—	—	—	—	—
Dryopteris dilatata A. Gray	—	—	—	—	—	—	72	—	—	—	—	—	—
Sambucus racemosa L.	—	—	—	16	—	5	32	—	—	—	—	—	—
Rubus idaeus L.	—	—	—	5	—	3	11	—	—	—	—	—	—
Prenanthes purpurea L.	—	—	—	5	—	—	7	—	—	—	—	—	—

Caractéristiques et différentielles d'ordre (*Quercetalia roboni-petraeae* Tx31) et d'alliance (*Quercion roboni-petraeae* Br.Bl. 32)

	1	a	b	c	d	2	3	4	a	b	c	d	e
Deschampsia flexuosa (L.) Trin.	87	97	100	86	86	97	100	100	100	100	100	100	100
Pteridium aquilinum (L.) Kuhn.	75	86	100	79	29	92	36	52	87	64	38	25	20
Melampyrum pratense L.	37	28	43	16	14	24	14	43	40	55	43	37	40
Carex pilulifera L.	50	24	43	11	29	23	39	18	53	9	9	12	—
Holcus mollis L.	12	—	14	53	86	27	—	53	33	18	76	100	20
Vaccinium myrtillus L.	—	10	14	11	—	6	50	2	7	—	—	—	—
Teucrium scorodonia L.	12	3	43	—	86	19	—	23	7	36	11	100	80
Galium saxatile L.	—	10	43	5	—	10	25	5	7	—	9	—	—
Hieracium umbellatum L.	—	—	—	—	57	8	—	15	—	27	5	50	40
Calluna vulgaris (L.) Hull	37	—	—	—	43	5	—	16	—	36	5	50	40
Hieracium sabaudum L.	—	—	—	5	—	2	—	6	7	9	5	—	—
Agrostis tenuis Sibth.	—	3	14	5	—	—	—	3	13	—	5	—	—
Hypericum pulchrum L.	12	—	—	—	—	7	4	5	—	—	—	37	40
Molinia caerulea L. Moench.	50	—	—	—	—	—	8	—	—	—	—	—	—
Carpinus betulus L.	37	—	—	95	14	30	—	33	—	9	90	—	—
Stellaria holostea L.	—	—	—	32	43	14	—	27	—	9	57	37	—
Festuca heterophylla Lam.	—	—	—	—	—	—	4	6	—	—	19	12	—
Silene nutans L.	—	—	—	—	—	—	—	13	7	—	—	100	—
Polypodium vulgare L.	—	—	—	—	—	—	8	10	—	—	—	72	100

Tableau 1. (cont.).

Espèces acidiclines													
Dryopteris carthusiana (Vill.) H.P. Fuchs	–	10	14	26	–	14	68	20	9	14	–	60	16
Lonicera periclymenum L.	75	24	57	68	71	47	–	40	36	67	50	40	50
Oxalis acetosella L.	–	–	–	–	–	–	29	–	–	–	–	–	–
Luzula pilosa Willd.	12	–	–	21	–	–	11	–	18	14	–	–	8
Athyrium filix-femina (L.) Roth.	–	–	–	–	–	6	14	–	–	–	–	–	–
Luzula sylvatica (Huds) Gaud.	37	–	–	5	–	2	4	–	–	9	–	40	6
Espèces de Fagetalia Pawl. 28 et Querco-Fagetea Br.Bl. et Vlieg. 37													
Hedera helix L.	37	–	–	–	–	6	–	7	–	33	–	–	13
Convallaria maialis L.	–	–	–	21	–	–	–	7	–	14	12	20	10
Polygonatum multiflorum (L.) All.	37	–	–	–	–	–	7	–	–	–	–	–	–
Anemone nemorosa L.	–	–	–	–	–	–	–	–	–	19	–	–	6
Autres espèces													
Rubus gp. fruticosus L.	87	28	57	68	43	45	50	53	–	67	25	–	40
Solidago virgaurea L.	12	–	29	21	71	18	11	13	27	29	12	–	20
Galeopsis tetrahit L.	–	3	14	11	–	6	14	–	–	–	25	–	3
Rubus idaeus L.	–	3	–	5	–	3	11	–	–	–	–	–	–
Bryophytes													
Hypnum cupressiforme L.	50	86	86	84	86	85	64	47	73	71	87	40	65
Dicranum scoparium Hedw.	62	69	71	53	100	68	72	67	73	48	100	80	66
Polytrichum formosum Hedw.	25	76	57	74	43	69	56	93	73	62	62	80	73
Rhytidiadelphus loreus Warnst.	–	38	14	26	–	27	72	47	9	19	–	20	22
Scleropodium purum (L.) Limpr.	37	21	14	5	43	18	11	27	45	24	62	40	34
Leucobryum glaucum Schimper	50	38	29	11	29	27	11	33	45	–	–	20	18
Rhytidiadelphus triquetrus WF.	37	10	–	26	–	13	4	40	27	33	12	60	33
Hylocomium splendens (Hedw.) Br. Eur.	37	7	14	5	29	10	8	20	27	24	12	20	22
Pleurozium schreberi Mitten	–	14	–	–	29	10	4	13	9	9	12	40	13
Dicranella heteromalla Sch.	12	3	–	16	–	6	11	7	9	9	–	–	6
Atrichum undulatum P. Beauv.	–	–	–	11	–	3	–	–	–	24	37	–	13
Thuidium tamariscinum Br. Eur.	–	–	–	11	–	3	–	–	–	5	–	–	2
Cladonia sp.	–	14	–	–	–	6	–	–	–	–	–	–	6

Tableau 2. Tableau synthétique de quelques groupements des *Quercetalia robori-petraeae.*

Tableau 2. (cont.).

Molinia coerulea
Potentilla erecta

Hieracium umbellatum
Cytisus scoparius
Juniperus communis

Lonicera periclymenum
Stachys officinalis
Deschampsia caespitosa
Luzula pilosa
Luzula sylvatica
Oxalis acetosella
Dryopteris dilatata
Luzula spicata
Luzula campestris

Carpinus betulus
Stellaria holostea
Polygonatum multiflorum
Festuca heterophylla
Melica uniflora
Endymion non scriptum

Quercus petraea
Fagus sylvatica
Hedera helix
Quercus robur
Corylus avellana
Anemone nemorosa
Convallaria majalis
Poa nemoralis

Rubus species
Solidago virgaurea
Betula pendula
Polypodium vulgare
Cladonia coniocraea
Populus tremula
Sorbus aucuparia

Ce groupement montre un certain nombre de variations (cf. Fig. 2). A côté du groupement typique (col. 2a) le plus répandu dans les conditions édaphiques moyennes s'observe déjà une phase dégradée, souvent ouverte où *Quercus petraea* domine (col. 2b), *Betula pendula, Castanea sativa* et *Corylus avellana* y sont très répandus; il en est de même de quelques espèces hélio-philes d'ourlets, manteaux ou pelouses acidiphiles (*Melampyrum pratense, Teucrium scorodonia, Hieracium sabaudum, Cytisus scoparius, Galium saxatile*, etc.). Sur sols plus épais et enrichis en éléments minéraux en bas de versant se rencontre une sous-association (col. 2c) oú pénètrent quelques espèces de l'alliance du *Carpinion betuli: Carpinus betulus* et *Stellaria holos-*

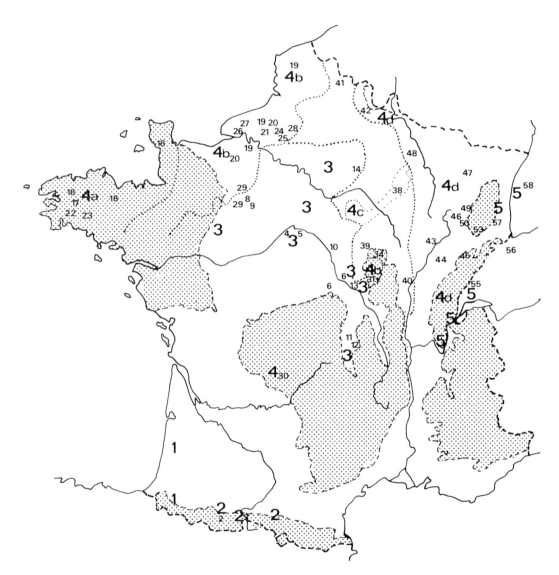

Fig. 3. Chorologie de quelques groupements des *Quercetalia robori-petraeae.*

Quercion pyrenaicae: 1 – *Blechno-Quercetum;*

Quercion robori-petraeae: 2 – *Teucrio-Quercetum,* 3 – *Peucedano-Quercetum,* 4 – *Fago-Quercetum:*

    4a – hyperatlantique (*Taxus, Ruscus, Blechnum* etc.),

    4b – atlantique à *Ilex aquifolium,*

    4c – atlantique atténué; subatlantique,

    4d – médioeuropéen: forme collinéenne, forme submontagnarde;

Genisto germanicae–Quercion: 5 – *Luzulo-Quercetum:* race subatlantique.

Pour les autres numéros se reporter au Tableau 1.

tea et des acidiclines comme *Luzula pilosa, Luzula sylvatica*. Enfin sur sol peu épais et en exposition chaude le groupement fait place à une chênaie sessiliflore de faible productivité (col. 2d), souvent ouverte, avec *Calluna vulgaris, Hieracium umbella-tum, Juniperus communis, Cytisus scoparius*, etc.

A partir de 750 m *Quercus petraea* disparait ou se raréfie fortement; la hêtraie montagnarde (col. 3) succède à la hêtraie-chênaie. La physionomie de ce groupement a été souvent modi-fiée par la réalisation de plantations anciennes (avec *Abies alba* ou *Picea abies*); la flore est peu perturbée sur les sols profonds. *Ilex aquifolium* est encore fréquent; *Rhytidiadelphus loreus* et *Blechnum spicant* montrent là leur fréquence la plus élevée. Mais ce groupement est surtout caractérisé par la présence de *Dryopteris dilatata, Sambucus racemosa, Rubus idaeus* et *Pre-nanthes purpurea* (rare). Les espèces acidiclines sont également plus répandues: *Dryopteris carthusiana, Oxalis acetosella* et *Athyrium filix-femina*.

Le Bas-Morvan est caractérisé par un quatrième type forestier (col. 4), se présentant sous la forme d'une chênaie–hêtraie ou d'une hêtraie–chênaie. *Ilex aquifolium* toujours présent ne joue plus qu'un rôle secondaire dans la strate arbustive. *Rhytidiadel-phus loreus* est plus rare et *Blechnum spicant* est absent. Le groupement typique (col. 4a) est accompagné de plusieurs vari-antes qui s'individualisent du fait des variations pédologiques et mésoclimatiques. Les bas de versant sont couverts par une sous-association (col. 4c) différenciée par *Carpinus betulus, Stellaria holostea, Festuca heterophylla* et *Holcus mollis*. Sur sols super-ficiels de plateau (col. 4b) *Quercus petraea* domine le peuple-ment généralement ouvert qui favorise le développement de *Calluna vulgaris, Hieracium umbellatum* et *Juniperus commu-nis*.

Les pentes fortes rocheuses sont occupées par une chênaie sessiliflore peu productive qui s'exprime sous deux formes selon l'exposition, une forme thermophile en adret, très ouverte, avec des espèces de lisière comme *Silene nutans* et une forme d'ubac à *Polypodium vulgare*.

A ce stade de l'étude il devient nécessaire de situer ces types forestiers par rapport aux divers groupe-ments de l'ordre des *Quercetalia robori-petraeae* décrits en France et dans les pays limitrophes. Nous ferons appel au Tableau 2 qui fournit une récapitu-lation des principaux travaux publiés jusqu'à ce jour sur ces forêts (cf. Rameau, 1981). Les données chorologiques sont parallélement prises en compte et synthétisées par la Figure 3.

Nous considérons que l'ordre des *Quercetalia ro-bori-petraeae* peut être subdivisé en trois alliances vicariantes, d'inégale importance géographique en France.

Le *Quercion pyrenaicae* est bien développé en Espagne nord-occidentale, par contre en France il est localisé au Pays Basque, au Piémont occidental des Pyrénées, aux Landes (où le plus souvent il a été détruit depuis longtemps, faisant place à des landes, ensuite replantées avec *Pinus pinaster*). (Comme diffé-rentielles largement répandues on peut citer *Blechnum spicant. Quercus pyrenaica, Erica vagans, Daboecia cantabrica, Eu-phorbia angulata, Asphodelus albus, Ruscus aculeatus, Rubia*

peregrina, Hypericum androsaemum; l'association la mieux représentée est le *Blechno-Quercetum*.)

Le *Quercion robori-petraeae* occupe l'essentiel du territoire français; les groupements du Morvan s'y rattachent.

Le *Genisto germanicae-Quercion petraeae* est le vicariant continental; de nombreuses espèces subatlantiques y disparais-sent progressivement (*Teucrium scorodonia, Holcus mollis, Hypericum pulchrum, Mespilus germanica, Hieracium laeviga-ta, Galium saxatile, Lonicera periclymenum*, etc.) alors que les espèces médioeuropéennes typiques y sont fréquentes (*Luzula luzuloides, Genista germanica, Cytisus nigricans*, etc.). En France cette alliance est représentée par le *Luzulo-Quercetum* observable en Alsace, dans le Jura méridional, sous une race où les subatlantiques précédemment citées sont encore présentes.

Le *Quercion robori-petraeae* du fait de son aire importante offre des variations géographiques très marquées. Le piémont des Pyrénées centrales et orientales héberge le *Teucrio-Querce-tum* (Tableau 2, col. 2–3) décrit par Lapraz (1966) en Catalogne; cette association montre fréquemment des transgressives des *Quercetalia pubescenti-petraeae*.

Le secteur ligérien qui possède un climat atlantique doux mais peu arrosé est bien caractérisé par le *Peucedano-Quercetum* (col. 4 à 14). Le plus souvent, la physionomie du groupement est sous la forme d'une chênaie sessiliflore ou d'une chênaie pédon-culée (fréquence des sols hydromorphes) et plus rarement d'une hêtraie-chênaie dans les terroirs légèrement plus arrosés (Fon-tainebleau). On y retrouve quelques éléments du *Quercion pyre-naicae* ayant migré jusque là, mais restant toujours dilués: *As-phodelus albus, Arenaria montana, Euphorbia angulata; Quer-cus pyrenaica* présent sur les buttes sableuses très sèches de Sologne est dans le *Betulo-Quercetum pyrenaicae* décrit par Braun-Blanquet (1967a, b) et *Ruscus aculeatus* est absent des forêts acidiphiles au profit des forêts mésoacidiphiles à méso-neutrophiles. L'association est bien individualisée par ailleurs par la fréquence de *Peucedanum gallicum, Pyrus cordata, Ser-ratula tinctoria* et *Sorbus torminalis*; des races régionales peu-vent être distinguées et en particulier celle du Pays de Fours qui correspond au premier type forestier morvandiau précédem-ment décrit (Tableau 2, col. 13).

Sur le reste de l'aire du *Quercion robori-petraeae* les forêts acidiphiles montrent une très grande homogénéité floristique qui nous conduit à ne retenir qu'une grande association le *Fago-Quercetum* s'exprimant par de grandes races régionales:
– races 'hyperatlantiques' avec *Taxus baccata, Plagiothecium undulatum, Ruscus aculeatus* (Angleterre, Bretagne, Cotentin: col. 15 à 18, = *Rusco-Fagetum* p.p.); ces forêts se présentent sous la forme de hêtraies-chênaies (avec des variantes hydro-morphes à *Quercus robur*, et des phases de dégradation à *Quer-cus petraea*);
– races atlantiques à *Ilex aquifolium, Betula pubescens, Rhy-tidiadelphus loreus*... (Bretagne, Normandie, Nord de la France: col. 19 à 32, = *Ilici-Fagetum* p.p.); il s'agit des hêtraies-chênaies à *Ilex aquifolium* faisant place parfois à des chênaies pédonculées hydromorphes ou à des chênaies sessiliflores lorsque le peuplement est dégradé ou le sol trop superficiel; les forêts acidiphiles du Haut-Morvan (col. 31 et 32) s'intègrent à cet ensemble, la race morvandelle ayant de plus un caractère sub-montagnard;
– races subatlantiques où *Ilex aquifolium* est encore présent mais très disséminé, caractère qui se retrouve pour *Sorbus tor-*

Tableau 3. Tableau synthétique des forêts du *Carpinion betuli* dans le Morvan.

Associations	1				2					3					
Sous-associations	a	b	c		a1	a2	b1	b2		a1	a2	b1	b2	c	
Nombre de relevés	4	2	7	13	16	10	5	19	52	13	18	4	5	4	44
Altitudes	850 à 900 m				300 à 750 m					250 à 600 m					
d.A. *Endymion non scriptum* (L.) Garcke	-	-	-	-	56	70	5	42	56	-	-	-	-	-	-
d.r. *Ilex aquifolium* L.	2	7	3	46	83	90	4	74	81	46	33	-	2	2	36
d.r. *Sambucus racemosa* L.	2	7	2	23	11	30	3	21	22	7	6	1	1	1	7
d.r. *Senecio fuchsii* Gmel.	2	2	2	46	11	20	2	11	15	7	-	1	1	1	-
Poa chaixii Vill.	-	-	-	-	6	-	17	5	6	54	28	1	1	1	28

Différentielles de sous-alliances : *Lonicero-Carpinenion* Rameau 80

Espèce	1a	1b	1c	1	2a1	2a2	2b1	2b2	2	3a1	3a2	3b1	3b2	3c	3
Rubus gr. *fruticosus* L.	3	2	4	69	72	100	4	89	81	85	56	3	4	4	73
Lonicera periclymenum L.	-	1	7	8	78	80	3	37	59	77	72	3	4	1	70
Luzula pilosa Willd.	1	1	8	8	50	60	3	21	48	69	61	1	5	7	64
Oxalis acetosella L.	3	7	4	61	44	70	4	63	57	-	33	7	2	3	25
Atrichum undulatum P. Beauv.	-	-	-	-	61	40	-	53	46	53	89	7	4	3	57
Dryopteris carthusiana (Vill) H.Fuchs	1	-	3	31	28	50	2	58	43	15	50	1	4	3	41
Polytrichum formosum Hedw.	-	3	2	-	39	40	2	32	35	53	67	7	7	7	45
Luzula sylvatica (Huds.) Gaud.	-	-	-	15	-	-	1	5	4	38	22	7	7	3	27

Différentielles de sous-alliances : *Daphno-Carpinenion* Rameau 80

Espèce	1a	1b	1c	1	2a1	2a2	2b1	2b2	2	3a1	3a2	3b1	3b2	3c	3
Ornithogalum pyrenaicum L.	-	-	-	-	-	-	-	-	-	-	6	1	4	1	14
Ligustrum vulgare L.	-	-	-	-	-	-	-	11	4	-	-	2	2	1	9
Evonymus europaeus L.	-	-	-	-	-	-	-	-	-	-	-	7	7	1	6
Cornus sanguinea L.	-	-	-	-	-	-	-	-	-	-	-	1	1	1	6
Scilla bifolia L.	-	-	-	-	-	-	-	-	-	-	-	1	2	1	6

Caractéristiques d'alliance : *Carpinion betuli* Oberd. 53

Espèce	1a	1b	1c	1	2a1	2a2	2b1	2b2	2	3a1	3a2	3b1	3b2	3c	3
Carpinus betulus L.	-	-	8	8	94	100	2	89	85	100	100	3	5	4	98
Stellaria holostea L.	1	1	1	8	78	80	4	47	65	54	67	3	4	3	61
Rosa arvensis Huds.	-	-	-	-	17	30	-	13	13	23	50	2	3	1	43
Potentilla sterilis Garcke	-	-	-	-	17	10	1	16	13	15	28	1	7	4	27
Prunus avium L.	-	-	-	-	11	20	-	11	13	15	17	1	1	1	16
Festuca heterophylla Lam.	-	-	-	-	-	-	1	-	-	62	39	-	-	7	39
Pulmonaria tuberosa Schrank.	-	-	-	-	-	-	-	-	-	23	11	2	3	3	18
Tilia cordata Miller	-	-	-	-	-	-	-	-	-	15	11	7	7	3	18

Caractéristiques et différentielles de Fagion sylvaticae (Tx. et Diem. 36)

Espèce	1a	1b	1c	1	2a1	2a2	2b1	2b2	2	3a1	3a2	3b1	3b2	3c	3
Cardamine heptaphylla (Vill.) O.E. Schul.	1	-	-	-	-	-	-	-	-	-	-	-	-	-	-
D. *Gymnocarpium dryopteris* (L.) Newn.	1	1	-	-	-	-	-	-	-	-	-	-	-	-	-
D. *Dryopteris dilatata* (Hoffm.) A. Gray	2	7	2	38	6	-	7	-	9	-	11	7	-	-	6

Espèces acidiphiles

Espèce	1a	1b	1c	1	2a1	2a2	2b1	2b2	2	3a1	3a2	3b1	3b2	3c	3
Holcus mollis L.	-	-	-	15	78	60	3	16	56	-	39	7	2	2	45
Pteris aquilinum (L.) Kuhn	-	-	-	15	78	20	3	11	46	33	11	-	-	1	20
Deschampsia flexuosa (L.) Trin.	1	5	2	38	56	50	7	5	31	85	22	7	1	1	30
Sorbus aucuparia L.	1	7	7	23	11	40	7	5	15	7	44	1	1	1	7

Espèces mésohygrophiles

Espèce	1a	1b	1c	1	2a1	2a2	2b1	2b2	2	3a1	3a2	3b1	3b2	3c	3
Athyrium filix-femina (L.) Roth.	2	-	6	77	22	40	3	79	35	-	11	1	5	2	16
Silene dioica (L.) Clairv.	-	7	-	8	20	20	3	26	26	33	33	2	4	3	25
Ajuga reptans L.	-	1	1	-	11	30	7	53	30	7	22	1	3	1	18
Deschampsia coespitosa (L.) Beauv.	-	-	-	-	-	20	-	26	13	15	44	7	2	1	32
Stachys sylvatica L.	-	7	7	8	-	20	-	42	19	-	-	7	2	1	5
Circaea lutetiana L.	-	7	1	-	-	20	-	32	15	-	11	-	1	1	7
Alnus glutinosa (L.) Gaertner	-	1	7	15	7	20	-	26	9	7	11	-	1	2	7
Stellaria nemorum L.	2	2	2	-	-	-	-	11	5	-	6	7	7	1	-
Carex brizoides L.	-	-	-	-	-	-	-	16	2	-	-	-	2	1	7
Lysimachia nemorum L.	-	-	8	8	-	-	-	-	6	-	-	-	-	1	-

Tableau 3. (cont.).

Caractéristiques d'ordre (Fagetalia sylvaticae Pawl. 28)

Espèce	%
Polygonatum multiflorum (L.) All.	39
Milium effusum L.	43
Galeobdolon luteum Huds.	39
Viola reichenbachiana Jordan et Reichenb.	45
Dryopteris filix-mas (L.) Schott	27
Euphorbia amygdaloides L.	36
Fraxinus excelsior L.	27
Ranunculus ficaria L.	27
Paris quadrifolia L.	9
Primula elatior (L.) Hill.	30
Asperula odorata L.	—
Acer pseudoplatanus L.	16
Melica uniflora Retz	9
Mercurialis perennis L.	9
Polystichum setiferum (Forsk.) Woynar	7
Ulmus glabra Hudson	—

Caractéristiques de classe (Querco-Fagetea Br.Bl. et Vlieg. 37)

Espèce	%
Quercus robur L.	68
Hedera helix L.	80
Fagus sylvatica L.	30
Corylus avellana L.	50
Poa nemoralis L.	55
Quercus petraea (Mat.) Liebl.	43
Anemone nemorosa L.	30
Moehringia trinerva (L.) Clairv.	32
Carex sylvatica Huds.	18
Vicca sepium L.	25
Brachypodium sylvaticum (Huds.) Beauv.	7
Crataegus laevigata (Poiret) D.C.	7

Espèces d'ourlet et de coupes

Espèce	%
Galeopsis tetrahit L.	20
Geranium robertianum L.	16
Epilobium montanum L.	9
Glechoma hederacea L.	16
Geum urbanum L.	14

Compagnes

Espèce	%
Solidago virgaurea L.	25
Viburnum opulus L.	16
Betonica officinalis L.	25
Betula pendula Roth.	18
Populus tremula L.	18
Crataegus monogyna Jacq.	20

Bryophytes

Espèce	%
Eurhynchium striatum (Scherb.) Schim.	80
Atrichum undulatum P. Beauv.	57
Polytrichum formosum Hedw.	45
Rhytidiadelphus triquetrus WF.	57
Mnium undulatum (L.) Hedw.	39
Thuidium tamariscinum Br. Eur.	48
Hypnum cupressiforme L.	9
Eurhynchium stockesii B.E.	11
Mnium affine Schw.	5
Rhytidiadelphus loreus Warnst.	5

Tableau 4. Tableau synthétique de quelques groupements du *Carpinion betuli.*

Groupements			
Nombre de relevés			

Ruscus aculeatus
Tamus communis
Rubia peregrina
Brachypodium sylvaticum
Luzula forsteri
Hypericum androsaemum
Pulmonaria affinis
Polystichum setiferum
Pulmonaria longifolia
Arum italicum
Iris foetidissima
Asphodelus albus
Helleborus viridis
Symphytum tuberosum
Potentilla montana

Carpinus betulus
Stellaria holostea
Rosa arvensis
Potentilla sterilis
Prunus avium
Festuca heterophylla
Campanula trachelium
Vinca minor
Tilia cordata
Pulmonaria tuberosa
Galium sylvaticum
Dactylis aschersoniana
Carex pilosa
Carex alba
Carex montana
Isopyrum thalictroides

Primula veris
Quercus pubescens
Melittis melissophyllum
Helleborus foetidus
Buxus sempervirens
Acer monspessulanus
Coronilla emerus
Lithospermum purpureo-caeruleum
Rubus caesius

Ilex aquifolium
Endymion non scriptum
Primula acaulis
Taxus baccata
Rhytidiadelphus loreus
Dryopteris tavellii

Poa chaixii
Carex umbrosa
Epipactis purpurata
Carex brizoides
Luzula luzuloides
Polygonatum verticillatum
Sambucus racemosa
Senecio fuchsii
Prenanthes purpurea
Festuca altissima

Oreopteris pyrenaicum
Ranunculus auricomus
Scilla bifolia
Narcissus pseudonarcissus

Anemone ranunculoides
Aconitum vulparia
Corydalis solida
Leucojum vernum
Lathraea squamaria
Corydalis cava

Lonicera periclymenum
Oxalis acetosella
Luzula pilosa
Dryopteris carthusiana
Luzula sylvatica
Dryopteris dilatata

Pteridium aquilinum
Holcus mollis
Teucrium scorodonia
Deschampsia flexuosa
Calamagrostis sativa
Viola reichenbachiana
Hypericum pulchrum
Carex pilulifera
Melampyrum pratense
Frangula alnus
Nespilus germanica
Veronica officinalis

Tableau 4. (cont.).

Cornus sanguinea
Ligustrum vulgare
Euonymus europaeus
Prunus spinosa
Lonicera xylosteum
Viburnum lantana
Cornus mas
Ribes alpinum
Daphne laureola
Daphne mezereum

Melica uniflora
Euphorbia amygdaloides
Viola reichenbachiana
Milium effusum
Lamium galeobdolon
Polygonatum multiflorum
Acer campestre
Fraxinus excelsior
Dryopteris filix-mas
Asperula odorata
Arum maculatum
Primula elatior
Ranunculus ficaria
Mercurialis perennis
Acer pseudoplatanus
Ranunculus nemorosus
Euphorbia dulcis
Scrophularia nodosa
Sanicula europaea
Neottia nidus-avis
Poa quadrifolia
Tilia platyphyllos
Asarum europaeum

Hedera helix
Fagus sylvatica
Corylus avellana
Anemone nemorosa
Quercus robur
Carex sylvatica
Vicia sepium
Poa nemoralis
Crataegus laevigata
Convallaria maialis
Lathyrus montanus
Carex digitata

Deschampsia cespitosa
Athyrium filix-femina
Circaea lutetiana
Veronica montana
Carex remota

Rubus gp. fruticosus
Crataegus monogyna
Brachypodium sylvaticum
Sorbus torminalis
Carex glauca
Viburnum opulus
Fragaria vesca
Solidago virgaurea
Geum urbanum
Betonica officinalis
Geranium robertianum
Ajuga reptans
Populus tremula

minalis (Champagne humide, Plaine de Saône: col. 34 à 41); la physionomie varie selon les régions (chênaie-hêtraie ou hêtraie-chênaie selon la pluviosité) et les caractères du sol (chênaie pédonculée de sols hydromorphes, chênaie sessiliflore sur sols superficiels; les forêts du Bas-Morvan constituent l'une de ces races (col. 34 à 37).

- races médioeuropéennes avec *Luzula luzuloides*, *Maianthemum bifolium* offrant des formes collinéennes et submontagnardes (passage vers le *Luzulo-Fageninon*) (Ardennes, Lorraine, Franche Comté, col. 42 à 53); les hêtraies-chênaies dominent avec présence là encore de chênaies pédonculées et de chênaies sessiliflores édaphiques.

L'interprétation de la hêtraie montagnarde est délicate: elle correspond très certainement à une forme appauvrie des hêtraies acidiphiles du Massif Central rattachées au *Luzulo nivae-Fagetum* (*Luzulo-Fagenion* ou *Ilici-Fagenion* pour Rivas Martinez, 1973). Il faut souligner la grande rareté des éléments des *Fagetalia sylvaticae*.

Les forêts mésoacidiphiles à neutrophiles

Bien qu'occupant une place réduite par rapport à la végétation acidiphile, la diversité floristique de ces forêts est beaucoup plus marquée traduisant des conditions écologiques bien différenciées (cf. Tableau 3 et Fig. 1c).

Dans les parties élevées du Haut-Morvan (altitude > 850 m) la hêtraie acidiphile s'interrompt sur les sols brunifiés plus riches en éléments minéraux pour faire place à une hêtraie mésoneutrophile relevant de l'*Asperulo-Fagenion* (col. 1a à 1c). Le peuplement arborescent est dominé par *Fagus sylvatica* auquel s'associent *Acer pseudoplatanus* et *Fraxinus excelsior*; *Quercus petraea* reste exceptionnel. La flore montagnarde reste dispersée et peu fournie (*Cardamine heptaphylla*, *Gymnocarpium dryopteris*, *Dryopteris dilatata*, etc.). Le cortège floristique comprend de nombreuses espèces neutrophiles et un lot assez important d'espèces acidiclines (*Dryopteris carthusiana*, *Oxalis acetosella*, *Polytrichum formosum*, *Lonicera periclymenum*, etc.).

A une altitude inférieure, les bas de versant et les vallées (Fig. 2b) (en dehors du lit majeur inondable où est installé le *Stellario-Alnetum*) sont occupés par une forêt mixte (col. 2a et 2b) où sont juxtaposés *Fagus sylvatica*, *Quercus robur*, *Quercus petraea*, *Carpinus betulus*, *Prunus avium*, *Fraxinus excelsior*, *Acer pseudoplatanus*, *Populus tremula*, etc. La physionomie de la strate arbustive est profondément marquée par l'abondance d'*Ilex aquifolium*; parmi les espèces remarquables on doit souligner la fréquence d'une atlantique: *Endymion non-scriptum* et de submontagnardes: *Sambucus racemosa* et *Senecio fuchsii*. Les espèces acidiclines forment un noyau fourni (*Lonicera periclymenum*, *Luzula pilosa*, *Oxalis acetosella*, *Atrichum undulatum*, *Dryopteris carthusiana*, *Polytrichum formosum*, *Luzula sylvatica*, etc.). En fonction de la richesse nutritive du sol, deux sous-associations peuvent être distinguées: une sous-association mésoacidiphile différenciée par *Holcus mollis*, *Pteris aquilinum*, *Deschampsia flexuosa* et *Sorbus aucuparia*; une sous-associa-

tion typique, comportant chacune selon le degré d'humidité du sol une variante mésophile et une variante mésohygrophile avec *Athyrium filix-femina*, *Silene dioica*, *Ajuga reptans*, *Deschampsia caespitosa*, *Stachys sylvatica*, *Circaea lutetiana*, *Alnus glutinosa*, etc.

Dans le Bas-Morvan, dans les mêmes conditions stationnelles et sur des placages de limons masquant le substrat granitique (Nord du Morvan = Avallonnais) (cf. Fig. 1a et 2b) se rencontre une autre association. La strate arborescente est surtout constituée par *Quercus robur* et *Quercus petraea* (*Fagus sylvatica* étant plus rare), avec *Carpinus betulus*, *Fraxinus excelsior*, *Prunus avium*, *Tilia cordata*, *Betula verrucosa*, *Populus tremula*, etc. *Ilex aquifolium* est moins fréquent et il ne joue plus qu'un rôle secondaire dans la strate arbustive. Une espèce médioeuropéenne, *Poa chaixii* se substitue à *Endymion non-scriptum*. Le noyau d'espèces acidiclines se retrouve ainsi que la structuration en sous-unités: sous-association mésoacidiphile; sous-association typique mésoneutrophile, possédant chacune une variante mésophile et une variante mésohygrophile; s'y ajoute une sous-association à *Polystichum setiferum* où *Tilia cordata* est abondant, sur pente forte et fraîche (suintements).

La position de ces deux associations par rapport aux groupements de l'alliance du *Carpinion betuli* se révèle très intéressante (Tableau 4) ceci malgré l'absence d'études phytosociologiques sur certaines régions; la synthèse chorologique est envisagée parallelement (Fig. 4).

Deux grands ensembles d'associations s'individualisent nettement dans le tableau synthétique. Le premier rassemble des associations décrites dans le Sud-Ouest de la France (climat atlantique doux à pluviosité souvent importante); les espèces différentielles sont nombreuses (*Ruscus aculeatus*, *Rubia peregrina*, *Blechnum spicant*, *Luzula forsteri*, *Hypericum androsaemum*, *Pulmonaria affinis*, *Polystichum setiferum*, *Pulmonaria longifolia*, *Arum italicum*, *Iris foetidissima*, *Asphodelus albus*, *Helleborus viridis*, *Symphytum tuberosum*, *Potentilla montana*, etc.) ce qui a conduit Lapraz (1963) à proposer une nouvelle alliance le *Rubio-Ruscion* vicariant du *Carpinion betuli*. L'idée d'alliance vicariance est à retenir mais en baptisant l'unité *Rubio-Carpinion*, nom plus conforme aux règles de la nomenclature syntaxonomique. Cette alliance qui occupe essentiellement le secteur atlantique aquitanien se subdivise encore en deux groupes d'associations: - un groupe calcicole avec l'*Isopyro-Quercetum*, le *Pulmonario affinis-Fagetum*, le *Viburno-Quercetum*, l'*Aceri monspessulanae-Fagetum*, le *Rubio-Fagetum* (col. 1 à 5); un groupe mésoacidiphile à mésoneutrophile avec l'*Androsaemo-Fagetum* et le *Periclymeno-Quercetum* (col. 6 à 9).

La distinction n'est pas toujours bien marquée entre ces deux groupes du fait de l'acidification assez marquée des sols (pluviosité élevée) et de l'amplitude très large de certaines espèces réputées calcicoles ou acidiphiles dans le nord-est de la France (compensations de facteurs du fait du climat?).

Le second ensemble correspond à l'alliance du *Carpinion* où les espèces du *Rubio-Carpinion* sont absentes ou présentes isolément et dans ce cas généralement diluées.

Nous avons proposé dans une étude précédente (Rameau, 1981) la création de deux sous-alliances: le *Lonicero-Carpinenion* différencié par la présence d'espèces acidiclines, voire de

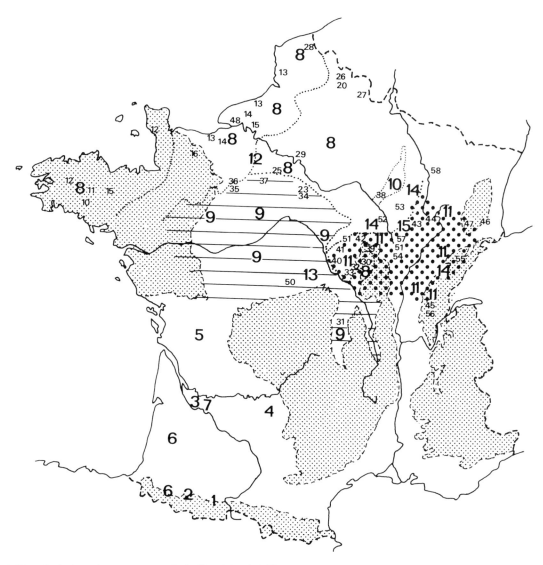

Fig. 4. Chorologie de quelques groupements du '*Carpinion betuli*'.
Alliance *Rubio–Carpinion:* 1 – *Isopyro*-Quercetum robori (1), 2 – *Pulmonario affinis–Fagetum* (2), 3 – *Viburno–Quercetum* (3), 4 –*Aceri monspessulanae–Fagetum* (4), 5 – *Rubio–Fagetum* (5), 6 – *Androsaemo–Fagetum* (6 à 8), 7 – *Periclymeno–Quercetum* (9);
Alliance *Carpinion betuli:* sous-alliance *Lonicero–Carpinenion:* 8 – groupe d'associations à *Endymion non-scriptum* (10 à 30), 9 – *Rusco–Carpinetum* (31 à 37), 10 – *Stellario–Carpinetum* (38), 11 – *Poo chaixii–Carpinetum* (39 à 47);
sous-alliance *Daphno–Carpinenion:* 12 – *Daphno–Fagetum* (48 et 49), 13 – *Lithospermo–Carpinetum* (50), 14 – *Scillo bifoliae–Carpinetum* (51 à 56), 15 – *Aconito vulpariae–Quercetum pedunculatae* (57 et 58).

Pour les autres numéros se reporter au Tableau 3.

quelques acidiphiles; le *Daphno–Carpinenion* défini par l'abondance des taxa calcicoles.

Nous ne considérerons ici que le *Lonicero–Carpinenion,* seul représenté au niveau du Morvan. Cette sous-alliance rassemble plusieurs associations ou 'groupes d'associations' affines.

Un premier ensemble de groupements apparait dans le Tableau 4 (col. 10 à 30), différenciés par la fréquence d'*Endymion non-scriptum, Ilex aquifolium, Primula acaulis,* etc. et localisés dans l'Ouest de la France (Bretagne, Normandie, Picardie et Nord); ils offrent des physionomies très variables avec

– des hêtraies-chênaies climatiques (avec ou sans *Carpinus betulus*) (= *Endymio–Fagetum, Rusco–Carpinetum* p. p. *Ilici–Fagetum* p.p.). Certaines de ces forêts se présentent parfois sous la forme de hêtraies pures. Ces futaies 'cathédrales' monospécifiques sont la création des forestiers: elles résultent d'un traitement qui a, peu à peu éliminé les essences secondaires. Sur le plan climacique il s'agit en réalité de *forêts mixtes* dominées par *Fagus sylvatica* mais où *Quercus* et *Carpinus betulus* souvent, sont bien représentés (de nombreux documents historiques le prouvent, et l'impossibilité de régénération qu'offrent actuellement ces 'hêtraies', témoigne de l'excès d'artificialisation dont elles ont fait l'objet. Plusieurs auteurs rangent ces forêts collinéennes dominées par *Fagus sylvatica* dans l'alliance du *Fagion*. Nous appuyant sur des critères floristiques et historiques (mise en place progressive des essences actuelles lors du postglaciaire: Bugnon *et al.*; 1981) nous estimons plus logique de les réunir aux chênaies-charmaies climatiques ou édaphiques au sein de l'alliance du *Carpinion betuli*.

– des chênaies-charmaies climatiques (avec *Quercus petraea* > *Quercus robur*) des régions peu arrosées, et des chênaies pédonculées-charmaies ou chênaies pédonculées-frênaies édaphiques rassemblées en un *Endymio-Carpinetum*, offrant aussi des variations floristiques selon les régions avec des races 'hyperatlantiques' avec *Ruscus aculeatus*, *Taxus baccata*, *Rhytidiadelphus loreus* (Bretagne, Cotentin, Angleterre) et des races atlantiques à *Ilex aquifolium* (par exemple Normandie et Picardie). La végétation comprend quelques espèces du *Rubio-Carpinion* (*Rubia peregrina*, *Blechnum spicant*, *Luzula forsteri*, *Hypericum androsaemum*, *Polystichum setiferum*, dont les migrations à partir du sud-ouest sont plus ou moins longues). Il serait sage sur le plan syntaxonomique de s'en tenir à deux associations: l'*Endymio-Fagetum* et l'*Endymio-Carpinetum* structurées en races, sous-associations, etc. Le groupement collinéen du Haut-Morvan s'identifie parfaitement aux forêts atlantiques de l'*Endymio-Carpinetum* sous une forme à caractère légèrement submontagnard.

Un deuxième ensemble de groupements réunis par la constance de *Ruscus aculeatus*, *Luzula forsteri*, caractérise le secteur ligérien: il s'agit du *Rusco-Carpinetum*, chênaie (charmaie) climatique (voire hêtraie-chênaie à charme climatique en quelques points mieux arrosés), chênaie pédonculée-charmaie édaphique (col. 31 à 37). L'aire du *Rusco-Carpinetum* (Fig. 4) arrive au contact du Morvan en Sologne Bourbonnaise, Pays de Fours, Puisaye (Fig. 1c). La Champagne humide héberge une chênaie-charmaie où *Fagus sylvatica* est dilué; la flore est privée des espèces atlantiques ou médioeuropéennes; ces forêts sont assimilables au *Stellario-Carpinetum* subatlantique (col. 38).

Enfin sur le Nord-Est de la France s'individualise une dernière grande association régionale médioeuropéenne avec *Poa chaixii*, *Carex umbrosa*, *Epipactis purpurata*, *Carex brizoides*, *Luzula luzuloides* (col. 39 à 47): le *Poo chaixii-Carpinetum*. Selon les caractères pédologiques, la physionomie oscille entre une hêtraie-chênaie-charmaie climatique et une chênaie pédonculée-charmaie édaphique sur sols plus ou moins engorgés. Les races sont nombreuses et les forêts mésotrophes du Bas-Morvan en représente une race se retrouvant sur les plateaux du Nivernais (Braque, 1978; cf. Fig. 1c et Fig. 4).

En conclusion, le Morvan est donc une région particulièrement intéressante du fait de cette réapparition au niveau du Haut-Morvan de deux groupements typiquement atlantiques (qui ont été définis sur la façade arrosée de la Manche): la race atlantique du *Fago-Quercetum* où *Ilex aquifolium* est omniprésent et l'*Endymio-Carpinetum* riche en *Endymion non-scriptum* et *Ilex aquifolium*. Ces deux associations se présentent sous une forme légèrement submontagnarde; leur développement dans le Morvan s'explique par la pluviosité très élevée qui y règne. Par ailleurs le Morvan se révèle être un carrefour phytogéographique exceptionnel comme le résument les Figures 1 b et 1c; on y observe en effet le contact entre le *Peucedano-Quercetum* ligérien et le *Fago-Quercetum*, le contact entre le *Rusco-Carpinetum* ligérien, l'*Endymio-Carpinetum* atlantique et le *Poo chaixii-Carpinetum* plus médioeuropéen et sur calcaire le contact entre le *Scillo-Carpinetum* bourguignon et le *Lithospermo-Carpinetum* du Berry.

Bibliographie

Bardat, J. & Frileux, P. N., 1981. Etude phytoécologique sur la végétation forestière du massif de Brotonne (Seine-Maritime). Documents Phytosociologiques, NS 5: 111–140.

Bolos, O. de, 1957. Datos sobre la vegetación de la vertiente septentrional de los Pireneos observaciones acerna de la zonación altitudinal en el Valle de Aran. Coll. Bot. 5: 465–514.

Braque, R., 1978. La forêt et ses problèmes dans le sud du Bassin Parisien (Berry-Nivernais). 943 pp. Thèse, Clermont-Ferrand.

Braun-Blanquet, J., 1932. Zur Kentnis nordschweizerischer Waldgesellschaften. Beih. Bot. Zentralblatt 49: 7–49.

Braun-Blanquet, J., 1967. Vegetationsskizzen aus dem Baskenland mit Ausblicken auf das weitere ibero-atlantikum II Teil. Vegetatio 14: 1–126.

Braun-Blanquet, J., 1967. La chênaie acidophile ibéro-atlantique (Quercion occidentale) en Sologne. Anales de Edafologia y agrobiologia: 53–87 Madrid.

Brunery, L., 1970. Les groupements forestiers de la région de Treignac (Corrèze), leur signification phytogéographique. Cahier des Naturalistes. Bull. Naturalistes Parisiens NS 26(1): 1–17.

Bugnon, F. & Rameau, J. C., 1973. L'Aconito vulpariae-Quercetum pedunculatae. Bull. Sci. Bourgogne 29: 5–16.

Bugnon, F. & Rameau, J. C., 1975. Les forêts acidiphiles du Morvan. Colloques Phytosociologiques 3: 44–52.

Bugnon, F. et al., 1981. Etudes sur les séries de végétation en Bourgogne: les types forestiers correspondant aux feuilles 34 (Dijon) et 41 (Autun) de la Carte de la Végétation. Bull. Soc. Bot. France 128: 7–20.

Clément, B. et al., 1975. Contribution à l'étude phytosociologique des forêts de Bretagne, Colloques Phytosociologiques 3: 53–72.

Comps, B. et al., 1980. Essai de synthèse phytosociologique sur les hêtraies collinéennes du domaine atlantique français. Documents Phytosociologiques, NS 5: 409–443.

Delelis, A. & Géhu, J. M., 1975. Apport à la phytosociologie de quelques forêts thermo-acidiphiles ligériennes et de leurs stades d'altération, Colloques Phytosocioloques 3: 83–114.

Dumé, G., 1975. Contribution à l'étude phytosociologique et écologique des forêts à chêne et à charme du Bassin Parisien au sens large. Thèse 3e cycle, Orsay. 92 pp.

Durin, L. & Géhu, J. M., 1963. Sur les hêtraies naturelles du nord-ouest de la France. Compte rendu Acad. Sci. 256: 3749–3751.

Durin, L. et al., 1967. Les hêtraies atlantiques et leur essaim climacique dans le nord-ouest et l'ouest de la France. Bull. Soc. Bot. du Nord de la France 20: 59–89.

Estrade, J. & Rameau, J. C., 1980. Premières observations sur les forêts riveraines des Vosges et du Morvan. Colloques phytosociologiques (sous presse).

Frileux, P. N., 1975. Contribution à l'étude des forêts acidiphiles de Haute Normandie. Colloques Phytosociologiques 3: 287–300.

Gaume, R., 1924a. Les associations végétales de la forêt de Preuilly (Indre et Loire). Bull. Soc. Bot. France 71, 24: 158–171.

Gaume, R., 1924b. Aperçu sur quelques associations végétales de la forêt d'Orléans (Loiret). Bull. Soc. Bot. France 24: 1194–1207.

Gaume, R., 1925a. La chênaie de chêne sessile de la forêt de Montargis (Loiret). Bull. Assoc. Naturalist. de la Vallée du Loing 8: 42–49.

Gaume, R., 1925b. Aperçu sur les groupements végétaux du plateau de Brie. Bull. Soci. Bot. France 72: 393–416.

Gaume, R., 1926. La flore de la forêt d'Orléans aux environs de Lorris (Loiret). Bull. Assoc. Naturalist. de la vallée du Loing 9: 101–115.

Géhu, J. M., 1961. Les groupements végétaux du Bassin de la Sambre française. Vegetatio 10, 2-6: 69–148. 161–208, 257–372.

Géhu, J. M., 1975. Aperçu sur les chênaies-hêtraies acidiphiles du sud de l'Angleterre. L'exemple de la New Forest. Colloques Phytosociologiques 3: 133–141.

Gruber, M. 1980. Le chêne sessile dans la vallée du Louron (Hautes Pyrénées). Bull. Soc. Histoire Natur. Toulouse 116 (1–2): 165–174.

Kiessling, P., 1983. Les chênaies du Jura Central suisse. Institut de Recherches Forestières 59, 3: 215–437.

Lapraz, G., 1963. La végétation de l'Entre-Deux-Mers: chênaies, châtaigneraies et charmaies mésophiles sur sol acide. Mémoire Soci. Sci. Physiques et Naturelles de Bordeaux 3: 1–36.

Lapraz, G., 1965. Note sur les chênaies-charmaies acidiphiles des Basses Vosges orientales entre Andlau et Ottrott (Poa chaixii-Carpinetum). Bull. Assoc. Philomatique d'Alsace Lorraine 12, 1: 41–57.

Lemée, G., 1937. Recherches écologiques sur la végétation du Perche. Thèse Sciences Naturelles, Paris. 392 pp.

Lemée, G., 1943. Etude sur la végétation et les sols des forêts de Randan et de Monpensier. Revue Sci. Natur. d'Auvergne 9: 69–81.

Lericq, R., 1965. Contribution à l'étude des groupements végétaux du Bassin français de l'Escaut. Thèse, Lille, 153 pp.

Malcuit, G., 1929. Contribution à l'étude phytosociologique des Vosges méridionales saônnoises: les associations de la Vallée de la Lanterne. Thèse: 211 pp.

Nègre, R., 1972. La végétation du Bassin de l'One – Pyrénées Centrales; 4e note Forêts. Veröff. Geobot. Inst. Zürich 49: 1–128.

Neuhäusl, R. & Neuhäuslová-Novotná, Z., 1966. Syntaxonomische Revision der azidophilen Eichen- und Eichenmischwälder im westlichen Teile der Tschechoslowakei. Folia Geobot. Phytotax. bohemaslovaca 1: 289–380.

Noirfalise, A., 1969. La chênaie mélangée à Jacinthe du domaine atlantique de l'Europe (Endymio-Carpinetum). Vegetatio 17: 131–150.

Oberdorfer, E., 1957. Süddeutsche Pflanzengesellschaften. Pflanzensoziol. 10: 1–564.

Quantin, A., 1935, L'évolution de la végétation à l'étage de la chênaie dans le Jura méridional. Thèse, Paris. Communication SIGMA 37. 382 pp.

Rameau, J. C., 1978. Notes sur le Carpinion mésotrophe du sud-est du Bassin Parisien et de la Bourgogne. Documents Phytosociologiques, NS 2: 353–363.

Rameau, J. C., 1981. Réflexions sur la synsystématique des forêts françaises de hêtre, chênes et charme. Application au système bourguignon. Bull. Soc. Bot. France 128: 3–4: 33–63.

Rameau, J. C. & Royer, J. M., 1975. Les forêts acidiphiles du sud-est du Bassin Parisien. Colloques Phytosociologiques 3: 319–340.

Rameau, J. C. & Timbal, J., 1979. Les groupements forestiers de fond de vallon des plateaux calcaires de Lorraine; étude phytosociologique. Documents Phytosociologiques, NS 4: 847–870.

Riomet, L. B. & Bournerias, M., 1952–1961. Flore de l'Aisne. Soc. Hist. Nat. Aisne. 356 pp.

Rivas Martinez, S., 1975. Observaciones sobre la sintaxonomia de los bosques acidofilos europas. Datos sobre las Quercetalia robori petraeae en la Peninsula iberica. Colloques Phytosociologiques 3: 1–10.

Roisin, P., 1967. Contribution à l'étude du domaine phytogéographique et des hêtraies atlantiques d'Europe. Thèse, Gembloux. 346 pp.

Timbal, J., 1975. Les rapports du Luzulo-Fagion et du Quercion robori petraeae dans le nord-est de la France. Colloques Phytosociologiques 3: 341–361.

Tutin, T. G. et al., 1964–1980. Flora Europaea. Cambridge University Press.

Vanden Berghen, C., 1968. Les forêts de la Haute Soule (Basses Pyrénées). Bull. Soc. Roy. Bot. Belge 102: 107–132.

Appendice

Localisations géographique et bibliographique des associations et sous-associations des différents tableaux

Tableau 1:

1 *Peucedano–Quercetum* Br. Bl. 67, race du Pays de Fours.
2 *Fago–Quercetum petraeae* Tx. 37, race atlantique morvandelle; a typique, b phase légèrement dégradée, c chênaie sessiliflore à *Calluna vulgaris*.
3 Hêtraie montagnarde à *Deschampsia flexuosa*.

4 *Fago–Quercetum petraeae* Tx. 37, race subatlantique morvandelle; a sous-association typique, b sous-association *carpinetosum,* c chênaie sessiliflore à *Calluna vulgaris,* d chênaie sessiliflore à *Silene nutans,* e chênaie sessiliflore à *Polypodium vulgare.*

Tableau 2:
Quercion pyrenaicae
1 Braun-Blanquet 67 (Espagne), Vanden Berghen 68, 2 relevés
(Pays Basque français).
Quercion robori-petraeae
Teucrio–Quercetum Lapraz 66: 2 Gruber 80 (Hautes-Pyrénées),
3 Nègre 72 (Pyrénées Vallée de l'One).
Peucedano–Quercetum Br. Bl. 67: 4 Braun-Blanquet 67: (So-
logne), 5 Géhu & Delelis 74 (Sologne – Centre du Bassin
Parisien), 6 Braque 78 (Berry, Nivernais), 7 Braque 78 (Ber-
ry, Nivernais);
'*Fago–Quercetum*': 8 Lemée 37 (Perche);
'*Quercetum roboris parisiense*': 9 Lemée 37 (Perche);
'*Querceto–Holcetum mollis*': 10 Rameau & Royer 74 (Puisaye,
Pays d'Othe), 11 Lemée 43 (Limagne);
'*Quercetum sessiliflorae occidentale*': 12 Lemée 43 (Limagne);
'*Quercetum roboris parisiense*': 13 Rameau 82 (Pays de Fours),
14 Bournerias (Laonnois), 15 Gaume 25 (Brie), 16 Gaume 24
(Indre et Loire), 17 Gaume 24 (Orléans), 18 Gaume 25
(Loiret), 19 Gaume 25 (Fontainebleau).
Fago–Quercetum Tx.55
races hyperatlantiques à *Taxus baccata, Ruscus aculeatus:* 20
Géhu 74 (New-Forest, Angleterre);
'*Rusco–Fagetum*' à *Vaccinium myrtillus:* 21 Géhu 74 (New-For-
est, Angleterre);
'*Rusco–Fagetum*': 22 Clément *et al.* 74 (Bretagne);
'*Vaccinio–Quercetum sessiliflorae taxetosum*': 23 Durin *et al.* 67
(Bretagne, Cotentin);
'*Rusco–Fagetum*' p.p. races atlantiques à *Ilex aquifolium:* 24
Durin *et al.* 67 (Picardie, Normandie);
'*Ilici–Fagetum luzuletosum*': 25 Durin *et al.* 67 (Picardie, Nor-
mandie);
'*Ilici–Fagetum vaccinietosum*': 26 Durin *et al.* 67 (Picardie,
Normandie); 'Forêts secondaires' 27 Clément *et al.* 74
(Bretagne);
'*Vaccinio–Quercetum*': 28 Clément *et al.* 74 (Bretagne);
'*Molinio–Quercetum*': 29 Frileux 74 (Normandie); '*Ilici–Fage-
tum*': 30 Frileux 74 (Normandie);
'*Mespilo–Quercetum*': 31 Bardat et Frileux 81 (Normandie);
'*Ilici–Fagetum*' 32 Bardat et Frileux 81 (Normandie);
'*Mespilo–Quercetum*': 33 Frileux 77 (Pays de Bray, Normandie), 34 Lemée 37 (Perche);
'*Quercetum occidentale ilicetosum*': 35 Brunerye 70 (Corrèze),
36 Rameau 82 (Morvan);
sous-association typique: 37 Rameau 82 (Morvan);
sous-association *carpinetosum:* 38 Rameau 82 (Morvan); hêtraie à *Deschampsia flexuosa,* race subatlantique: 39 Rameau
82 (Morvan);
sous-association typicum: 40 Rameau 82 (Morvan); sous-association *carpinetosum:* 41 Rameau 82 (Morvan);
chênaie à *Silene nutans:* 42 Rameau 82 (Morvan);
chênaie à *Polypodium vulgare;* 43 Rameau & Royer 74 (Champagne);

Fago–Quercetum campanense: 44 Rameau 74 (Plateaux bour-
guignons), 45 Rameau (inédit) (Plaine de la Saône), 46 Lericq 65 (Bassin de l'Escaut);
races médioeuropéennes: 47 Géhu 61 (Bassin de la Sambre), 48
Rameau 74 (Plateau de Haute-Saône), 49 Rameau &
Schmitt 80 (Vallée de l'Ognon, Doubs), 50 Rameau (inédit)
(Jura du Nord), 51 Rameau & Royer 74 (Pays d'Amance et
d'Apance), 52 Timbal 74 (Lorraine);
'hêtraies acidiphiles à Canche flexueuse': 53 Timbal 74 (Ar-
gonne);
'hêtraies acidiphiles': 54 Timbal 74 (Basses Vosges gréseuses);
hêtraie acidiphile à Myrtille: 55 Timbal 74 (Bases Vosges gréseuses);
hêtraies acidiphiles à Callune: 56 Timbal 74 (Bases Vosges gréseuses);
chênaies sessiliflores: 57 Timbal 74, (Vosges);
hêtraies acidiphiles de l'étage montagnard inférieur des Vosges:
58 Malcuit 29 (Vosges).
Quercetum sessiliflorae
Genisto germanicae–Quercion petraeae
Luzulo–Quercetum Knapp 42
races subatlantiques: 59 Quantin 35 (Jura méridional), 60 Kiessling 81 (pied du Jura suisse), 61 Braun-Blanquet 32 (Suisse);
Quercetum medioeuropaeum: 62 Issler 36 (Vosges);
'chênaie sessiliflore': 63 Oberdorfer 57 (Forêt Noire), 64 Ober-
dorfer 57 (Forêt Noire), 65 Oberdorfer 57 (Forêt Noire);
race médioeuropéenne:
Luzulo–Quercetum genistetosum variante chaude: 66 Neuhäusl
& Neuhäuslová-Novotná 66 (Tchécoslovaquie);
Luzulo–Quercetum genistetosum: 67 Neuhäusl & Neuhäuslová-
Novotná, 66 (Tchécoslovaquie);
Luzulo–Quercetum typicum variante chaude: 68 Neuhäusl &
Neuhäuslová-Novotná 66 (Tchécoslovaquie);
Luzulo–Quercetum typicum: 69 Neuhäuśl & Neuhäuslová-No-
votná 66 (Tchécoslovaquie).

Tableau 3:
1 Hêtraie montagnarde à *Asperula odorata,* a sous-association
typique, b sous-association *stellarietosum nemori,* c sous-association *deschampsietosum flexuosae;*
2 *Endymio–Carpinetum* Noirfalise 68, 2a sous-association
holcetosum, 2a$_1$ variante mésophile, 2a$_2$ variante méso-hy-
grophile; 2b sous-association typique, 2b$_1$ variante mésophile,
2b$_2$ variante mésohygrophile;
3 *Poo chaixii–Carpinetum* (Oberd. 57) Rameau 78, 3a sous-
association *holcetosum,* 3a$_1$ variante mésophile, 3a$_2$ variante
mésohygrophile; 3b sous-association typique, 3b$_1$ variante
mésophile, 3b$_2$ variante mésohygrophile; 3c sous-association
polystichetosum setiferi.

Tableau 4:
Alliance *Rubio–Carpinion*
1 *Isopyro–Quercetum robori* Tx. et Diemont 36, O. de Bolos
57 – Pyrénées
2 *Pulmonario affinis–Fagetum* Comps *et al.* 81 – Pyrénées
3 *Viburno–Quercetum* Lapraz 63 – Entre Deux Mers
4 *Aceri monspessulano–Fagetum* Comps *et al.* 81 – Quercy
5 *Rubio–Fagetum* Roisin 67 – Charentes
6 *Androsaemo–Fagetum,* sous-ass. typique, Comps *et al.* 81 –
Pyrénées occidentales, Landes

7 *Androsaemo–Fagetum* sous-ass. à *Quercus petraea*, Comps *et al.* 81 – Pyrénées occidentales et Landes

8 *Androsaemo–Fagetum* sous-ass. à *Quercus pubescens*, Comps *et al.* 81 – Pyrénées occidentales

9 *Periclymeno–Quercetum* Lapraz 63 – Ėntre Deux Mers
Alliance *Carpinion betuli*
sous-alliance *Lonicero–Carpinenion*

10 *Rusco–Melico–Fagetum taxetosum* Clément *et al.* 74 – Bretagne

11 *Rusco–Melico–Fagetum typicum* Clément *et al.* 74 – Bretagne

12 *Rusco–Fagetum* Durin *et al.* 67 — Bretagne

13 *Ilici–Fagetum melicetosum* Durin *et al.* 67 p.p. – Picardie, Normandie

14 *Ilici–Fagetum* Durin *et al.* 67 p.p. – Picardie, Normandie

15 *Endymio–Fagetum typicum* Durin *et al.* 67 – Bretagne, Normandie

16 *Endymio–Fagetum circeaetosum* et *dryopteridetosum* Durin *et al.* 67 – Bretagne, Normandie

17 *Endymio–Fagetum dryopteridetosum*, Bardat et Frileux 81 – Normandie

18 *Endymio–Fagetum primuletosum*, Bardat et Frileux 81 – Normandie

19 *Endymio–Fagetum holcetosum*, Bardat et Frileux 81 – Normandie

20 'hêtraie à *Endymion non-scriptum*', Lericq 65 – Escaut

21 'hêtraies neutromésophiles et acides', Picard *et al.* 72 – Forêt de Bellème (Orne)

22 'hêtraies neutromésophiles et hêtraies acides à *Melica*', Timbal 71 – Normandie

23 'hêtraies neutromésophiles et hêtraies acides à *Melica*', Timbal 77 – Fontainebleau

24 '*Melico–Fagetum typicum*', Roisin 68 – Picardie, Normandie

25 *Endymio–Fagetum*, Timbal 72 – Compiègne

26 'chênaie–frênaie mélangée atlantique', Lericq 65 – Escaut

27 *Querco–Carpinetum endymietosum*, Géhu 59 – Sambre

28 *Endymio–Carpinetum*, Noirfalise 68 – Flandre

29 'chênaie–charmaie', Dumé 75 – Centre du Bassin Parisien

30 *Endymio–Quercetum* Tx. et Diem. 36, Rameau 82 – Morvan

31 '*Querco–Carpinetum occidentale*' Lemée 43 – Limagne

32 '*Hyperico–Carpinetum*' Braque 78 – Pays de Fours, Berry

33 '*Melico–Fagetum*' Braque 78 – Berry, Pays de Fours

34 '*Asperulo–Fagion*', Dumé 75 – Fontainebleau

35 '*Querco–Carpinetum occidentale*' Lemée 37 – Perche

36 '*Querco–Fagetum*' Lemée 37 – Perche

37 *Rusco–Carpinetum* Noirfalise 68 – Sud ouest du Bassin Parisien

38 *Stellario–Carpinetum*, Rameau 82 – Champagne

39 *Poo chaixii–Carpinetum* (Oberd. 57) Rameau 78, Rameau 82 –Morvan

40 '*Oxalido–Carpinetum*' Braque 78 – Nivernais

41 '*Luzulo sylvaticae–Fagetum*', Braque 78 – Nivernais

42 *Poo chaixii–Carpinetum*, Rameau 78 – Auxois

43 *Poo chaixii–Carpinetum*, Rameau 78 – Bassigny

44 *Poo chaixii–Carpinetum*, Rameau 78 – Pays d'Amance et Apance

45 *Poo chaixii–Carpinetum*, Beaufils 81 – Jura central

46 *Poo chaixii–Carpinetum*, Lapraz 65 – Vosges alsaciennes

47 *Poo chaixii–Carpinetum*, Estrade 82 – Massif Vosgien
Sous-alliance *Daphno–Carpinenion*

48 *Daphno–Fagetum* Durin et Géhu 63, Bardat et Frileux 81 – Normandie

49 *Daphno–Fagetum* Durin *et al.* 67 – Normandie, Picardie

50 '*Lithospermo–Carpinetum*' Oberd. 57, Braque 78 – Berry, Limagne

51 *Scillo bifoliae–Carpinetum* Rameau 74 – Bourgogne

52 *Scillo bifoliae–Carpinetum* Rameau 74 – Sud-Est du Bassin Parisien

53 *Scillo bifoliae–Carpinetum* Rameau 74 – race submontagnarde

54 *Scillo bifoliae–Carpinetum* sous-association thermo-xérophile Rameau 74 – Bourgogne

55 *Scillo bifoliae–Carpinetum*, Rameau 75 – Jura du Nord

56 *Scillo bifoliae–Carpinetum*, Beaufils 81 – Jura central

57 *Aconito vulpariae–Quercetum pedunculatae* Bugnon et Rameau 73 – Bourgogne

58 *Aconito vulpariae–Quercetum pedunculatae*, Rameau et Timbal 79 – Lorraine

Geographical variation in associations of juniper scrub in the central European plain*,**

J. J. Barkman
Biological Station, 9418 PD Wijster, The Netherlands

Keywords: Characteristic cover value, Denmark, *Dicrano-Juniperetum,* Federal Republic of Germany, Geographical variation, German Democratic Republic, *Helichryso–Juniperetum,* Netherlands, Poland, Presence degree, Scrub community, Soil type, Sweden, Vicariants

Abstract

Dense scrub of *Juniperus communis* is assigned to many associations, ungrazed lowland juniper scrub on level, dry, poor, acid sandy soil in NC Europe to two vicarious associations, the *Helichryso–Juniperetum* in Poland and the *Dicrano–Juniperetum* occupying an area from E Germany (Rügen) and S Sweden (Blekinge) to the central Netherlands. The former is characterized by 20 differential species, mostly xerophytic herbs, the latter by 25 species. The shares of cryptogams are 20% and 44% respectively. The *Dicrano–Juniperetum* is clearly differentiated into a NE group of two subassociations (vicariants) and a SW group of six subassociations. The differential species and the area of distribution of the eight subassociations are discussed. The two associations, as well as the two subassociation groups of *Dicrano–Juniperetum* are found in similar habitats and are obviously controlled by climate alone. The eight subassociations of the *DJ*, however, also differ, at least in part, by soil factors.

Finally, the geographical variation of the constancy (presence degree) and abundance (cover degree) of a number of species within the two vicarious associations throughout the European area is given (Tables 1, 2). It is shown that within a narrowly delimited habitat and group of plant communities the geographical indicator values of species may become much greater than if only their geographical occurrence is considered, irrespective of habitat.

Stands of juniper scrub (*Juniperus* spp.) occur throughout Europe. Stands of *Juniperus communis* ssp. *communis* are lacking only in the Mediterranean lowlands, in the alpine zone and in the arctic and subarctic regions. Hence they occur in a wide range of climates. They also occur on a wide variety of soils, including acid and calcareous sands, loam, marl, pure chalk as well as granitic and gneissic rocks, both wet and dry. *Juniperus communis* is, however, restricted to well-aerated mineral soils, poor in nitrogen and phosphorus.

As a consequence of the wide ecological range of *Juniperus communis* its stands show an equally wide variety of undergrowth types. Extra ecological and floristic variation is added by the occurrence on different slope aspects and by the fact that some stands are heavily grazed, others are not. Juniper scrubs therefore belong to a great many associations and even to different classes, viz. *Vaccinio–Piceetea, Querco–Fagetea, Rhamno–Prunetea* and *Festuco–Brometea.*

This paper deals only with the geographical variation in lowland ungrazed juniper scrub on level, dry, poor, acid soil, mostly sand. This type is closely allied to the *Dicrano–Pinion.* I have studied it in the

* Communication No. 234 of the Biological Station Wijster. Communication No. 74 of the Department of Plant Ecology of the Agricultural University, Wageningen.
** Nomenclature follows Heukels-van Oosstroom, Flora van Nederland, 19 ed. (1977); Smith, The Moss Flora of Britain and Ireland (1976); Landwehr, Atlas Nederlandse Levermossen (1980); Wirth, Flechtenflora (1980).

Netherlands, the German Federal Republic (N part), the German Democratic Republic (N part), Denmark, S Sweden, and Poland (N and E part). On the basis of 149 vegetation relevés a clear division can be made in an eastern association, the *Helichryso–Juniperetum* nov. ass. (n.p.), occurring from Warsaw to at least the Russian border (*HJ*), and a western association, the *Dicrano–Juniperetum* nov. ass. (n.p.), occurring in S, DK, DDR, D and NL (*DJ*). For the delimitation of their character (faithful) species these associations have to be compared carefully with all other associations of the NC European plains, including other juniper associations. But we can indicate the differential species.

The *Helichryso–Juniperetum* has 20 differential species against the *Dicrano–Juniperetum* of which six species do not occur in western Europe at all: *Cytisus nigricans, Gypsophila fastigiata, Koeleria glauca, Peucedanum oreoselinum, Scabiosa ochroleuca* and *Viscaria vulgaris*. The other 14 species do occur in western Europe, but not or rarely in the vicarious *Dicrano–Juniperetum: Artemisia campestris, Carex ericetorum, C. hirta, Genista tinctoria, Helichrysum arenarium, Potentilla argentea, Solidago virgaurea, Thymus serpyllum, Veronica spicata* and the cryptogams *Brachythecium albicans, B. velutinum, Cetraria islandica, Cladonia furcata* and *Rhacomitrium canescens*. The majority of these species are xerophytic, in accordance with the dry continental climate of E Poland. In addition there is a great number of differential species of macrofungi (Barkman, unpublished).

The *Dicrano–Juniperetum* has 25 differential species against the *HJ*, all of which do occur in E Poland, however. These species are: *Agrostis vinealis, Campanula rotundifolia, Carex pilulifera, Chamaenerion angustifolium, Deschampsia flexuosa, Hieracium umbellatum, Luzula campestris, Polypodium vulgare, Potentilla erecta, Rubus fruticosus* s.l., *Sambucus nigra, Senecio sylvaticus, Stellaria media, Urtica dioica*, and the cryptogams: *Barbilophozia barbata, Cladonia coccifera, C. gracilis, C. portentosa, Dicranum scoparium, Hypnum jutlandicum, Lecidea granulosa, Lophocolea bidentata, Plagiothecium curvifolium, Pseudoscleropodium purum* and *Ptilidium ciliare*. In accordance with the more humid climate of the area of the *DJ* the share of cryptogams among the differential species (11 out of 25) is higher than in the *HJ* (5 out of 20). In addition, the *DJ* has a number of differential fungi (Barkman, unpublished).

The *Dicrano–Juniperetum* itself also shows a clear geographical variation. There is a north eastern type (group of subassociations) (Denmark and S Sweden) and a southwestern type (NW Germany, Netherlands), with 2 and 6 subassociations respectively.

As they differ mainly in geographical (climatological) respect, they might also be considered vicariants and subvicariants, but in this case these notions are synonymous with subassociations, as there is no other possible way of subdivision of the *DJ*. Besides, the SW subassociations also differ in soil type, at least partly.

A. NE group (vicariant)
Differential species: *Carex arenaria, Galium verum, Luzula campestris, Potentilla erecta, Quercus petraea, Ribes uva-crispa, Stellaria graminea, Viola canina* and the moss *Rhytidiadelphus triquetrus*. There are two subassociations (subvicariants):
A1. *DJ knautietosum*
S Sweden (Blekinge).
Differential species (with regard to A2, B and *HJ*): *Agrostis tenuis*, Fragaria vesca, Geranium robertianum, Geum urbanum, Knautia arvensis, Rosa canina*, Veronica officinalis* and the mosses *Plagiomnium affine** and *Rhodobryum roseum**.
Differential species with regard to A2 and B (also frequent in *HJ*): *Hieracium pilosella**, and the cryptogams *Ceratodon purpureus** and *Cladonia arbuscula*.
Differential species with regard to A2 (also frequent in B): *Hieracium laevigatum* and the cryptogams *Atrichum undulatum, Barbilophozia barbata* and *Cladonia uncialis*.
A2. *DJ majanthemetosum*
N and W Sjaelland (Seeland), Jylland (Jutland), N Schleswig-Holstein.
Differential species with regard to A1 and B: *Majanthemum bifolium* and the moss *Pseudoscleropodium purum**. Differential species with regard to A1 (also frequent in B): *Empetrum nigrum, Frangula alnus, Galium saxatile, Holcus lanatus*, Lonicera periclymenum, Molinia caerulea** and the cryptogams *Cladonia glauca** and *Lophozia ventricosa*.
The only juniper scrub in the G.D.R. that be-

longs to the *Dicrano–Juniperetum*, viz. the one on Fährinsel near Hiddensee (Rügen), a small island in the Baltic, is exactly intermediate between the Swedish and the Danish subassociation: this scrub has 6 out of the 16 differential species of the *DJ knautietosum* (marked above with an asterisk), as well as 4 out of the 10 differential species of the *DJ majanthemetosum* (asterisk, see above).

B. SW group (vicariant)
NW Germany, Netherlands.

Differential species: *Dryopteris carthusiana, D. dilatata, Erica tetralix, Genista pilosa* and the bryophytes: *Barbilophozia lycopodioides* s.l. (incl. *B. hatcheri* which I regard conspecific), *Campylopus paradoxus, C. pyriformis, Dicranella heteromalla, Leucobryum glaucum, Orthocaulis kunzeanus* and *Polytrichum formosum*. Notice that 9 of the 11 differential species are cryptogams (cf. NE group with 1 out of 9). There are six subassociations.

B1. *DJ myrtilletosum*
Lüneburger Heide, Hümmling (Niedersachsen), Haltern (Westfalen), Auf Kölmich (Eiffel).

Differential species: *Campanula rotundifolia, Dryopteris dilatata, Galium saxatile, Oxalis acetosella, Rubus ideaus, Trientalis europaea, Vaccinium myrtillus* and the mosses: *Atrichum undulatum, Hylocomium splendens, Plagiomnium affine, Plagiothecium curvifolium* and *Rhodobryum roseum*. Also 17 species of fungi.

B2. *DJ senecietosum sylvatici*
Province of Drenthe (NE Netherlands)

Differential species: *Hieracium umbellatum, Lonicera periclymenum, Polypodium vulgare, Senecio sylvaticus* and the mosses: *Dicranoweisia cirrhata* (terrestrial!), *Eurhynchium (Kindbergia) praelongum* and *Pseudoscleropodium purum*. In addition 17 species of fungi.

B3. *DJ solanetosum nigri*
Twente, Achterhoek and adjacent parts of Germany (E Netherlands, W Westphalia).

Differential species: *Moehringia trinervia, Solanum nigrum* and the lichen *Cladonia uncialis*. No differential fungi.

B4. *DJ caricetosum arenariae*
River dunes along river Vecht (NE Netherlands).

Differential species: *Carex arenaria, Solanum dulcamara* and the cryptogams: *Cladonia foliacea* and *Dicranum spurium*. No differential fungi.

B5. *DJ lophozietosum ventricosae*
Salland (E Netherlands)

Differential species: *Erica tetralix, Polygonum convolvulus, Ribes uva-crispa* and the cryptogams *Cladonia pityrea* and *Lophozia ventricosa*. No differential fungi.

B6. *DJ orthocauletosum kunzeanae*
Veluwe (central Netherlands).

Differential species: *Aira praecox* and the hepatics *Gymnocolea inflata* and *Orthocaulis (Lophozia) kunzeanus*. No differential fungi.

It is interesting to notice that only the first two subassociations of the SW group have differential fungi (17 species each), whereas the others have none, except in a negative sense: fungus species drop out, when we go from B1 to B6, i.e. from NE to SW. This is, incidentally, just the continuation of a more general phenomenon in juniper scrub, since those in Denmark and particularly in S Sweden are richer in fungi than the Dutch and W German stands.

The six subassociations of the SW group each inhabit a small continuous area, with the exception of the *DJ myrtilletosum*, which is very discontinuous. However, the further we go south, the higher the altitude we find it: in the Hümmling at about 30 m, near Haltern at 100 m, in the Eiffel between 400 and 500 m above sea level.

The floristic differences between the two vicarious associations are probably largely due to climate. The same applies to the NE and SW vicariant of the *DJ*. Within these vicariants, however, edaphic factors also play a role: The S Swedish stands are often found on weathered granitic rocks, the Danish on sand.

Within the SW vicariant the subvicariants differ as follows: 1, 5 and 6: on hills pushed up by glaciers; preglacial coarse sand with gravel, very dry; 2: postglacial cover sands: fine sands, often overlying a compact layer of boulder clay (moist); 3: as 2, without boulder clay in the subsoil. However, groundwater table often fairly high; 4. medium fine, very dry cover sands and river dune sands.

In Table 1 I have presented a choice of differential species of the associations, the two vicariants and the 8 subassociations. The figures refer to the presence degrees of the species. In this way clinal variation of the geographical behaviour of species within one (group of) association(s) may be examined.

70

Table 1. Sociological range of some species in juniper scrub on poor sandy soil in the central European plain.

Presence degrees (in %)	Dicrano–Juniperetum								HJ
	SW vicariant						NE vicariant		
	Vel	Sall	Vecht	Tw	Dr	Lün	DK	S	PL
	DJ orthocauletosum kunzeanae	DJ lophozietosum ventricosae	DJ caricetosum arenariae	DJ solanetosum nigri	DJ senecietosum sylvatici	DJ myrtilletosum	DJ majanthemetosum	DJ knautietosum	Helichryso–Juniperetum
Orthocaulis kunzeanus	50	29	8	7	24	25	8	–	–
Lophozia ventricosa	30	64	31	7	21	13	17	–	–
Carex arenaria	10	14	54	7	2	6	66	50	–
Solanum nigrum	–	14	–	31	2	6	–	–	–
Senecio sylvaticus	10	36	31	21	60	43	33	17	–
Hylocomium splendens	10	–	–	10	33	70	84	100	17
Vaccinium myrtillus	20	29	8	3	2	100	17	67	–
Campylopus pyriformis	70	79	38	66	50	25	8	–	–
Genista pilosa	20	29	–	28	29	20	–	–	–
Erica tetralix	–	64	31	48	29	31	–	–	–
Dryopteris carthusiana	40	57	8	41	41	50	8	17	–
Leucobryum glaucum	60	71	8	24	31	50	–	–	–
Lonicera periclymenum	–	–	8	17	33	–	25	–	–
Galium saxatile	20	57	8	3	55	95	84	–	–
Empetrum nigrum	70	–	–	3	45	25	84	–	–
Majanthemum bifolium	–	7	–	–	–	13	50	17	–
Knautia arvensis	–	–	–	–	–	–	–	66	17
Fragaria vesca	–	–	–	–	–	–	8	100	33
Geranium robertianum	–	–	–	–	–	–	8	83	–
Galium verum	–	–	–	–	–	–	33	100	–
Quercus petraea	–	–	–	–	–	–	25	33	–
Rhytidiadelphus triquetrus	–	–	–	–	–	–	50	66	–
Hieracium pilosella	10	14	–	3	24	20	33	100	84
Ceratodon purpureus	20	21	31	28	33	45	58	100	84
Cladonia arbuscula	50	–	31	41	17	6	17	84	84
Brachythecium velutinum	–	14	–	3	–	–	17	25	50
Thymus serpyllum	10	7	–	–	–	–	8	17	67
Carex hirta	–	–	–	–	–	–	–	17	50
Rhacomitrium canescens	–	–	–	–	–	–	–	–	84
Helichrysum arenarium	–	–	–	–	–	–	–	–	67
Carex ericetorum	–	–	–	–	–	–	–	–	50

In Table 2 not only the presence degree (P), but also the characteristic cover value (C) has been given for a choice of species within the Dicrano–Juniperetum. The latter value refers to the average real cover (in % of the sample plot surface) in those relevés of a vegetation type, in which a species occurs. Multiplication of C and P therefore yields the total cover value per type.

Table 2. Sociological range of some species in vicarious subassociations of the *Dicrano–Juniperetum*. P = presence degree in %, C = characteristic cover value (average real cover percentage in the relevés in which a species is present) $r = 0$, $+ = 0.5$, $1 = 1$, $2m = 3$, $2a = 8$, $2b = 18$, $3 = 35$, $4 = 60$, $5 = 85$).

		Vel orth.	Sall loph.	Vecht car.	Tw sol.	Dr sen.	Lün myrt.	DK maj.	S knaut.
Agrostis vinealis	P	80	64	92	76	79	63	50	67
	C	8	6	1	5	5	0.4	3	7
Dicranum scoparium	P	100	100	100	100	98	100	100	100
	C	19	11	14	17	11	6	1	3
Barbilophozia barbata	P	100	79	54	38	62	81	8	50
	C	12	5	2	4	6	4	1	0.7
Barbilophozia lycopodioides	P	30	21	31	17	29	36	–	–
	C	0.7	0.5	0.5	2	0.9	7	–	–
Rhodobryum roseum	P	–	7	15	10	17	25	17	100
	C	–	0.1	0.3	0.7	0.6	6	2	4
Lophocolea bidentata	P	30	˒ 50	15	10	64	63	67	33
	C	0.3	3	0.5	0.5	3	9	8	1
Brachythecium rutabulum	P	40	36	15	21	88	87	92	100
	C	0.5	0.9	4	0.3	3	9	18	5
Carex arenaria	P	10	14	54	7	2	6	66	50
	C	8	4	7	4	0.5	0.5	11	6
Deschampsia flexuosa	P	80	79	62	48	67	100	100	100
	C	0.7	14	7	10	20	40	34	31
Mnium affine	P	–	–	–	–	5	25	33	100
	C	–	–	–	–	2	1	3	4

Obviously P and C sometimes show parallel variation, as is the case with *Carex arenaria*, but more often they do not. The presence (constancy) of *Deschampsia flexuosa* and *Dicranum scoparium* for instance is very high throughout the whole area, but the former has a very low cover degree in the Veluwe and a high cover in the NE (Lüneburg, Denmark and Sweden), whereas *Dicranum scoparium* shows the opposite behaviour.

Barbilophozia barbata has about the same presence degrees in the *DJ caricetosum arenariae, DJ senecietosum sylvatici* and the *DJ knautietosum arvensis*, but its cover degree varies from 0.7 to 5.7. As to cover degree *Barbilophozia lycopodioides* has its optimum in the *DJ myrtilletosum*, but this is hardly reflected by a higher presence degree.

Summary and conclusions

1. Juniper scrubs in Europe belong to widely different vegetation types.

2. Even the ungrazed scrubs on poor, acid, dry mineral soils display a remarkable geographical variation from W to E Europe, including two vicarious associations and eight vicarious subassociations.

3. Geographical variation can be observed both on a large scale and on a small scale.

4. The small-scale (local) variation can be partly of climatic, partly of edaphic origin. The distinction between subvicariants and subassociations becomes obscure at this level and in the case of *Dicrano–Juniperetum* the two even happen to be synonymous.

5. Within a given plant association individual species may show a geographic variation in frequency that is not reflected in their general frequency. This variation may be clinal or discontinuous. Within a narrowly delimited syntaxon the geographical indicator value of species may therefore be much greater than in case its general occurrence is considered. Compare the increased ecological indicator value of species if examined within a single syntaxon.

6. Geographical variation of the phytosociological behaviour of a plant species can be expressed in graphs or tables showing the variation of its presence degree or its abundance (cover degree). These two variations are by no means always parallel.

Accepted 19.1.1984.

Données synchorologiques sur la végétation littorale européenne

J.-M. Géhu[1] & J. Franck[2]
[1]*Université de Paris V;* [2]*Université de Lille II;* [1,2]*Centre de Phytosociologie de Bailleul, F-59270, France*

Keywords: Atlantic Coast, Dunes, Europe, France, Geosigmetum, Heathland, Littoral, Salt marsh, Synendemism, Synvicariance.

Abstract

Four aspects of the synchorology of plant communities along the European coastline are treated: (1) with examples from sand dunes (*Agropyreta* and *Ammophileta*) and salt marshes (*Puccinellieta maritimae*) synvicariance is elucidated; (2) The increase in synvicariance towards the inner sand dunes is discussed as a result of an increasing effect of autochthonous climatic features; (3) with examples from sand dunes and cliffs the concept of synendemism is elucidated; (4) synvicariance at the landscape (geosigmetum) is discussed with examples from the French coastline.

Introduction

Notre propos est d'attirer l'attention sur quelques faits de synchorologie concernant la végétation du littoral européen.

L'exposé se déroulera en quatre points:
- le premier point concerne la vicariance géographique d'associations de dunes et de prairies salées à l'échelle de l'Europe;
- le deuxième point démontre sur la côte atlantique française l'accélération des phénomènes de géosynvicariance dans les dunes internes soumises à une accentuation des caractères autochtones du climat local;
- le troisième point illustre par quelques exemples la notion de synendémisme côtier;
- le quatrième point développe la notion de vicariance paysagère en étudiant la substitution géographique des *geosigmeta* sur le littoral français de la Belgique à l'Espagne.

Exemple de Géovicariance d'associations de dunes et de prés salés sur le littoral européen

Nous illustrerons ce premier point par quatre exemples de catégories de communautés végétales, deux de dunes, deux de prés salés.

Pour chaque catégorie de communautés (ou groupe d'associations) on donnera un tableau phytosociologique synthétique* précisant les différences floristiques entre les unités et une carte de distribution visualisant les substitutions d'ordre géographique.

Les dunes

On traitera ici des associations de la zone des dunes embryonnaires (*Agropyreta juncei*) et de celles de la zone des dunes blanches (*Ammophileta*).

Les associations à Agropyrum junceum *des dunes embryonnaires*

Il existe sur le littoral européen trois associations à *Agropyrum junceum* de dunes initiales. Le Tableau synthétique 1 en précise les différences floris-

---

* Chaque tableau a été établi en fonction de documents publiés ou inédits archivés à la Station de Phytosociologie de Bailleul où ils peuvent être consultés.

Vegetatio 59, 73–83 (1985).
© Dr W. Junk Publishers, Dordrecht. Printed in the Netherlands.

74

Tableau 1. Agropyreta juncei.

Numéros des associations:	1	2	3
Nombre de relevés:	121	433	50
Agropyron junceum	V	V	V
Elymus arenarius	IV		
Honckenya peploides	III	II	
Eryngium maritimum	+	IV	IV
Euphorbia paralias		IV	III
Calystegia soldanella		IV	II
Sporobolus pungens			IV
Echinophora spinosa			IV
Medicago marina			III
Matthiola sinuata			III
Anthemis maritima			III
Ammophila arenaria	III	III	I
Otanthus maritimus		+	I
Pancratium maritimum		R	I
Galilea mucronata			+
Cakile maritima	II	II	III
Salsola kali	I	I	II
Crithmum maritimum		I	I
Polygonum maritimum		r	II
Euphorbia peplis			I

1 = *Elymo–Agropyretum*
2 = *Euphorbio–Agropyretum*
3 = *Agropyretum–Mediterraneum*

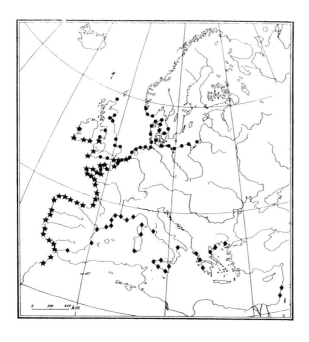

Fig. 1. Carte 1: ● *Elymo–Agropyretum,* ★ *Euphorbio–Agropyretum,* ◆ *Agropyretum–Mediterraneum.*

tiques. La Carte 1 (Fig. 1) indique la distribution géographique de ces trois associations géovicariantes qui exploitent la même zone écologique du littoral.
- L'*Elymo–Agropyretum* a une dispersion nord-atlantique, ouest baltique;
- L'*Euphorbio–Agropyretum* a une aire thermo-atlantique, du sud des Iles-Britanniques au Maroc;
- L'*Agropyretum* méditerranéen se rencontre tout autour de la Méditerranée.

Les associations des dunes blanches, mobiles à Ammophila arenaria

On peut distinguer sur le littoral européen quatre associations à *Ammophila arenaria* de dunes blanches, plus ou moins meubles (Tableau 2, Fig. 2):

Tableau 2. Ammophileta arenariae.

Numéros des associations:	1	2	3	4
Nombre de relevés:	223	238	100	40
Ammophila arenaria	V	V	V	V
Elymus arenarius	V	+		
Lathyrus maritimus	III			
Sonchus arvensis	III	r		
Ammocalamagrostis baltica	II			
Festuca arenaria	II			
Eryngium maritimum	I	IV	IV	IV
Euphorbia paralias		IV	V	IV
Calystegia soldanella		IV	III	I
Festuca juncifolia		II		
Galium arenarium		II		
Otanthus maritimus		+	IV	
Pancratium maritimum		r	III	II
Echinophora spinosa				V
Anthemis maritima				V
Sporobolus pungens				II
Cutandia maritima				II
Agropyron junceum	II	II	II	II
Medicago marina		+	II	V
Matthiola sinuata		I		II
Crucianella maritima			II	III
Cakile maritima	I	I	I	I
Salsola kali		r	+	+
Honckenya peploides	+	+		
Carex arenaria	+	+		
Hypochoeris radicata	+	+		
Polygonum maritimum			I	I

1 = *Elymo–Ammophiletum*
2 = *Euphorbio–Ammophiletum*
3 = *Otantho–Ammophiletum*
3 = *Echinophoro–Ammophiletum*

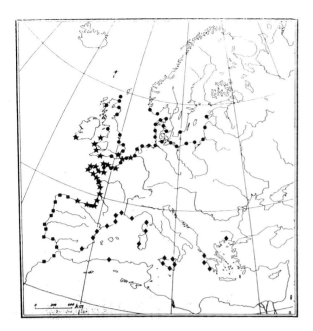

Fig. 2. Carte 2: ● *Elymo-Ammophiletum,* ★ *Euphorbio-Am-mophiletum,* ■ *Otantho-Ammophiletum,* ◆ *Echinophoro-Am-mophiletum.*

– L'*Elymo-Ammophiletum* possède une disper-sion nord-atlantique, sud-baltique;
– L'*Euphorbio-Ammophiletum* présente une ré-partition atlantique, du sud des Iles Britanniques au nord de l'Espagne;
– L'*Otantho-Ammophiletum* dans lequel l'*Am-mophila arenaria* est déjà représenté le plus sou-vent par sa variété *arundinacea* est davantage thermophile et cantonné à l'ouest ibérique et au nord-ouest marocain;
– L'*Echinophoro-Ammophiletum* a une distribu-tion ouest méditerranéenne du Péloponèse au Magreb et à l'Espagne.

Les prairies salées

Il ne sera envisagé ici que le cas des prairies de *Puccinellia maritima (Puccinellieta maritimae)* et des subcuvettes à *Plantago maritima (Plantagineta maritimi)* du schorre.

Les associations de prés salés à *Puccinellia mariti-ma*

Il s'agit dans l'ensemble de prairies denses et dominées par *Puccinellia maritima.* Vers le sud-ouest elles peuvent avoir une signification primaire,

ou secondaire (dégradation par pâturage de l'*Hali-mionetum*).

Sur la base des combinaisons floristiques le Ta-bleau synthétique 3 permet la distinction de 3 associations sur le littoral nord-ouest européen.

La Carte 3 (Fig. 3) precise la distribution géogra-phique de trois *Puccinellieta* vicariants.
– Le *Stellario humifusae-Puccinellietum mariti-mae* ass. nov. est strictement atlantique-subarc-tique (Islande, nord Scandinavie);
– L'*Astero-Puccinellietum maritimae* ass. nov. à dispersion nord-atlantique-baltique s'étend du nord des Iles Britanniques à la Scandinavie;
– L'*Halimiono portulacoidis-Puccinellietum ma-ritimae* Géhu 76, plus atlantique s'observe du nord-ouest de l'Espagne aux Iles Britanniques et à l'Allemagne. Il est enrichi en espèces thermo-philes qui s'estompent progressivement vers le nord-est.

Tableau 3. Puccinellieta maritimae.

Numéros des associations:	1	2	3
Nombre de relevés:	14	87	201
Puccinellia maritima	V	V	V
Stellaria humifusa	II		
Halimione pedunculata		II	
Halimione portulacoides			III
Limonium vulgare			II
Spartina anglica			II
Arthrocnemum perenne			I
Plantago maritima	II	III	I
Triglochin maritimum	II	II	II
Armeria maritima	I	+	+
Cochlearia officinalis	+	+	
Aster tripolium		III	V
Spergularia media		II	II
Spergularia salina		II	+
Glaux maritima		I	+
Limonium humile		+	
Cochlearia anglica			+
Agrostis stolonifera	II	+	+
Atriplex hastata	+	+	II
Festuca rubra	I	+	
Suaeda maritima		III	III
Salicornia gpe *europaea*		II	II
Salicornia gpe *dolichostachya*		II	I

1 = *Stellario humifusae-Puccinellietum maritimae*
2 = *Astero-Puccinellietum maritimae*
3 = *Halimiono portulacoidis-Puccinellietum maritimae*

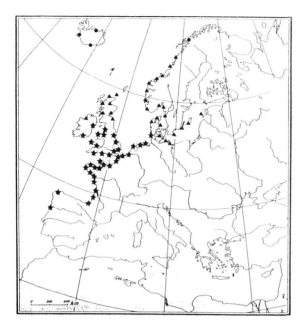

Fig. 3. Carte 3: ● *Stellario–Puccinellietum,* ▲ *Astero–Puccinellietum,* ★ *Halimiono–Puccinellietum.*

Tableau 4. Plantagineta maritimae.

	I	2
Numéros des associations:	I	2
Nombre de relevés:	42	58
Plantago maritima	V	V
Limonium vulgare		V
Halimione portulacoides		IV
Arthrocnemum perenne		II
Puccinellia maritima	V	IV
Spergularia media	III	IV
Aster tripolium	III	V
Armeria maritima	V	IV
Glaux maritima	V	II
Triglochin maritimum	I	V
Festuca littoralis	I	III
Limonium humile	+	+
Cochlearia officinalis	+	
Cochlearia anglica		I
Salicornia div. sp.	I	III
Suaeda maritima	+	III
Atriplex hastata		I

1 = *Glauco–Plantaginetum*
2 = *Plantagini–Limonietum*

Les subcuvettes du schorre à Plantago maritima

Sur les schorres très plats ou à morphologie de subcuvette, le ralentissement du ressuyage du substrat après le flot de marée se traduit toujours dans la végétation du pré salé par l'apparition d'une prairie plus rase, plus ouverte d'où régresse fortement *Puccinellia maritima* mais où se développent de façon optimale des espèces comme *Plantago maritima.* Le phénomène est observable aussi bien sur les côtes atlantiques européennes qu'américaines.

Sur les côtes d'Europe il est possible de distinguer deux associations principales de subcuvettes La (Tableau 4, Fig. 4):

- Le *Glauco–Plantaginetum maritimae* ass. nov. est cantonné sur les côtes nord-atlantiques de l'Ecosse et de la Scandinavie moyenne, tandis que le *Plantagini–Limonietum* Westhoff possède une aire plus méridionale, du sud de la France au sud de l'Angleterre et à L'Allemagne du nord-ouest. Pour être complet il faudrait aussi mentionner dans le cadre de ces végétations de subcuvettes à plantain maritime, le *Cochleario–Plantaginetum* Géhu 76, endémique nord-finistérien.

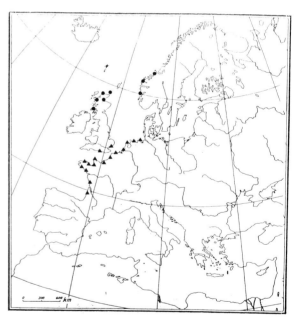

Fig. 4. Carte 4: ● *Glauco–Plantaginetum,* ▲ *Plantagini–Limonietum.*

Accélération de la géosynvicariance dans les dunes internes à climat autochtone

Le phénomène de l'accélération de la géosynvicariance des associations dunaires internes sous

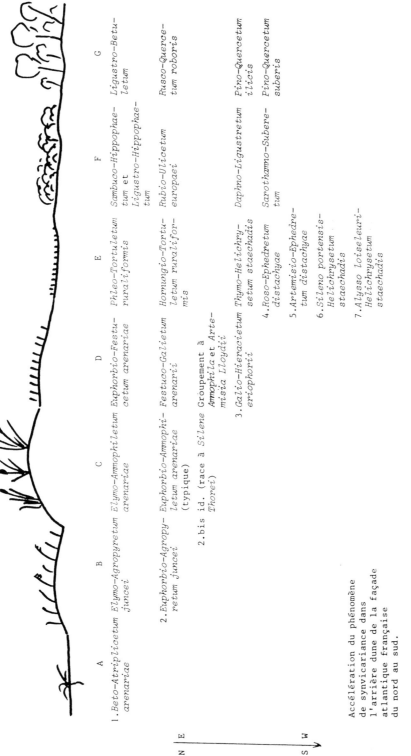

Fig. 5. *Schéma 1*: Accélération du phénomène de synvicariance dans l'arrière dune de la façade atlantique française du nord au sud.

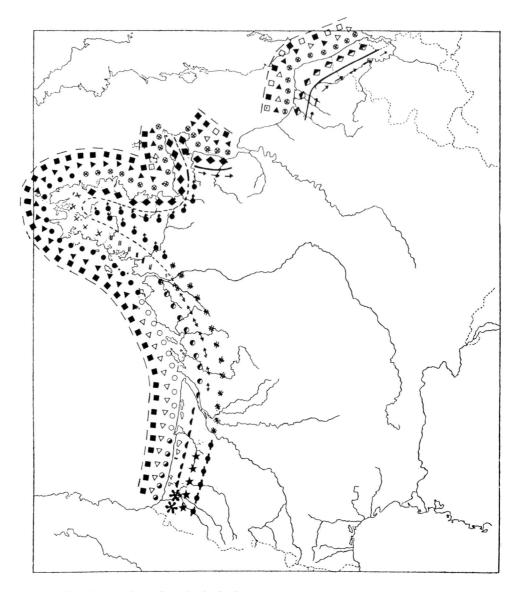

Fig. 6. Carte 5: Accélération des phenomènes de géovicariance dans les transects dunaires atlantiques français. Répartition des associations zone par zone.

l'effet de l'accentuation des conditions microclimatiques à caractère autochtone est facile à mettre en évidence par la comparaison de transects dunaires relevant les associations présentes de la plage à la forêt sur sable et effectués en divers secteurs des côtes atlantiques de France (Fig. 5, Fig. 6).

Sur le versant maritime des systèmes dunaires (hauts de plages, dunes embryonnaires, dunes blanches à oyats), soumis à l'influence directe du climat allochtone venu de la mer, le phénomène de substitution des groupements, du nord au sud de la façade atlantique est peu marqué: une seule association au niveau des hauts de plages, le *Beto–Atri-*

plicetum arenariae, avec cependant une variation plus thermophile à *Salsola soda* au sud de la Loire.

Deux associations dans la zone des dunes embryonnaires, l'*Elymo–Agropyretum* vers le nordest, l'*Euphorbio–Agropyretum* ailleurs.

Deux associations dans la zone des dunes blanches à oyats, l'*Elymo–Ammophiletum* au nord-est, l'*Euphorbio–Ammophiletum* à l'ouest, plus une remarquable race à *Silene thorei* de cette dernière association en Aquitaine-Charentes.

Par contre dans l'intérieur des dunes, et sur les versants continentaux oú se développent les caractères autochtones du climat et par suite de l'accélé-

ration du phénomène de géosynvicariance qui en résulte, il existe un nombre beaucoup plus élevé d'associations:
- quatre dans la zone des gazonnements à fétuques, en retrait des Ammophilaies (dunes grises);
- sept dans la zone interne des dunes noires à cryptogames.

Au-delà, en raison même de la structure plus complexe de la végétation qui interfère sur le microclimat, le nombre des substitutions redevient, plus faible et se stabilise à quatre (zone des fourrés et des forêts littorales sur sables dont il existe quatre grands types sur les côtes atlantiques françaises).

Toutes proportions gardées l'intéressante notion de climat allochtone et autochtone, brillamment développée par Carbiener (1966) en montagne vosgienne est donc transposable à l'interprétation des systemes de végétation dunaire. A condition édaphique sensiblement égale, elle est un des éléments les plus intéressants d'explication de la géosynvicar-

iance dunaire et de son accélération de la plage aux dépressions sèches internes.

Notion de synendémisme littoral

Nous avons précédemment montré sur le littoral français (Géhu, 1978) que diverses associations côtières, notamment de dunes, de falaises, et pour une moindre part de prés salés possédaient une aire très réduite et pouvaient être considérées comme synendémiques du littoral français, voire d'une portion restreinte de celui-ci.

Pour ne reprendre que quelques associations citées ci-dessus, le *Galio–Hieracietum eriophorii* n'existe qu'entre Arcachon et Biarritz et l'*Alysso loiseleuri–Helichrysetum staechadis* n'apparaît que sur quelques dizaines de kilomètres vers l'embouchure de l'Adour.

Toujours dans la dune sont des synendémiques françaises les associations C_2 bis, D_2, D_2 bis, D_3, E_2,

Dunes embryonnaires	—	*Beto–Atriplicetum laciniatae* (A)
	□	*Elymo–Agropyretum juncei* (B1)
	■	*Euphorbio–Agropyretum juncei* (B2)

Dunes blanches	△	*Elymo–Ammophiletum arenariae* (C1)
	▲	*Euphorbio–Ammophiletum arenariae* (C2)
	▽	*Euphorbio–Ammophiletum* race à *Silene thorei* (C2 bis)

Dunes grises	⊗	*Euphorbio–Festucetum arenariae* (D1)
	●	*Festuco–Galietum arenarii* (D2)
	○	Gre à *Artemisia lloydii* (D2 bis)
	◕	*Galio–Hieracietum eriophori* (D3)

Dunes noires	◈	*Phleo–Tortuletum ruraliformis* (E1)
	◆	*Hornungio–Tortuletum ruraliformis* (E2)
	×	*Thymo–Helichrysetum staechadis* (E3)
	‖	*Roso–Ephedretum distachyae* (E4)
	◑	*Artemisio–Ephedretum distachyae* (E5)
	◣	*Sileno–Helichrysetum staechadis* (E6)
	*	*Alysso–Helichrysetum staechadis* (E7)

Fourrés	—	*Ligustro–Hippophaetum rhamnoides* (F1)
	----	*Rubio–Ulicetum europaei* (F2)
	↕	*Daphno–Ligustretum vulgaris* (F3)
	★	*Sarothamno–Quercetum suberis* (F4)

Forêts	↑	*Ligustro–Betuletum pubescentis* (G1)
	♠	*Rusco–Quercetum roboris* (G2)
	⋇	*Pino–Quercetum ilicis* (G3)
	♦	*Pino–Quercetum suberis* (G4)

Tableau 5. Ulici maritimi–Ericeta vagantis.

		1	2	3
	Numéros des associations:	1	2	3
	Nombre de relevés	15	49	24
Car.	*Erica vagans*	V	V	V
	Ulex europaeus ssp. *maritimus*	V	V	V
D_1	*Smilax aspera*	III		
	Leucanthemum crassifolium	III		
	Potentilla montana	II		
	Vincetoxicum officinale	II		
	Daucus gummifer	II		
	Tamus communis	II		
	Lithodora diffusa	+		
D_2	*Erica ciliaris*		III	
	Daucus gadeceaui		II	
	Plantago recurvata		II	
	Simaethis planifolia		II	
	Scorzonera humilis		II	
	Asparagus prostratus		I	
D_3	*Schoenus nigricans*	+		V
	Carex flacca			IV
	Betonica officinalis			III
	Brachypodium sylvaticum			III
	Sanguisorba officinalis			III
	Pedicularis sylvatica			III
	Filipendula vulgaris			III
	Scilla verna			III
	Genista anglica			II
	Viola lactea			II
	Orchis elodes			II
	Carex panicea			II
	Agrostis setacea			II
	Erica tetralix			II
	Genista pilosa			I
All.	*(Ulicion maritimi)*			
	Ulex gallii fo. *humilis*	+	II	III
	Dactylis glomerata var. *marina*	II	II	II
	Festuca rubra spp. *pruinosa*	III	II	I
	Lotus corniculatus var. *crassifolius*	II	II	II
	Juncus maritimus		+	I
U.S.	*Potentilla erecta*	I	III	V
	Erica cinerea	+	V	V
	Calluna vulgaris	+	II	I
	Sieglingia decumbens		III	+
Comp	*Rubia peregrina*	V	II	+
	Centaurea nemoralis	II	I	I
	Cirsium filipendulum	II	III	
	Brachypodium pinnatum	V	III	
	Rubus sp.	III		II
	Pteridium aquilinum	I		I
	Molinia coerulea		II	IV
	Viola riviniana		II	III
	Teucrium scorodonia		+	I
Autres espèces		3	8	9

E_3, E_4, E_5, E_6, E_7, F_2, F_3, F_4, G_3 et G_4 du Schéma 1 (Fig. 5).

A propos de la notion de synendémisme il est intéressant de préciser que la combinaison d'espèces même relativement banales ou non rares peut créer un groupement ou une association de haute signification biogéographique, à caractère de préciosité réelle.

Dans le cadre de cet exposé nous attirons plus spécialement l'attention sur le cas de trois associations strictement endémiques de landes littorales (sur falaises) à *Erica vagans* ne possédant chacune qu'une aire ponctuelle mais appartenant au même groupe d'association (*Ulici maritimi–Ericeta vagantis*) (Tableau 5).

L'une de ces landes est cantonnée aux falaises argileuses du pays basque français vers St-Jean-de-Luz, Biarritz (Géhu & Géhu-Franck, 1981); une autre n'existe qu'au sommet des grandes falaises de schistes des côtes ouest de Belle-Ile et de Groix (Géhu & Géhu-Franck, 1975), au sud de la Bretagne; la troisième enfin est cantonnée sur les falaises du cap Lizzard en Cornouailles britannique (Bridgewater, 1980).

Vicariance paysagère de géosigmeta

Sous l'impulsion de R. Tüxen l'application des méthodes phytosociologiques à l'étude des paysages a fait l'objet de nombreux travaux. Un colloque de notre association internationale y a été consacré (Tüxen, 1978).

Pour terminer cet exposé nous souhaitons simplement rappeler que les paysages phytosociologiques sont également sujets à variations chorologiques et influencés par les phénomènes synchorologiques.

Le Tableau 6 permet ainsi de déceler dans le paysage des dunes sèches du littoral atlantique français, l'existence de trois *géosigmeta* (dont quatre sous-géosigmeta) de la Belgique à l'Espagne. La liste quantifiée des associations constitutives y apparaît.

Figure 7 precise la répartition de ces *géosigmeta* et de leurs sous-unites. Les sables méditerranéens sont occupés par un quatrième géosigmetum, bien distinct des trois premiers, mais qui n'est pas étudié ici.

Tableau 6. Geosigmeta dunaires atlantiques.

Geosigmeta	1							2a					2b									3a									3b						
Côtes	Manche orientale							Massif armoricain														Centre									Sud-ouest						
Régions	Flandre-Picardie							Contenin Côtes-du-nord					Finistère					Morbihan				Charentes						Gironde			Landes et				Adour		
Numéros de relevés:	1	2	3	4	5	6	7	8	9	10	11	12	13	14	15	16	17	18	19	20	21	22	23	24	25	26	27	28	29	30	31	32	33	34	35	36	37
Nombre d'associations:	10	8	10	10	11	11	7	11	7	10	9	7	8	7	7	7	7	7	6	5	6	7	7	7	7	7	7	7	8	7	8	8	6	7	8	6	7

Beto-Atriplicetum laciniatae	/+ /1 /+ … /+ /+ … /1 /+ /1 /+ … /1 /+ … /+ /1 /+ … /+ /+ … /1 … /+
Euphorbio-Agropyretum juncei	/+ /1 /1 /1 … /+ /+ /+ .r … /2 /1 /1 … /1 /1 /1 /1 … /1 /1 /2 … /1 /2 /1 /1 … /1 /1 /1 /2 … /1 /1 /2
Euphorbio-Ammophiletum arenariae	/2 03 03 03 03 /2 03 … /1 /1 /1 /1 /2 … /2 /1 /1 .+ /1 … /1 /1 /1 /r … /2 /1 /2 … /2 /2 /2 /2 … /2 /2 /2 /2 … /2 /2 .+ .r
Euphorbio-Festucetum arenariae	.+ .+ 02 /1 .+ … .+
Festuco-Galietum maritimi	.+ .+ … .+ … /+ /+
Elymo-Agropyretum juncei	.r … .+ .r .+
Elymo-Ammophiletum arenariae	.+ .+ .+ .1 … .+ .1
Tortulo-Phleetum arenarii	02 02 02 01 02 03 02 … 02 02 02 02 02
Sambuco-Hippophaetum	04 04 02 05 03 02 .+ … 05 04 05 03 /1
Ligustro-Hippophaetum	01 .+ 02 01 .r .+ 04
Claytonio-Anthriscetum	/1 .+ 01 /1 .+ /+ /1 .+ … .r .+
Ligustro-Betuletum	.+ 02 01 .r /1 03
Ulici-Geranietum sanguinei	.+ .+ .+ .+ .+ … .+ .+ … .+ .+ … /1 .+
Rubio-Ulicetum	.1 02 .r /+ .+ … .+ .+ .+ … /1 .+
Hornungio-Tortuletum	02 02 02 02 .+ .+ .1 … .+ .+ .+ 01 .1
'Mielles' à Koeleria albescens et Linum angustifolium	05 04 05 03 /1 … .+ .+ .+ .1
Festuco-Galietum arenarii	02 02 /1 /1 /+ /1 /2 03 … .+ /+ /+ … /1 /1 /2 /2 02 02 /1 … 03 03 03 03 02 /1
Thymo-Helichrysetum	04 04 02 /1 /2 /1 /1 /1 … .+ /+ … 04 .+ /2 03 03 03 02 /1 … 03 02 02 02 02 03 /1
Roso-Ephedretum	05 05 03 04 03 03
Ourlet à Cistus salviaefolius	.+ … /+
Corynephoretum à Tuberaria guttata	/+ /+ /+ +W … /+ /+ /+ /+ … /+ /+ /+
Gpt à Ammophila et Artemisia lloyd.	.+ .r … .+ .1 .+ … .1 .1 .+ … .+ .r
Pino-Quercetum ilicis	/1 /1 /2 /2 02 02 /1 … 04 .+ /2 02 02 03 02 /1
Hieracio eriophori-Galietum arenar.	04 .+ /2 03 03 03 02 /1
Pino-Quercetum suberis	.+ 04 04 03 /1 … .+ 04 04 03 /1
Artemisio-Ephedretum	/1 .+ /1 /2 /1 … /1 .+ /1 /2 /1
Daphno-Ligustretum	02 02 02 02 .+ … 02 /2 .1 .+
Sileno-Helichrysetum	/1 .+ /1 … /1 .r .+
Scopario-Sarothamnetum	02 02 02 02 … .r .+
Alysso-Helichrysetum	02 04 04
Sarothamno-Suberetum	/+ /1 .5

Forêts sur dunes	Ligustro-Betuletum	Rusco-Quercetum	Pino-Quercetum ilicis	Pino-Quercetum suberis
	Zone de la forêt caducifoliée		Zone des forêts littorales sempervirentes	

.... race à *Silene thorei* de l'*Euphorbio-Ammophiletum* et race à *Salsola soda* du *Beto-Atriplicetum*; le 1er symbole correspond à la forme du paysage dans l'espace, 0: spatial, / : linéaire, . : ponctuel. Le chiffre quantifie l'espace occupé par chaque groupement dans le paysage étudié: 5 = 75 à 100%, 4 = 50 à 75%, 3 = 33 à 50%, 2 = 20 à 33%, 1 = 10 à 20%, + = 5 à 10%, r = 1 à 5%.

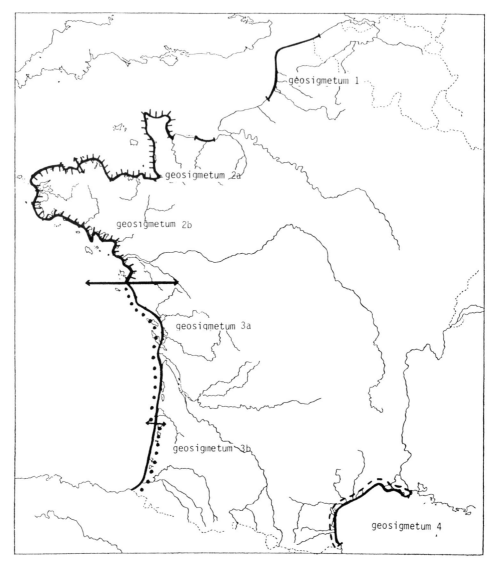

Fig. 7. Carte 6: Paysages dunaires.

Bibliographie

Bridgewater, P., 1980. Phytosociological studies in the British heath formation I. Phytocoenologia 8: 191–235.

Carbiener, R., 1966. La végétation des Hautes-Vosges dans ses rapports avec les climats locaux, les sols et la géomorphologie. Ms. 109 pp. Thèse, Orsay.

Géhu, J.-M. & Géhu-Franck, J., 1975. Apport à la connaissance phytosociologique des landes du littoral de Bretagne. Coll. phytosociol. 2, Les Landes, Lille 1973, pp. 193–212. Cramer, Vaduz.

Géhu, J.-M., 1976. Approche phytosociologique synthétique de la végétation des vases salées du littoral atlantique français. Coll. phytosociol. Les Vases salées, Lille 1975, pp. 395–462.

Cramer, Vaduz.

Géhu, J.-M., 1977. Le concept de sigmassociation et son application à l'étude du paysage végétal des falaises atlantiques françaises. Vegetatio 34: 112–125.

Géhu, J.-M., 1978a. Les phytocoenoses endémiques des côtes françaises occidentales. Bull. Soc. Bot. France 125: 199–208.

Géhu, J.-M., 1978b. Les sigmassociations de la xérocere des dunes atlantiques françaises de Dunkerque à Biarritz. In: R. Tüxen (ed.), Ber. Intern. Symp. I.V.V. Rinteln, pp. 77–82. Cramer, Vaduz.

Géhu, J.-M. & De Foucault, B., 1978. Les pelouses à Tortula ruraliformis des dunes du nord-ouest de la France. Coll. phytosociol. 6, Lille (1977), pp. 269–273. Cramer, Vaduz.

Géhu, J.-M., et al., 1979. Etude phytocoenotique analytique et globale de l'ensemble des vases et prés salés et saumâtres de la façade atlantique française. Min. Envir. 1 514 pp. Neuilly-sur-Seine.

Géhu, J.-M. & Géhu-Franck, J., 1981. Aperçu phytosociologique sur les falaises d'Hendaye et de St-Jean-de-Luz (Pays basque). Doc. phytos N.S.V.: 363–374.

Géhu, J.-M., 1982. La végétation des plages de sables et des dunes des côtes françaises. Ms. 61 pp. Bailleul.

Tüxen, R., (éd.), 1978. Assoziationskomplexe (Sigmetea). Ber. Intern. Symp. I.V.V. Rinteln, 1977. Cramer, Vaduz.

Accepted 19.1.1984.

Liens entre chorologie et différenciation de quelques associations du Mesobromion erecti d'Europe occidentale et centrale*

J. M. Royer

Laboratoire de Taxonomie Expérimentale et Phytosociologie, Faculté des Sciences, La Bouloie, 25030 Besançon Cedex, France

Keywords: Calcareous grassland, Central Europe, Chorological differentiation, Ecological differentiation, *Festuco-Brometea*, Marl grassland, *Mesobromion erecti,* Semi-dry grassland, Western Europe

Abstract

Chorological phenomena are very important for the differentiation of the *Mesobromion erecti* associations in western and central Europe. We recognized several associations of grasslands specific of calcareous brown-soils (on marl) in a range from the southwest of France to Switzerland and the Boulonnais, which correlated with a floristic impoverishment. Beyond the north and the east of France, the characteristic species disappear: the associations of marl grasslands cannot be differentiated.

Chorological phenomena are also important for the differentiation of the semi-dry grasslands in a range from England to Germany, with four greater associations (*Cirsio-Brometum, Festuco-Brachypodietum, Festuco lemanii-Brometum, Gentiano-Koelerietum*) and several local communities and minor associations.

Three association groups are recognized in the *Mesobromion erecti:* in addition to the northwest European and the centralwest European association groups (Willems, 1980), we distinguish a southwest European group very rich in subatlantic and submediterranean species.

Introduction

Les pelouses calcaires de l'Europe occidentale et centrale appartiennent à la classe *Festuco-Brometea* et à l'órdre *Brometalia erecti.* Ce dernier, d'origine subatlantique-sub-méditerranéenne, s'oppose à l'ordre *Festucetalia vallesiacae* d'origine continentale. On distingue habituellement au sein des *Brometalia* d'une part le *Xerobromion* sur sols peu profonds et en climat chaud, d'autre part le *Mesobromion erecti,* alliance plus mésophile. Les associations relevant du *Mesobromion erecti* sont répandues dans toute l'Europe occidentale et centrale, depuis l'Irlande et le sud de la Suède jusqu'au nord de l'Espagne, aux Apennins et aux Alpes orientales (Wolkinger & Plank, 1981).

*Nomenclature d'espèces d'après Flora Europaea, par Tutin *et al.* (1964–1980); nomenclature syntaxonomique d'après Royer (1978, 1982) et Oberdorfer (1978).

Les phénomènes chorologiques jouent un grand rôle dans la différenciation des associations du *Mesobromion erecti,* tout comme les phénomènes écologiques. Nous envisagerons très succinctement le cas des seconds, puis nous développerons celui des premiers.

Les principales variations constatées au sein du *Mesobromion erecti,* liées à des facteurs écologiques, sont les suivantes (Fig. 1):

– sur sols profonds, passage progressif à *l'Arrhenatherion elatioris* et optimum pour *Onobrychis viciifolia, Salvia pratensis, Campanula glomerata . . .*

– sur sols acidifiés, passage progressif au *Nardion* et au *Calluno-Genistion,* avec *Chamaespartium sagittale, Calluna vulgaris, Danthonia decumbens, Antennaria dioica . . .*

– sur sols bruns calcaires des roches marneuses, passage progressif au *Molinion coeruleae* et optimum pour *Blackstonia perfoliata, Tetragonolobus maritimus . . .*

Vegetatio 59, 85–96 (1985).

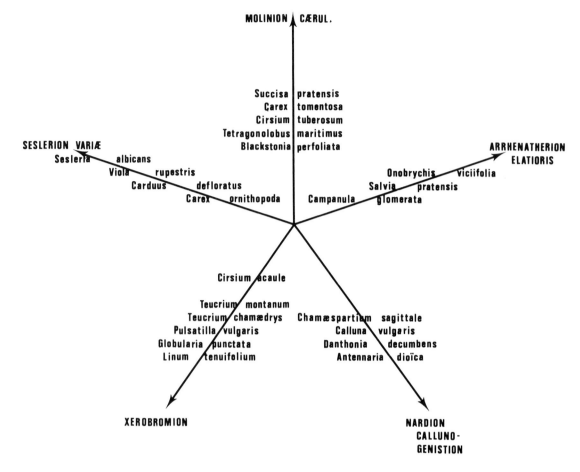

Fig. 1. Principales variations écologiques au sein du *Mesobromion erecti.*

– sur sols superficiels, passage au *Xerobromion* et optimum pour *Teucrium chamaedrys, Teucrium montanum, Pulsatilla vulgaris, Cirsium acaule . . .*

– sur rendzines, passage au *Seslerion variae,* en climat montagnard (*Seslerio–Mesobromenion*) a- vec *Sesleria albicans, Viola rupestris, Thesium alpi- num, Carex ornithopoda . . .*

Tout comme Willems (1980), nous ne retiendrons pas comme facteur essentiel l'activité humaine (pâ- turage et fauchage) considérée comme détermi- nante par Muller (1966) et Oberdorfer (1978).

Dans les régions où l'alliance *Mesobromion erec- ti* est bien développée, cette variation écologique conduit à la séparation d'associations distinctes, par exemple en Bourgogne (Royer, 1973, 1978, 1982), dans le Jura français (Pottier-Alapetite, 1943), dans le Jura Suisse (Zoller, 1974), dans la

région parisienne (Bournerias, 1979). Par contre dans d'autres régions la variabilité plus faible ne conduit qu'à des sous-associations, par exemple en Allemagne du sud (Oberdorfer, 1978). La Figure 2 résume ces données pour 3 régions différentes: Bourgogne, Jura Suisse, Schwäbische Alb (Jura souabe): Cette figure permet par ailleurs de com- prendre le rôle fondamental de la chorologie qui permet la distinction d'associations ou groupe- ments différents en fonction de la région considé- rée.

Dans le cadre de ce travail nous développerons deux cas où les facteurs chorologiques jouent un rôle essentiel, celui des associations propres aux sols bruns calcaires marneux (passage au *Molinion coeruleae*) et celui des associations propres aux sols superficiels (passage au *Xerobromion*).

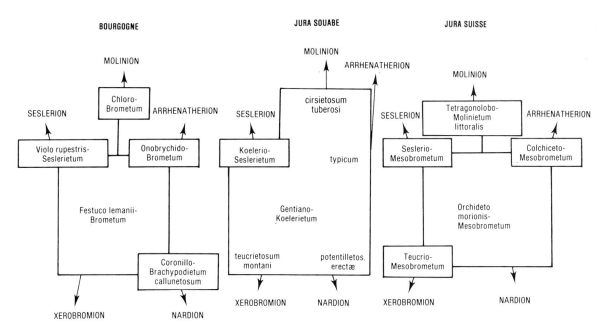

Fig. 2. Influence de la chorologie sur la différenciation d'associations ou de sous-associations au sein du *Mesobromion erecti.*

Influences chorologiques au niveau des associations marnicoles

Ces associations sont caractérisées par une dizaine d'espèces assez constantes provenant du *Molinion coeruleae* d'une part (*Tetragonolobus maritimus, Cirsium tuberosum, Silaum silaus, Succisa pratensis, Carex panicea, Carex tomentosa*), d'espèces particulières au moins régionalement d'autre part (*Blackstonia perfoliata, Molinia coerulea* ssp. *arundinacea, Senecio erucifolius, Peucedanum cervaria, Plantago maritima* ssp *serpentina, Centaurium erythraea . . . .*).

L'analyse du Tableau 1 permet de dégager les principales influences chorologiques conduisant à la distinction d'associations particulières selon les régions.

La richesse floristique maximum est atteinte dans le sud-ouest de la France, par exemple dans les Charentes où nous avons défini récemment une association provisoire à *Ophrys scolopax* et *Carex flacca* (colonne 2) bien différenciée par de nombreuses espèces méditerranéennes et méditerranéo-atlantiques (*Ophrys scolopax, Ophrys fusca, Serapias lingua, Linum salsoloïdes* ssp. *suffruticosum, Carduncellus mitissimus, Dorycnium pentaphyllum* ssp. *pentaphyllum, Polygala calcarea . . .*).

Tableau 1. Associations marnicoles à *Blackstonia perfoliata* et *Tetragonolobus maritimus.*

Numéro des associations	1	2	3	4	5	6	7	8	9	10	11	12	13	14
Nombre de relevés	4	5	L	30	45	11	5	14	10	17	10	10	10	19
Espèces caractéristiques du groupe d'associations														
Blackstonia perfoliata	2	5	+	5	5	5	3	5	1	.	2	1	1	.
Peucedanum cervaria	.	2	+	5	4	.	.	.	1	2	.	2	1	.
Inula salicina	.	.	+	1	2	1	.	.	1	1	1	1	1	.
Agrostis stolonifera	.	.	.	2	3	2	2	.	1	3	.	.	.	.
Molinia caerulea ssp. *arundinacea*	.	2	.	2	3	.	.	.	3	5	4	5	5	.
Cirsium tuberosum	2	2	.	1	1	.	.	.	5	4	.	5	5	3
Silaum silaus	.	1	.	1	1	5	.	.	5	2	2	1	5	.

Tableau 1. (continué)

Numéro des associations	1	2	3	4	5	6	7	8	9	10	11	12	13	14
Nombre de relevés	4	5	L	30	45	11	5	14	10	17	10	10	10	19
Senecio erucifolius	.	.	+	1	2	4	.	.	4	2	1	3'	.	.
Carex tomentosa	.	1	+	1	1	1	1	.	3	1	.	.	3	.
Succisa pratensis	.	1	+	1	1	.	.	3	4	4	4	2	5	.
Tetragonolobus maritimus	1	.	+	.	3	1	3	1	5	4	5	4	5	4
Carex panicea	.	.	.	.	1	1	.	.	4	2	.	1	4	.
Plantago maritima ssp. *serpentina*	.	.	.	.	1	.	.	.	.	.	3	.	.	.

Espèces méditerranéennes

Ophrys scolopax	3	5												
Serapias lingua	2	1												
Dorycnium pentaphyllum ssp. *pentaphyllum*	1	2												
Ophrys fusca	.	2												
Linum suffruticosum ssp. *salsoloïdes*	.	3												
Carduncellus mitissimus	.	3	+											

Espèces méditerranéo-atlantiques et subméditerranéennes

Polygala calcarea	.	5	+	1	.	1	3	1	1	.	.	.	.	.
Coronilla minima	.	3	+	1	1	.	.	.	.	.	.	.	.	.
Thesium humifusum	.	5	+	1	.	.	1	1	1	.	.	.	.	.
Odontites jaubertiana	.	.	+											
Seseli montanum	1	3	.	2	2	.	5	.	2	1	2	.	.	.

Espèces médioeuropéennes

Gentianella germanica	.	.	.	1	3	1	1	3	4	2	4	1	2	4
Gentianella ciliata	.	.	.	2	1	.	.	.	2	.	3	.	1	3
Aster amellus	.	.	.	1	2	.	.	.	.	1	.	3	.	.

Autres espèces fréquentes

Bromus erectus	4	5	+	5	5	4	4	2	5	5	4	5	5	5
Brachypodium pinnatum	3	2	+	3	5	5	5	5	2	5	2	3	2	5
Carex flacca	3	5	+	5	5	5	5	5	4	5	5	4	4	5
Festuca gr. *duriuscula*	4	5	+	4	3	4	5	5	4	4	4	5	4	5
Cirsium acaule	.	5	+	3	5	1	3	5	4	4	2	2	1	5
Briza media	3	5	+	4	5	5	4	5	5	4	3	4	4	5
Lotus corniculatus	4	2	+	4	5	5	5	5	5	5	5	3	4	5
Linum catharticum	3	3	+	3	4	5	2	5	4	4	5	5	3	4
Euphorbia brittingeri	.	2	+	2	2	.	.	.	2	3	1	4	4	2
Hippocrepis comosa	1	5	+	5	4	2	4	2	3	3	3	5	4	4
Koeleria pyramidata	2	.	+	2	2	.	3	3	4	2	4	2	3	4
Sanguisorba minor	3	.	.	4	5	5	5	5	2	4	2	5	1	5
Asperula cynanchica	.	5	.	4	4	1	5	3	.	4	1	5	.	2
Prunella grandiflora	.	.	+	2	4	.	2	.	3	5	4	5	5	4
Scabiosa columbaria	2	2	+	2	2	1	4	3	.	1	1	2	1	4
Plantago media	2	2	+	2	3	3	.	2	1	4	1	3	1	4
Leontodon hispidus	2	3	+	3	4	3	3	3	5	4	3	2	2	5
Carlina vulgaris	1	4	+	4	4	3	2	4	2	3	1	2	.	4
Pimpinella saxifraga	1	.	.	1	3	1	1	5	.	5	2	4	2	2
Ophrys insectifera	.	2	+	1	2	.	3	2	1	1	.	1	2	.
Carex caryophyllea	.	1	.	1	4	.	2	2	1	2	2	3	1	3
Leucanthemum vulgare	.	3	.	.	1	4	.	4	3	2	2	1	.	.
Genista tinctoria	.	2	+	5	4	5	.	1	3	4	3	1	2	2
Gymnadenia conopsea	.	1	+	3	5	.	1	5	5	5	1	5	4	2
Orobanche gracilis	.	.	+	.	3	5	.	.	.	5	1	1	1	.
Thymus gr. *serpyllum*	2	5	+	2	4	3	5	5	1	2	2	2	1	.
Orchis militaris	1	1	+	1	1	1	4	.	4	.	.	1	1	.

L'association à *Ophrys scolopax* et *Carex flacca* présente des affinités évidentes avec l'*Orchido-Brometum* (colonne 1) décrit par Braun-Blanquet & Susplegas (1937) à partir de quatre relevés provenant des Corbières et des Cévennes (Braun-Blanquet, 1952). Cette association encore mal connue est homologue des associations marnicoles septentrionales (présence de *Cirsium tuberosum, Tetragonolobus maritimus, Blackstonia perfoliata*). Il en est de même pour d'autres groupements comme le *Bromo-Cirsietum tuberosi* de Catalogne (de Bolos, 1967) et le groupement à *Catananche coerulea* du Quercy (Verrier, 1979), situés dans la même région.

Dans le centre de la France une seule association est reconnue, publiée à partir d'une simple liste, le *Blackstonieto-Senecietum erucifolii* (Braque & Loiseau, 1972). Ce groupement original (colonne 3) est dépourvu d'espèces médioeuropéennes, est riche en espèces méditerranéo-atlantiques (avec encore *Carduncellus mitissimus*) et possède une espèce particulière, *Odontites jaubertiana* qui remplace *Odontites lutea*.

A l'est de la France les pelouses marneuses se rapportent au *Chloro perfoliatae-Brometum* (colonne 4) pour ce qui concerne les collines de Bourgogne et de Champagne méridionale (Royer, 1973, 1978). Cette association contient à la fois des espèces médioeuropéennes et des espèces subatlantiques-subméditerranéennes. Elle est pauvre en éléments du *Molinion*. Par contre *Peucedanum cervaria* est ici particulièrement exhubérant, de même que *Centaurium erythraea*. Dans le sud du Jura, à l'étage collinéen, nous relevons un groupement très affine du *Chloro-Brometum* (colonne 5), quoique appauvri en espèces méditerranéo-atlantiques et possédant quelques espèces propres comme *Orobanche gracilis, Tetragonolobus maritimus* (Royer, non publié).

Au nord-ouest de la France ainsi que dans la région parisienne, les auteurs ont fréquemment mentionné la spécificité de la pelouse marnicole à *Blackstonia perfoliata* et *Carex flacca* (Bournérias, 1979). Toutefois peu de relevés sont publiés actuellement, excepté pour le Perche (colonne 6) par Lemée (1937) qui le premier a défini cette association, et pour le Laonnois (colonne 7) d'où proviennent quelques relevés de Prelli (1968). Par rapport aux précédentes, il y a appauvrissement en espèces méditerranéo-atlantiques et en espèces marnicoles, les espèces médioeuropéennes étant par ailleurs fort

mal représentées (absence de *Gentianella ciliata, Aster amellus, Euphorbia brittingeri*...). Vers le nord et le nord-est de la France, les associations marnicoles s'appauvrissent encore. L'association à *Dactylorhiza fuchsii* et *Brachypodium pinnatum* (colonne 8) des marnes du Boulonnais décrite par Géhu (1959) s'apparente au groupement précédent et représente l'extrême irradiation vers le nord des associations marnicoles.

A l'est de la France en altitude (étages collinéen supérieur et montagnard), ainsi qu'en Suisse apparaissent des végétations marnicoles différentes (colonnes 9 à 13), presque dépourvues d'espèces méditerranéo-atlantiques, bien pourvues en espèces médioeuropéennes. Certaines espèces marnicoles trouvent là leur optimum (*Molinia coerulea* ssp. *arundinacea, Tetragonolobus maritimus, Succisa pratensis, Carex panicea, Cirsium tuberosum*) alors que d'autres se raréfient (*Blackstonia perfoliata, Peucedanum cervaria*). De nouvelles espèces différencient cet ensemble notamment *Trifolium montanum, Parnassia palustris, Carex montana, Potentilla erecta*. Cette variante montagnarde et médioeuropéenne de pelouse marnicole est représentée localement dans les zones froides du Plateau de Langres (colonne 9, groupement à *Tetragonolobus* et *Carex tomentosa*) à proximité de marais tuffeux (Rameau & Royer, 1978). Elle est surtout développée au niveau du Jura central (colonne 10, association à *Plantago serpentina* et *Tetragonolobus*) où elle est étudiée par Pottier-Alapetite (1943), du Jura du sud où nous l'avons relevée (colonne 11, non publié) et de l'Aargau (colonnes 12 et 13). Dans cette dernière région, Zoller (1954) définit un *Tetragonolobo-Molinietum littoralis* très représentatif de ce type de pelouse. On retrouverait plus à l'est, vers Zürich des groupements équivalents comme semble l'indiquer la liste incomplète du *Molinietum littoralis* (Braun-Blanquet & Moor, 1938) qui comprend à côté de *Molinia coerulea* ssp. *arundinacea* (= *M. littoralis*), *Peucedanum cervaria, Cirsium tuberosum*, des espèces spéciales comme *Polygala chamaebuxus* et *Festuca amethystina*.

Au-delà du nord et de l'est de la France et du nord de la Suisse, les associations marnicoles autonomes disparaissent (Korneck, 1974; Oberdorfer, 1978; Witschel, 1980). Il ne subsiste localement dans ces groupements que quelques espèces du *Molinion coeruleae* comme *Cirsium tuberosum*, permettant

90

de différencier des sous-associations comme le *Gentiano-Koelerietum cirsietosum tuberosi* du Jura Souabe (colonne 14).

En conclusion, il apparaît que les groupements les plus typés se situent dans le Midi de la France. Il pourrait s'agir du berceau chorologique des associations marnicoles, leur richesse floristique s'amenuisant en direction du nord et de l'est, l'apport médioeuropéen ou montagnard ne compensant pas totalement les pertes en espèces méditerranéennes. La Figure 3 résume la répartition de ces différents groupements.

Influences chorologiques au niveau des associations mésoxérophiles

Moins typées que les associations marnicoles, les associations mésoxérophiles sont souvent mal sé-

parées des autres associations du *Mesobromion erecti* et remplacées alors par des variantes ou des sous-associations. On y rencontre un ensemble d'espèces xérophiles qui affectionnent également les associations du *Xerobromion* comme *Pulsatilla vulgaris, Teucrium chamaedrys, Teucrium montanum, Linum tenuifolium, Globularia punctata, Potentilla tabernaemontani, Genista pilosa, Coronilla minima, Odontites lutea* . . . voire des espèces spécifiques du *Xerobromion* comme *Fumana procumbens, Helianthemum apenninum* et *Ononis pusilla*. Ces diverses espèces pénètrent dans une moindre mesure les associations du *Seslerio-Mesobromenion*.

L'analyse du Tableau 2 permet de dégager l'importance du facteur chorologique au niveau de ces associations mésoxérophiles. Nous n'avons réuni dans ce tableau que les informations provenant de

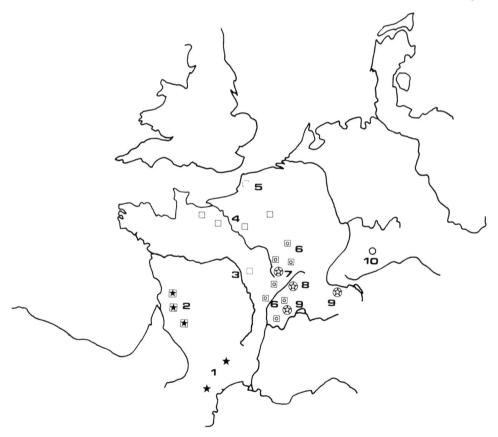

Fig. 3. Associations marnicoles du *Mesobromion erecti.*

Légende: 1, *Orchido-Brometum;* 2, Gpt. à *Ophrys scolopax* et *Carex flacca;* 3, *Blackstonieto-Senecietum erucifolii;* 4, *Chloro perfoliatae-Caricetum flaccae;* 5, Ass. à *Orchis fuchsii* et *Brachypodium pinnatum;* 6, *Chloro perfoliatae-Brometum;* 7, Gpt. à *Tetragonolobus* et *Carex tomentosa;* 8, Ass. à *Plantago serpentina* et *Tetragonolobus;* 9, *Tetragonolobo-Molinietum littoralis;* 10, *Gentiano-Koelerietum cirsietosum tuberosi;* étoile = influence subméditerranéenne; carré = influence subatlantique-subméditerranéenne; rond = influence médioeuropéenne; étoile blanche = influence 'montagnarde'.

Tableau 2. Associations mésoxérophiles du *Mesobromion.*

Numéros des associations	1	2	3	4	5	6	7	8	9	10	11	12	13	14	15	16	17	18	19	20	21	22	23	24	25	26	27	28	29
Nombre de relevés	24	5	18	23	5	5	12	20	20	30	51	10	35	14	4	26	8	35	12	65	16	37	13	68	12	12	54	22	15
Espèces boréo-atlantiques																													
Thymus praecox ssp. *arcticus*	5	3	5	3	5	5	1	2	.	.	.	.	.	.	.	.	.	.	.	.	.	.	.	.	.	.	.	.	.
Gentianella amarella	2	2	3	1	5	3	1	1	.	.	.	.	.	.	.	.	.	.	.	.	.	.	.	.	.	.	.	.	.
Filipendula vulgaris	2	2	2	1	.	.	.	.	.	.	.	.	.	4	2	1	.	.	.	.	.	.	.	.	.	.	.	.	.
Espèces subatlantiques-subm. et subméditerranéennes																													
Polygala calcarea	1	.	1	.	2	2	3	3	3	3	3	2	3	.	2	1	1	1	2	1	4	.	.	.	.	1	.	.	.
Thesium humifusum	.	.	.	.	.	4	1	3	1	.	1	3	4	.	.	1	5	3	2	3	4	2	.	.	.	.	.	.	.
Seseli montanum	.	.	.	.	.	.	1	.	.	1	.	3	5	4	4	4	5	5	4	5	3	5	5	.	.	.	.	.	.
Thymus praecox ssp. *praecox*	.	.	.	.	.	.	3	5	.	3	5	5	.	4	4	4	4	3	4	5	5	4	5	5	.	.	1	.	.
Genista pilosa	.	.	.	.	.	.	.	.	1	.	.	2	.	.	1	1	2	2	2	5	3	1	2	1	1	1	.	.	.
Coronilla minima	.	.	.	.	.	.	1	.	.	.	.	2	.	1	1	5	4	2	.	.	.	1	.	.	.	.	.	.	.
Odontites lutea	.	.	.	.	.	.	.	.	.	.	.	1	2	3	.	.	1	2	1	3	3	.	.	.	.	.	.	.	.
Prunella laciniata	.	.	.	.	.	.	.	.	.	.	.	1	.	1	4	.	1	.	1	1	2	2	.	2	2	.	1	.	.
Veronica prostrata ssp. scheereri	.	.	.	.	.	.	.	.	.	.	.	1	2	.	.	1	.	.	1	.	1	2	.	1	.	.	.	.	.
Linum leonii	.	.	.	.	.	.	.	.	.	.	.	.	2	.	1	4	.	1	.	.	1	.	.	.	.	3	.	.	.
Fumana procumbens	.	.	.	.	.	.	.	.	.	2	2	.	.	.	.	3	1	1	.	1	.	.	.	.	.	.	.	.	.
Festuca hervieri	.	.	.	.	.	.	.	.	.	.	.	.	.	2	3	1	4	1	.	.	.	.	4	.	.	.	.	.	.
Helianthemum apenninum	.	.	.	.	.	.	.	.	.	.	.	.	.	1	3	.	1	.	1	.	.	.	4	2	.	.	.	.	.
Carduncellus mitissimus	.	.	.	.	.	.	.	.	.	.	.	.	.	4	4	1	.	.	.	.	.	.	.	.	.	.	.	.	.
Ranunculus gramineus	.	.	.	.	.	.	.	.	.	.	.	.	.	2	3	.	.	.	.	.	.	.	.	.	.	.	.	.	.
Cytisus decumbens	.	.	.	.	.	.	.	.	.	.	.	.	.	.	.	.	1	2	4	4	2	1	.	.	.	.	.	.	.
Carex hallerana	.	.	.	.	.	.	.	.	.	.	.	.	.	4	2	.	.	2	.	2	.	.	.	.	.	.	.	.	.
Espèces médioeuropéennes et médioeuropéennes-subméditerranéennes																													
Pulsatilla vulgaris	.	.	4	.	.	.	1	1	1	1	1	5	3	.	2	1	2	3	4	2	4	3	.	1	.	3	4	1	3
Teucrium chamaedrys	.	.	.	.	.	.	3	4	1	1	2	5	5	4	3	3	2	5	2	4	5	5	5	5	4	1	2	2	.
Linum tenuifolium	.	.	.	.	.	.	.	.	.	2	1	2	.	.	2	3	4	3	3	5	4	1	3	1	.	4	.	.	.
Globularia punctata	.	.	.	.	.	.	1	1	1	.	1	2	1	5	1	1	2	4	2	3	.	2	1	4	4	3	.	3	.
Teucrium montanum	.	.	.	.	.	.	1	2	2	1	1	3	4	1	4	1	4	4	1	3	3	3	4	3	3	.	1	.	.
Potentilla tabernaemontani	.	.	.	.	.	.	.	.	3	.	.	2	4	3	5	2	5	2	4	3	4	5	5	5	5	5	5	3	5
Prunella grandiflora	.	.	.	.	.	.	.	.	2	.	.	3	2	.	3	1	1	3	5	3	5	3	.	3	2	5	1	4	3
Euphorbia cyparissias	.	.	.	.	.	.	.	.	.	.	1	5	2	4	1	5	4	1	.	1	5	5	3	4	4	.	4	2	4
Gentianella germanica	.	.	.	.	.	.	.	3	4	3	1	3	.	.	.	2	.	1	1	4	.	.	.	1	1	2	3	1	5
Espèces médioeuropéennes strictes																													
Gentianella ciliata	.	.	.	.	.	.	.	.	.	.	.	.	.	.	.	.	1	.	.	1	5	.	.	1	1	4	1	1	5
Carlina acaulis	.	.	.	.	.	.	.	.	.	.	.	.	.	.	.	.	.	.	.	2	.	.	.	.	1	2	5	.	.
Festuca rupicola	.	.	.	.	.	.	.	.	.	.	.	.	.	.	.	.	.	.	.	.	.	.	.	.	.	.	.	.	3
Autres espèces fréquentes																													
Festuca gr. duriuscula	4	5	5	4	5	5	4	5	5	5	5	5	5	5	.	.	5	2	5	5	5	5	5	1	5	3	4	4	3
Cirsium acaule	5	3	5	4	5	5	3	4	5	5	5	2	5	.	4	4	4	4	5	5	5	1	2	2	4	4	5	.	5
Briza media	5	4	5	5	4	5	5	5	4	5	3	.	2	4	4	3	2	5	5	4	1	1	4	4	4	4	2	5	5
Lotus corniculatus	5	5	5	5	4	3	4	3	5	5	5	4	4	1	4	4	1	3	4	5	4	3	4	5	5	4	3	5	5
Bromus erectus	5	5	5	3	5	5	2	2	1	1	2	3	2	5	4	5	4	5	5	5	5	5	5	5	5	5	5	5	1
Hippocrepis comosa	2	.	3	1	5	5	4	5	4	1	3	4	5	4	4	5	3	4	5	4	5	5	4	5	5	5	5	3	2
Scabiosa columbaria	4	5	5	1	4	4	2	3	3	4	5	4	5	3	3	3	5	5	5	4	5	5	2	3	3	5	4	3	5
Helianthemum gr. nummularium	5	5	4	4	5	4	1	4	3	2	3	5	5	.	2	2	1	3	2	4	5	5	4	4	5	2	4	5	2
Brachypodium pinnatum	5	5	4	5	5	.	5	5	5	5	5	3	5	1	3	5	3	5	5	5	5	5	3	3	4	4	3	5	4
Sanguisorba minor	5	5	4	5	5	5	5	.	5	5	5	3	5	1	3	5	5	4	2	4	5	5	5	3	5	5	5	5	4
Asperula cynanchica	5	1	3	2	5	5	4	4	5	3	5	5	5	1	2	3	3	4	3	4	5	5	2	5	5	.	5	3	1

92

Tableau 2. (continué)

Numéros des associations	1	2	3	4	5	6	7	8	9	10	11	12	13	14	15	16	17	18	19	20	21	22	23	24	25	26	27	28	29
Nombre de relevés	24	5	18	23	5	5	12	20	20	30	51	10	35	14	4	26	8	35	12	65	16	37	13	68	12	12	54	22	15
Carex flacca	5	5	5	4	5	5	5	4	5	5	4	1	4	.	.	4	4	4	5	4	5	5	1	3	4	.	5	4	4
Pimpinella saxifraga	3	4	4	2	3	1	4	5	3	4	4	3	5	.	.	4	2	1	.	1	5	5	1	.	4	.	2	3	5
Anthyllis vulneraria	2	3	3	1	4	4	2	4	1	2	2	1	3	3	2	3	3	3	2	2	2	5	5	3	4	5	.	1	4
Linum catharticum	4	4	4	3	5	5	4	5	5	4	5	4	.	5	.	1	3	2	4	4	5	5	.	3	3	4	4	2	5
Carlina vulgaris	4	3	3	1	5	1	2	3	4	3	3	2	3	.	.	2	3	3	.	3	5	5	1	3	.	3	4	2	4
Koeleria pyramidata	1	5	4	1	4	3	.	.	5	5	5	2	5	5	1	3	4	3	3	4	5	5	4	2	4	5	4	3	5
Plantago media	3	3	3	2	1	1	.	3	1	2	2	.	.	.	.	3	2	1	1	3	2	.	.	1	4	4	4	2	5
Carex caryophyllea	4	2	4	2	5	5	.	.	3	.	.	4	.	4	.	.	1	3	2	3	5	4	2	5	4	2	2	5	4
Leontodon hispidus	4	5	3	4	3	1	1	.	5	5	4	2	.	.	.	2	4	2	4	2	5	.	.	1	.	4	4	1	5
Centaurea scabiosa	1	3	3	1	.	.	2	2	1	1	2	2	5	.	.	1	2	2	2	2	5	5	3	1	2	1	3	3	3

France et des régions voisines, notamment Suisse et Allemagne du sud ainsi que quelques données provenant d'Angleterre, à titre comparatif. Trois groupes d'associations sont discernables à la premiè-re lecture du tableau:

– groupe nord-ouest européen, défini par Willems (1980) pour les groupements anglais (colonnes 1 à 6),

– groupe sud-ouest européen, inédit, rassemblant surtout des groupements français (colonnes 7 à 24),

– groupe centre-ouest européen, défini par Willems (1980) essentiellement ici pour des groupements allemands (colonnes 25 à 32).

Le groupe d'associations nord-ouest européen regroupe l'ensemble des associations mésoxérophiles et surtout mésophiles du *Mesobromion erecti* répandues en Irlande, Angleterre, Danemark et Suède (Willems, 1980). Outre *Thymus praecox* ssp. *arcticus,* et *Gentianella amarella,* ce groupe est caractérisé par *Dactylorhiza fuchsii, Centaurea nigra, Euphrasia nemorosa,* . . . Les espèces subatlantiques manquent presque toutes (sauf *Polygala calcarea*) alors que les diverses espèces mésoxérophiles d'affinités médioeuropéennes sont absentes excepté *Pulsatilla vulgaris,* rarissime. Nous n'avons pris à titre d'exemple que le *Cirsio–Brometum* anglais représenté ici par diverses variantes définies par Shimwell (1968, 1971), association qui apparaît comme l'une des plus mésoxérophiles du groupe.

Le groupe d'associations centre-ouest européen regroupe l'ensemble des associations mésoxérophiles et mésophiles du *Mesobromion erecti* depuis le Benelux jusqu'à la Suisse, la Bavière et l'Allemagne du nord (Willems, 1980). Les groupements mésoxérophiles inclus dans cet ensemble sont riches en espèces médioeuropéennes et subméditerranéennes-médioeuropéennes absentes du groupe nord-ouest européen comme *Teucrium chamaedrys, Teucrium montanum, Euphorbia cyparissias, Gentianella germanica, Gentianella ciliata, Carlina acaulis* . . . Ils sont généralement dépourvus d'espèces à tendance subatlantique. Nous avons pris ici à titre d'exemple le *Teucrio–Mesobrometum* du Jura Suisse (colonne 25) décrit par Zoller (1954), ainsi que diverses variantes regionales du *Gentiano–Koelerietum* d'Allemagne du sud, provenant des travaux de Korneck (1974) et Oberdorfer (1978). Willems (1980) a regroupé le *Gentiano–Koelerietum* avec le *Mesobrometum erecti* et a isolé une association mésoxérophile nommée *Antherico–Brometum* pour cette même région. Nous n'avons pas assez d'éléments pour conclure, aussi avonsnous transitoirement conservé le *Gentiano–Koelerietum.*

L'intérêt de notre travail réside essentiellement en la mise en évidence d'un troisième groupe d'associations, distinct des deux précédents et dont la définition est basée sur la présence de nombreuses espèces subatlantiques-subméditerranéennes et subméditerranéennes. L'influence chorologique est donc déterminante pour l'identification de ce groupe d'associations sud-ouest européen défini par des espèces comme *Polygala calcarea, Thesium humifusum, Veronica prostrata* ssp. *scheereri, Linum leonii, Carduncellus mitissimus, Seseli montanum, Festuca hervieri* (subatlantiques-subméditerranéennes), *Coronilla minima, Fumana procumbens, Carex hallerana* (subméditerranéennes),

Thymus praecox ssp. *praecox, Odontites lutea, Prunella laciniata, Cytisus decumbens*... Dans ce groupe d'associations répandu de la Normandie au Jura et au Quercy s'observent surtout des associations mésoxérophiles (seules mentionnées ici), généralement très riches en *Teucrium chamaedrys, Teucrium montanum, Potentilla tabernaemontani, Globularia punctata*...

Les influences chorologiques permettent de distinguer cinq grandes associations au sein de ce groupe:

– le *Festuco-Brachypodietum* au nord-ouest de la France (colonnes 7 à 11),

– un *Mesobromion* xérophile au centre du Bassin Parisien qui est encore à étudier en détail (colonnes 12, 13 et 16),

– un groupement à *Carduncellus mitissimus* et *Ranunculus gramineus* en Berry et Quercy (colonnes 14 et 15),

– le *Festuco lemanii-Brometum* en Bourgogne, Champagne et Lorraine (colonnes 17 à 22),

– un *Mesobromion* xérophile propre au Jura français (colonnes 23 et 24).

Le *Festuceto-Brachypodietum* se situe à la charnière du groupe d'association nord-ouest européen et sud-ouest européen. En effet il contient encore *Thymus praecox* ssp. *arcticus, Gentianella amarella, Dactylorhiza fuchsii, Centaurea nigra*, mais aussi *Polygala calcarea, Thesium humifusum, Thymus praecox* ssp. *praecox*, ainsi que *Teucrium montanum, Teucrium chamaedrys*. Il s'agit cependant d'une association très pauvre en éléments mésoxérophiles. Nous avons rassemblé ici des données provenant de Normandie, notamment du Perche (Lemée, 1937), des environs de Caen (Lemée, 1932), de la vallée de la Bresle (De Blangermont & Liger, 1964), du Pays de Bray (Frileux, 1977) ainsi que de Picardie (Boullet, 1981).

Au centre du Bassin Parisien, le *Mesobromion* mésoxérophile s'enrichit sensiblement en espèces subméditerranéennes constituant un groupement original qui reste à définir. Nous avons repris ici les données concernant le Valois (Jovet, 1949), le Laonnois (Prelli, 1968) et la Beauce (Maubert, 1978).

Plus au sud, en Berry et en Quercy notamment le *Xerobromion* domine et le *Mesobromion* se réfugie sur les sols les plus épais. De nouvelles espèces à affinités méditerranéennes apparaissent comme *Ranunculus gramineus, Carduncellus mitissimus*, alors que les espèces mésophiles présentes dans les autres associations examinées ici régressent et disparaissent, notamment *Carex flacca, Carlina vulgaris, Pimpinella saxifraga, Plantago media, Leontodon hispidus*. Ce groupement original à *Carduncellus* et *Ranunculus gramineus* est connu du Berry (Maubert, 1978) et du Quercy (Verrier, 1979).

Le *Festuco lemanii-Brometum* du centre-est, du nord-est français est également très riche en espèces d'affinités méditerranéennes *(Cytisus decumbens,* et de façon plus sporadique *Carex hallerana, Helianthemum apenninum, Cytisus supinus* var. *gallicus, Fumana procumbens*...). Il renferme par ailleurs la presque totalité des espèces submediterranéennes-subatlantiques du groupe d'associations, de même que l'ensemble du lot mésoxérophile médioeuropéen. Le tableau rassemble les données concernant la Champagne crayeuse (Royer, 1973, 1978), la Bourgogne occidentale (Royer, 1978), le Barséquanais (Royer, 1982), la Haute-Marne et la Bourgogne orientale (Royer, 1973, 1978), la Lorraine (Haffner, 1960).

Enfin, sur la bordure du Jura français s'observent des groupements proches du *Xerobromion* et rappelant par ailleurs le *Festuco lemanii-Brometum* et le *Gentiano-Koelerietum*. Notamment les espèces subatlantiques manquent ou se raréfient dans cette région (*Polygala calcarea, Thesium humifusum*). La colonne 23 est relative au *Bromo-Caricetum halleranae* de la région de Besançon (Pottier-Alapetite, 1943) et la colonne 24 à un groupement équivalent du Jura du sud (Royer, non publié).

La Figure 4 résume la distribution des principales associations et sous-associations mésoxérophiles analysées ici. La limite des trois groupes d'associations reste à préciser de même que l'extension du groupe sud-ouest européen vers le sud, notamment au niveau des montagnes (Alpes, Pyrénées, Massif Central).

En conclusion, il apparaît que la chorologie est un facteur prédominant pour la séparation des associations au sein de l'alliance *Mesobromion erecti*. Un axe chorologique apparaît privilégié, celui qui contourne le Massif Central à l'ouest, apportant de nombreuses espèces méditerranéennes ou méditerranéo-atlantiques au sein de la végétation des pelouses.

94

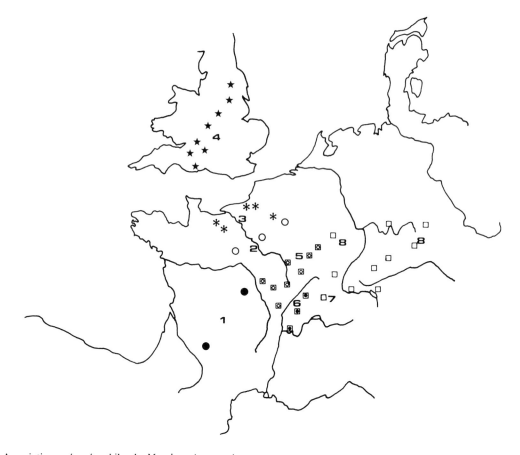

Fig. 4. Associations mésoxérophiles du *Mesobromion erecti.*

Légende: 1, Gpt. à *Carduncellus* et *Ranunculus gramineus;* 2, *Mesobromion* xérophile; 3, *Festuco–Brachypodietum;* 4, *Cirsio–Brometum;* 5, *Festuco lemanii–Brometum;* 6, *Bromo–Caricetum halleranea;* 7, *Teucrio–Mesobrometum;* 8, *Gentiano–Koelerietum;* voir aussi Figure 3.

Bibliographie

Blangermont, C. de & Liger, J., 1964. Végétation des pelouses crayeuses de la vallée de la Bresle (Seine-Maritime). Revue Soc. Sav. Haute-Normandie, Sciences, 36: 29–47.

Bolos, O. de, 1967. Comunidades vegetales de las Comarcas proximas al litoral situadas entre los Rios Llobregat y Segura. Mem. Real. Academia Ciencas y Artes Barcelona 38, 1. 268 pp.

Boullet, P., 1981. Les pelouses calcaires et leur appauvrissement thermophile entre Seine et Somme. DEA, Lille. 108 pp.

Bournerias, M., 1979. Guide des groupements végétaux de la région parisienne. CDU et SEDES, Paris. 509 pp.

Braque, R. & Loiseau, J. E., 1972. Contribution à l'étude de la flore et de la végétation du Centre de la France. Revue Sciences Nat. Auvergne 38: 27–33.

Braun-Blanquet, J., 1952. Les groupements végétaux de la France méditerranéenne. CNRS, Montpellier. 298 pp.

Braun-Blanquet, J. & Moor, M., 1938. Prodromus der Pflanzengesellschaften. Fasc. 5. Verband des Bromion erecti. Montpellier. 64 pp.

Braun-Blanquet, J. & Susplegas, J., 1937. Reconnaissance phytosociologique dans les Corbières. SIGMA, Comm. 61, Saint-Dizier. 16 pp.

Frileux, P. N., 1977. Les groupements végétaux du Pays de Bray (Seine-Maritime et Oise). Caractérisation, Ecologie, Dynamique. Thèse Univ. Rouen. 207 pp.

Géhu, J. M., 1959. Les pelouses calcaires de la 'Cuesta' boulonnaise. Bull. Soc. Et. Sc. Angers, NS 2: 205–221.

Haffner, P., 1960. Pflanzensoziologische und pflanzengeographische Untersuchungen im Muschelkalkgebiet des Saarlandes mit besonderer Berücksichtigung der Grenzgebiete von Lotharingen und Luxemburg. Natursch. Landschaftspfl. Saarland 2: 66–164.

Jovet, P., 1949. Le Valois. Phytosociologie et phytogéographie. SEDES, Paris. 389 pp.

Korneck, D, 1974. Xerothermvegetation in Rheinland-Pfalz und Nachbargebieten. Schriftenr. Vegetationsk., Bad-Godesberg, 7. 196 pp.

Lemée, G., 1932. Etudes phytogéographiques sur les plaines jurassiques normandes. Bull. Soc. Bot. France 79: 637–650.

Lemée, G., 1937. Recherches écologiques sur la végétation du Perche. Thèse Univ. Paris. 389 pp.

Maubert, P., 1978. Contribution à l'étude phytosociologique des pelouses calcicoles du Bassin Parisien. Thèse Orsay. 84 pp.

Muller, Th., 1966. Die Wald-, Gebüsch-, Saum, Trocken- und Halbtrockenrasengesellschaften des Spitzberges. Natur. u Landschaft. Bad.-Württ. 3. Ludwigsburg: 278–475.

Oberdorfer, E., 1978. Süddeutsche Pflanzengesellschaften. 2. Fischer, Jena. 355 pp.

Pottier-Alapetite, G., 1943. Recherches phytosociologiques et historiques sur la végétation du Jura central et sur les origines de la flore jurassienne. SIGMA, Comm. 81, Tunis. 333 pp.

Prelli, R., 1968. Contribution à l'étude des pelouses calcaires du Laonnois. DEA, Orsay. 80 pp.

Rameau, J. C. & Royer, J. M., 1978. Les Moliniaies du Plateau de Langres. Coll. Phytosoc. 5, Lille, 1976. Cramer, Vaduz: 269–288.

Royer, J. M., 1973. Essai de synthèse sur les groupements végétaux de pelouses, éboulis et rochers de Bourgogne et Champagne méridionale. Ann. Scient. Univ. Besançon 3. 13: 157–316.

Royer, J. M., 1978. Nouvelles données sur le Mesobromion Br.-Bl. et Moor em. Oberd. 49 de Bourgogne et Champagne. Doc. Phytosoc. NS 2: 393–399.

Royer, J. M., 1982. Etude phytosociologique des pelouses du Barséquanais, du Barsuraubois, du Tonnerrois et de l'Est-Auxerrois. Bull. Soc. Sc. Hist. Natur. Yonne, Auxerre 113: 217–247.

Shimwell, D. W., 1968. The phytosociology of calcareous grasslands in the British Isles. Thesis Univ. Durham. 340 pp.

Shimwell, D. W., 1971. Festuco-Brometea Br.-Bl. et R. Tx. 1943 in the British Isles: the phytogeography and phytosociology of limestone grasslands. Vegetatio 23: 1–60.

Tutin, T. G., et al. 1964–1980. Flora Europaea. Cambridge University Press.

Verrier, J. L., 1979. Contribution à la systématique et à la synécologie des pelouses sèches à thérophytes d'Europe. Thèse Orsay. 204 pp.

Willems, J. H., 1980. Observations on north-west european limestone grassland communities. VI. Phytosociological and geographical survey of Mesobromion communities in North-West Europe. Thesis Utrecht.

Witschell, M., 1980. Xerothermvegetation und dealpine Vegetations-Komplexe in Südbaden. Beihefte Veröffent. Natursch. und Landschaft. Baden-Wurtt. 17: 212 pp.

Wolkinger, F. & Plank, S., 1981. Les pelouses sèches en Europe. Collection Sauvegarde de la Nature 21. Conseil de l'Europe, Strasbourg. 70 pp.

Zoller, H., 1954. Die Typen der Bromus erectus-Wiesen des Schweizer Juras. Beitrage zur geobot. Landesaufn. Schweiz 33. 305 pp.

Accepted 19.1.1984.

Appendice

Localisation géographique et bibliographique des associations et sous-associations des tableaux synoptiques.

Tableau 1:
1. *Orchido–Brometum* Br.-Bl. 37 – Corbières, Causses – Braun-Blanquet (1952).
2. Gpt à *Ophrys scolopax* et *Carex flacca* – Charentes, Périgord – Royer, inédit.
3. *Blackstonieto–Senecietum erucifolii* Braque et Loiseau 72 – Nièvre – Braque et al. (1972).
4. *Chloro perfoliatae–Brometum* Royer 73 nom. inv. – Champagne, Bourgogne – Royer (1978).
5. *Chloro perfoliatae–Brometum* Royer 73 nom. inv. – Sud Jura – Royer, inédit.
6. *Chloro perfoliatae–Caricetum flaccae* Lemée 37 – Perche – Lemée (1937).
7. *Chloro perfoliatae–Caricetum flaccae* Lemée 37 – Laonnois – Prelli (1968).
8. Ass. à *Orchis fuchsii* et *Brachypodium pinnatum* Géhu 59 – Boulonnais – Géhu (1959).
9. Gpt à *Tetragonolobus* et *Carex tomentosa* – Plateau de Langres – Rameau & Royer (1978).
10. *Mesobromion* marnicole montagnard – Jura du sud – Royer, inédit.
11. Ass. à *Plantago serpentina* et *Tetragonolobus* P. Al. 43 – Jura central – Pottier-Alapetite (1943).
12. *Tetragonolobo–Molinietum littoralis asperuletosum* Zoller 54. Aargau – Zoller (1954).
13. *Tetragonolobo–Molinietum littoralis tofieldietosum* Zoller 54. Aargau – Zoller (1954).
14. *Gentiano–Koelerietum* Knapp 42 ex. Bornk. 60 *cirsietosum tuberosi* Muller 78. Jura souabe–Oberdorfer (1978).

Tableau 2:
1-6. *Cirsio–Brometum* Shimwell 68 – Angleterre – Shimwell (1968) (Tableau 17) 1: *typicum* – 2: *Carex* variante – 3: *astragaletosum* – 4: *brachypodietosum* – 5: *typicum* – 6: *Onobrychis – Serratula* variante.
7. *Festuco–Brachypodietum* De Litardière 28 – Perche – Lemée (1937).
8. *Festuco–Brachypodietum* De Litardière 28 – Région de Caen – Lemée (1932).
9. *Festuco–Brachypodietum* De Litardière 28 – Vallée de La Bresle – de Blangermont & Liger (1964).
10. *Festuco–Brachypodietum* De Litardière 28 – Pays de Bray – Frileux (1977).
11. *Festuco–Brachypodietum* De Litardière 28 – Oise – Boullet (1981).
12. *Mesobromion* xérophile – Laonnois – Prelli (1968).
13. *Mesobromion* xérophile – Valois – Jovet (1949).
14. Groupement à *Carduncellus* et *Ranunculus gramineus* – Quercy – Verrier (1979).
15. Groupement à *Carduncellus* et *Ranunculus gramineus* – Berry – Maubert (1978).
16. *Mesobromion* xérophile – Beauce – Maubert (1978).
17-22. *Festuco lemanii–Brometum* Royer 73 nom. inv. corr.
17. *ononidetosum* Royer (73) 78 – Champagne crayeuse – Royer (1978).
18. *festucetosum hervieri* (Lericq 72) Royer 73 – Yonne – Royer (1978).
19. *cytisetosum gallici* Royer 78 – Barséquanais – Royer (1982).
20. *typicum* Royer 73 corr. 78 – Haute-Marne, Bourgogne – Royer (1978).

21. *typicum* Royer 73 corr. 78 – Lorraine – Haffner (1960).

22. variante – Lorraine – Haffner (1960).

23. *Bromo–Caricetum halleranae* P. Al. 43 – Jura central – Pottier-Alapetite (1943).

24. *Mesobromion* xérophile – Jura du sud – Royer, inédit.

25. *Teucrio-Mesobrometum* Zoller 54 – Jura suisse – Zoller (1954).

26–32. *Gentiano-Koelerietum* Knapp 42 ex Bornk. 60.

26. race de Sud Eifel – Korneck (1974).

27. race de Kalk Eifel – Korneck (1974).

28. race de Bavière du sud – Lang in Oberdorfer (1978).

29. race du Jura franconien – Korneck in Oberdorfer (1978).

30. race du Main – Korneck in Oberdorfer (1978).

31. race du Jura franconien – Korneck in Oberdorfer (1978).

32. race du Jura souabe – Muller *et al.* in Oberdorfer (1978).

Chorologische Phänomene in Wasserpflanzengesellschaften Mitteleuropas*

W. Pietsch

8027 Dresden, Am Tälchen 16, D.D.R.

Keywords: Border of distribution area, Central Europe, Floristic-sociological structure, Human activities, Hydrochemical composition, Ionic content, *Juncetalia bulbosi*, Oceanic species

Abstract

On the basis of three selected examples chorological phenomena in waterplant communities of central, east and southern Europe are represented. Species of the Atlantic-oceanic floral element of the order *Juncetalia bulbosi*, like *Pilularia globulifera*, *Eleocharis multicaulis* and *Deschampsia setacea* are growing in markedly secondary sites near their eastern border, i.e. in the Lusatian district of central Europe, especially in sand, gravel and clay pits as well as in residue lakes of opencast mines. As a result of intensive recent human activities numerous new sites are being established now causing an expansion of their distribution area. With decreasing oceanity from the central part to the eastern border of the area a change in the floristic-sociological structure of the relevant plant communities and in the hydrochemical composition of the respective water bodies takes place. In western Europe the species grow in soft water poor in minerals, in the eastern part (central Europe) in mineral-rich, hard water with calcium and sulphate.

At present *Ceratophyllum submersum* has a high frequency in the northeastern area in Mecklenburg. The sites are morainic lakes (kettle holes) and small waters and ponds near settlements, which are enriched in minerals and nitrates in consequence of the intensive agricultural use of inorganic fertilizers. Moreover the waters are alkaline and rich in minerals, especially bicarbonate and possess a high range of total hardness.

The example of eight different sites of *Aldrovanda vesiculosa* illustrates a striking change of its diagnostic value in its disjunct north–south distribution. In the northern part of central Europe *Aldrovanda* is found in the communities of the *Utricularietea intermedio–minoris*, both in waters with a low acidity rich in sulphate and in alkaline waters rich in bicarbonate (*Sphagno–Utricularion* and *Scorpidio–Utricularion* respectively). In southern central Europe and in southeastern Europe *Aldrovanda* occurs in *Lemnetea* and *Potametea* communities. In northeastern Poland *Aldrovanda* is found together with *Hydrilla verticillata*, *Chara tomentosa*, *Myriophyllum verticillatum* and *Scorpidium scorpioides*.

Einleitung

Durch die Tätigkeit des Menschen können außerhalb des eigentlichen Arealzentrums Standortsverhältnisse geschaffen werden, die bestimmten Arten und ihren Gesellschaften optimale Wuchsbedingungen ermöglichen. So kommt es dann an der äußersten Arealgrenze zur großflächigen Entwicklung von Vegetationsverhältnissen, in denen die Arten sich durch hohe Individuenzahlen und Biomasseleistungen auszeichnen und zu einer Erweiterung des Verbreitungsareals beitragen. Einmal sind es Vorgänge des Sand-, Kies-, Ton- und Braunkohlenabbaus und der Melioration, die ein spezifisches Rohbodensubstrat und eine bestimmte hydrochemische Beschaffenheit der Wasserkörper der entstehenden Pionierstandorte schaffen. Am Bei-

* Die Sippennomenklatur richtet sich nach Rothmaler *et al.* (1970).

Vegetatio 59, 97–109 (1985).
© Dr W. Junk Publishers, Dordrecht. Printed in the Netherlands.

spiel von Vertretern des atlantischen Florenelementes der *Juncetalia bulbosi* Pietsch 71 wird das gehäufte Auftreten an der östlichen Arealgrenze als chorologisches Phänomen betrachtet. Zum anderen ist es der Einfluß der Landwirtschaft, der als Folge des erhöhten Einsatzes mineralischen Düngers, zu einer Anreicherung der Gewässer mit Mineralstoffen führt. Ein gehäuftes Auftreten von *Ceratophyllum submersum* in Mettelmecklenburg an der nordöstlichen Arealgrenze wird untersucht.

In einem dritten Beispiel wird das unterschiedliche soziologische und ökologische Verhalten einer disjunkt verbreiteten Art, *Aldrovanda vesiculosa*, in Mittel-, Ost- und Südosteuropa untersucht. Neben der floristisch-soziologischen Struktur wird auch die jeweilige hydrochemische Beschaffenheit der miteinander zu vergleichenden Siedlungsgewässer besprochen und am Beispiel von absoluten und relativen Ionenfelddiagrammen nach der Methode von Maucha (1932) dargestellt. Bei den Diagrammen des absoluten Ionengehaltes handelt es sich um einfache Kreisdiagramme, in denen die Größe der Kreisfläche dem Gesamtsalzgehalt des jeweiligen Gewässers entspricht.

Chorologische Phänomene

Für die Einschätzung des chorologischen Verbreitungsspektrums der zu besprechenden Wasserpflanzen wurden die Verbreitungskarten von Meusel, Jäger & Weinert (1964) sowie von Jäger (1968) zugrunde gelegt. Außerdem wurden die von Jäger (1964) für *Ceratophyllum submersum* und *Aldrovanda vesiculosa* mitgeteilten Verbreitungskarten bei der Betrachtung des jeweiligen Arealbildes mit herangezogen.

Arten des atlantischen Florenelementes und ihre Gesellschaften an der Ostgrenze ihres Verbreitungsareals

Während der letzten zwei Jahrzehnte ist ein gehäuftes Auftreten von Vertretern des atlantischen Florenelementes an der Ostgrenze ihres Verbreitungsareals, in der Altmark und vor allem aber in der Lausitz festzustellen (Pietsch, 1978, 1979). Temperat-euozeanische und meridional-ozeanische Arten, wie *Pilularia globulifera, Eleocharis multicaulis, Apium inundatum, Potamogeton polygoni-*

folius und *Deschampsia setacea* sowie ihre in der Ordnung der *Juncetalia bulbosi* Pietsch 71 zusammengefaßten Gesellschaften, wie *Pilularietum globuliferae, Eleocharitetum multicaulis, Carici (serotinae)–Deschampsietum setaceae* und die *Utricularia minor–Potamogeton polygonifolius*-Ges., bestimmen das atlantische Florenbild der Lausitz. Bereits vor der Jahrhundertwende war die Lausitz für das Auftreten atlantisch-subatlantisch verbreiteter Arten bekannt (Barber, 1893, Drude, 1902). Gräbner bezeichnete deshalb die Lausitzer Niederung als eine pseudatlantische Exklave. Mit zunehmender Intensivierung des Braunkohlenbergbaus in der Lausitz wurde ein großer Teil der bisher bekannten Standorte vernichtet. *Eleogiton fluitans, Apium inundatum, Deschampsia setacea* und *Hypericum elodes* galten bereits seit Ende der dreißiger Jahre als ausgestorben. *Eleocharis multicaulis, Pilularia globulifera* und *Potamogeton polygonifolius* hatten sich nur noch an den Randzonen des einstigen zentralen Verbreitungsgebietes erhalten. Während der letzten 15 Jahre ist nun wiederum ein gehäuftes Auftreten von *Pilularia globulifera, Apium inundatum* und in etwas geringerer Intensität auch von *Deschampsia setacea* zu beobachten. *Pilularia globulifera* und *Juncus bulbosus* siedeln an den neuen Standorten in solchen Massenbeständen, daß an eine Gefährdung im Moment nicht zu denken ist.

Das chorologische Phänomen liegt darin, daß außerhalb des eigentlichen Verbreitungszentrums des atlantischen Florenelementes an der östlichen Arealgrenze unter Abnahme der Ozeanitätsbindung, durch das Vorherrschen des ostdeutschen Binnenlandklimas charakterisiert, eine ungewöhnliche Häufung atlantischer Arten und ihrer Gesellschaften anzutreffen ist.

Die Vertreter der *Juncetalia bulbosi* besiedeln in der Lausitz Pionierstandorte von saurer (azidophiler) und nährstoffarmer (oligotropher) Beschaffenheit mit geringem Pufferungsvermögen des Bodensubstrates und des Wasserkörpers. Es handelt sich überwiegend um Arten einer Erstbesiedlungsvegetation an Standorten spezifischer edaphischer Verhältnisse. Diese werden im wesentlichen durch die Aziditätsverhältnisse, den Trophiegrad und das Pufferungsvermögen ihrer Standorte sowie durch spezielle mikroklimatische Verhältnisse in der bodennahen Luftschicht bestimmt. Die ostwärts zunehmende Winterkälte dürfte das weitere Vordringen der atlantischen Arten nach dem östlichen Mitteleuropa hin verhindern. Der Verlauf der $-1\,^\circ$C-Januar-Isotherme in Verbindung mit der Verteilung des Feuchtigkeitsgehaltes hängt mit dem östlichen Verbreitungsareal des atlantischen Florenelementes eng zusammen.

Soziologisches Verhalten

Am Beispiel von je 10 soziologischen Aufnahmen werden für *Pilularia globulifera* (Tab. 1) und *Eleocharis multicaulis* (Tab. 2) die floristisch-soziologische Struktur der Vorkommen der Altmark und der Lausitz dargestellt. Gegenüber dem Arealzentrum in W und NW Europa zeigt sich eine gewisse Verarmung an Charakterarten, so z.B. an *Hypericum elodes, Baldellia ranunculoides* und *Ranunculus ololeucos*, während *Eleogiton fluitans* nur noch in Gräben des Jeggauer Moores in der Altmark reichlich entfaltet ist. Abgesehen von den Vorkommen in den Fischteichen, fehlt den zahlreichen neuen Sekundärstandorten des Sand-, Kies-, Ton- und Braunkohlenabbaus *Littorella uniflora*. Dafür hat *Eleocharis acicularis* eine größere Bedeutung für die Zusammensetzung der Vegetationsverhältnisse erlangt. Gegenüber den Standorten des Arealzentrums, an denen es mehrfach zu Überlagerungen der *Littorella*-reichen Vegetation mit den

Arten der eigentlichen azidophilen Ufervegetation der *Juncetalia* kommt, zeigt sich in der Lausitz, aufgrund extremer edaphischer Verhältnisse, eine deutliche standörtliche Trennung zwischen der *Littorella*- und *Eleocharis acicularis*-reichen Vegetation der *Littorelletalia* und der *Eleocharis multicaulis*- und *Juncus bulbosus*-reichen Vegetation der *Juncetalia bulbosi*. *Pilularia globulifera* und *Eleocharis multicaulis* erreichen das Optimum ihrer Vorkommen in Ausbildungen der *Juncetalia bulbosi*, die eine charakteristische im Wasser flutende und den gesamten Wasserkörper ausfüllende Vegetation bilden, der jegliche isoetiden Arten fehlen.

Das *Pilularietum globuliferae* läßt sich an seiner östlichen Arealgrenze in eine typische Subass. (Tab. 1, Aufn. 1 bis 4), durch dominantes Auftreten von *Pilularia globulifera* und *Juncus bulbosus*, vereinzelt auch von *Eleogiton fluitans* gekennzeichnet, in eine Subass. v. *Myriophyllum heterophyllum* (Aufn. 5 u. 6) innerhalb von Sand- und Kiesgruben, in eine Subass. v. *Eleocharis acicularis* (Aufn. 7 u. 8) im flachen sandig-fraktionierten Litoralbereich von Sandgruben und Tagebauseen sowie in Gräben mit *Eleocharis acicularis* und *Apium inundatum* als Differentialarten, sowie in eine Subass. v. *Sparganium minium* (Aufn. 9 u. 10) untergliedern.

Innerhalb des *Eleocharitetum multicaulis* unterscheiden wir neben einer typischen Subass. (Tab. 2, Aufn. 1 bis 3), durch dominantes Auftreten von *Eleocharis multicaulis* und *Juncus*

Tab. 1. Pilularia globulifera-reiche Gesellschaften an der östlichen Arealgrenze in der Altmark und der Lausitz.

	1	2	3	4	5	6	7	8	9	10
No. der Aufnahme	1	2	3	4	5	6	7	8	9	10
Größe der Aufnahmefläche m²	200	200	400	240	120	120	80	200	200	200
Gesamtdeckung in %	100	100	100	100	100	100	95	100	100	100
Gesamtartenzahl	3	9	11	11	8	8	9	8	11	6
Wassertiefe in cm	120	90	80	110	80	120	50	65	40	45
C-Ass.:										
Pilularia globulifera	5.5	5.5	5.5	4.5	4.5	3.4	3.4	3.4	4.5	4.5
OC-KC Juncetalia bulbosi										
u. *Littorelletea uniflori:*										
Juncus bulbosus	2.3	3.4	2.3	4.5	4.5	3.4	3.4	3.4	3.4	2.3
Apium inundatum	.	.	.	+.3	.	.	1.3	2.3	+.3	.
Eleocharis acicularis	.	.	.	.	.	.	3.4	3.4	+.3	.
Eleogiton fluitans	.	2.3	.	1.3	.	.	.	(+.3)	.	.
Potamogeton polygonifolius	.	1.1	.	.	.	.	.	+.1	.	.
Potametea-Arten:										
Potamogeton natans	.	+.1	+.1	1.1	1.1	+.1	+.1	+.1	+.1	.
Callitriche palustris	.	.	+.1	1.1	1.3	+.2	.	.	+.3	1.1
Polygonum amphibium	+.1	.	+.1	.	+.1	.	+.1	.	+.1	.
Elodea canadensis	.	.	+.1	+.1	.	+.1	.	.	.	.
Myriophyllum heterophyllum	.	.	.	.	3.4	4.5	.	.	.	.
Phragmitetea-Arten:										
Eleocharis palustris	.	+.2	+.2	+.3	+.2	+.2	.	+.2	.	.
Equisetum fluviatile	.	1.3	+.1	.	.	+.1	+.1	.	+.1	+.3
Glyceria fluitans	.	+.2	.	+.2	.	.	+.1	.	+.1	.
Sparganium emersum	.	.	+.3	+.1	.	.	+.3	+.3	.	.
Phragmites australis	.	+.2	+.2	.	+.2	.	.	.	.	.
Alisma plantago-aquatica	.	.	.	+.1	.	.	.	.	+.1	+.1
weitere Arten:										
Sparganium minimum	.	.	+.1	.	.	.	.	.	2.3	3.4

Tab. 2. *Eleocharis multicaulis*-reiche Gesellschaften an der östlichen Arealgrenze in der Lausitz.

No. der Aufnahme	1	2	3	4	5	6	7	8	9	10
Größe der Aufnahmefläche m²	200	120	160	20	40	60	80	120	200	160
Gesamtdeckung in %	100	100	95	90	90	95	95	100	100	100
Gesamtartenzahl	3	5	7	10	14	13	9	9	12	12
Wassertiefe in cm	60	45	50	30	25	25	45	40	30	25
C-Ass.:										
Eleocharis multicaulis	5.5	5.5	4.5	3.4	3.4	4.5	4.5	3.4	3.4	4.5
OC-KC: Juncetalia bulbosi										
u. *Littorelletea uniflori:*										
Juncus bulbosus	3.4	3.4	4.5	4.5	4.5	3.4	2.3	2.3	3.4	2.3
Potamogeton polygonifolius	.	.	+.1	.	+.1	.	+.1	+.1	.	.
Potametea-Arten:										
Nymphaea alba	.	+.1	1.1	.	+.1	1.1	+.1	+.1	.	.
Nymphaea candida	+.1	.	+.1	+.1	+.1	.	1.1	.	.	.
Potamogeton natans	.	+.1	+.1	.	+.1	+.1	.	+.1	.	.
Charetea-Arten:										
Chara globularis	.	.	.	.	.	.	3.4	4.5	.	.
Chara vulgaris	.	.	.	.	.	.	2.3	1.1	.	.
Utricularietea-Arten:										
Utricularia minor	.	.	.	+.3	1.3	3.1	.	.	+.3	+.1
Sparganium minimum	.	.	.	+.1	2.3	1.1	.	.	.	.
Utricularia intermedia	.	.	.	.	3.4	2.3	.	.	+.3	.
Utricularia ochroleuca	.	.	.	.	2.3	3.4	.	.	.	+.3
Scheuchzerio–Caricetea-Arten:										
Drosera intermedia	.	+.1	+.1	3.4	+.1	1.1	+.1	+.1	1.1	1.1
Hydrocotyle vulgaris	.	.	.	+.1	+.1	1.1	+.1	.	1.1	1.1
Agrostis canina	.	.	.	+.2	+.2	+.2	.	.	2.3	1.3
Carex lasiocarpa	.	.	.	+.1	.	1.2	.	+.2	+.2	+.2
Eriophorum angustifolium	.	.	.	3.1	+.1	.	.	.	1.1	1.1
Carex rostrata	.	.	.	.	.	+.2	.	.	+.2	+.2
Rhynchospora fusca	.	.	.	.	.	.	.	.	4.5	3.4
Carex nigra	.	.	.	.	.	.	.	.	+.2	+.2

bulbosus gekennzeichnet, noch eine Subass. v. *Utricularia minor* (Aufn. 4 bis 6) mit einer Var. v. *Drosera intermedia* und einer typischen Variante der elektrolytärmsten Siedlungsgewässer, sowie eine Subass. v. *Chara globularis* (Aufn. 7 u. 8) etwas karbonatreicherer aber ausgesprochen phosphatarmer Standorte. Außerdem sind großflächig Durchdringungen mit dem *Rhynchosporetum* anzutreffen (Aufn. 9 u. 10).

Dagegen hat sich *Deschampsia setacea* in ihrem diagnostischen Wert auffällig verändert. Charakterisiert sie in W Europa vor allem Ausbildungen des *Eleocharitetum multicaulis* von Heideteichen und Heidemooren (Schoof-van Pelt, 1973, Dierßen, 1975), so besiedelt sie in der Lausitz sandig-fraktionierte Bodensubstrate, wie Teichböden oder flache Litoralbereiche von Tagebauseen, denen *Eleocharis multicaulis* fehlt (Pietsch, 1973). *Deschampsia* bildet an der Mehrzahl der Standorte mit Elementen der *Scheuchzerio–Caricetea* eine Vegetation, die von uns bereits früher als *Carici (serotinae)–Deschampsietum setaceae* beschrieben wurde (Pietsch, 1968).

Ökologisches Verhalten

Am Beispiel von absoluten und relativen Ionenfelddiagrammen werden Unterschiede und Gemeinsamkeiten in der ökologisch-hydrochemischen Beschaffenheit der *Pilularia globulifera-* (Tab. 3) und *Eleocharis multicaulis*-reichen Siedlungsgewässer (Tab. 4) des Arealzentrums und der an der östlichen Arealgrenze dargestellt. Sehr eindeutig zeigen sich die Verschiedenheiten in den unterschiedlich großen Kreisflächen der Diagramme für den absoluten Ionengehalt (Gesamtsalz- bzw. Elektrolytgehalt in mg/l). Die Beispiele 1 und 2 der *Pilularia*-reichen Siedlungsgewässer in W Frankreich und den Niederlanden besitzen die kleinsten Kreisflächen, diejenigen der Altmark und der Lausitz (Beispiele 3 und 4) weisen dagegen eine um 4 bis 8 mal größere Kreisfläche auf. Ebenso besitzen die *Eleocharis multicaulis*-reichen Siedlungsgewässer Irlands (Tab. 4, Beispiel 1) und NW Frankreichs (Beispiel 2) die kleinsten und diejenigen der Altmark (Beispiel 5) und der Lausitz (Beispiel 6) um 3- bis 4-fach größere Kreisflächen.

Betrachten wir dagegen die relativen Ionenfelddiagramme, so lassen alle aufgeführten Beispiele sowohl des Arealzentrums als auch der östlichen Vorpostenstandorte eine gewisse Gemeinsamkeit erkennen. Alle Siedlungsgewässer zeichnen sich durch einen auffällig geringen Anteil an gebundener Kohlensäure (HCO_3) aus. Es handelt sich bei allen um Calcium-Sulfatgewässer vom bikarbonatarmen Typ, im Bereich der Meeresküsten des Atlantik sowie der Nord- und Ostsee durch einen erhöhten Anteil an Natrium und Chlorid ausgezeichnet. Spielen Natrium und

Tab. 3. Absoluter und relativer Ionengehalt verschiedener Siedlungsgewässer von *Pilularia globulifera* L. in West- und Mitteleuropa.

Standorte	Hydrochemische Kenngrößen										Gesamt-ionen
	Na$^+$	K$^+$	Ca^{++}	Mg^{++}	Fe$^{2+/3+}$	SO$_4$	Cl	HCO$_3$	NO$_3$	PO$_4^{3-}$	
Beispiele:											
1. mg/l			1,9				19,3				116,0
2: mg/l			18,4				34,5				277,0
3: mg/l	12,6	3,7	48,2	7,2	0,01	94,4	36,2	15,3	5,8	0,068	232,5
mval-%	15,0	2,6	66,2	16,2	0,01	54,0	27,9	6,9	2,6	0,06	
4a: mg/l	10,2	3,2	60,5	9,7	0,02	157,7	30,1	4,3	6,1	0,068	282,7
mval-%	10,2	1,9	69,3	18,5	0,23	77,9	19,5	1,6	2,2	0,05	
4b: mg/l	6,8	3,4	123,6	21,2	1,1	296,2	61,1	8,5	5,9	0,072	529,9
mval-%	3,6	1,1	74,1	20,7	0,5	75,2	21,1	1,7	1,2	0,03	

Chlorid an den Standorten des Arealzentrums eine wesentliche Rolle, so sind es Calcium und Sulfat an denen der östlichen Arealgrenze. Deutlich sichtbar ist bei den Kreisdiagrammen die Abnahme des hellen Kreissektors als Anteil von Natrium und Kalium, sowie die Zunahme des schwarz ausgewiesenen Sektors für den Calciumgehalt mit weiterer Entfernung vom Arealzentrum. Eine gewisse Ausnahme bilden die Siedlungsgewässer von *Eleocharis multicaulis* in unmittelbarer Nähe der polnischen Ostseeküste (Beispiel 7), übrigens das letzte Vorkommen dieser Art an der Ostgrenze überhaupt. Als Folge der unmittelbaren Nähe des Meeres ist hier der Anteil des Chlorids gegenüber des Sulfates wiederum auffällig erhöht. Generell läßt sich feststellen, daß die Vertreter der *Juncetalia bulbosi* an der östlichen Arealgrenze extrem saure bis saure, ausgesprochen kalkarme, aber calcium- und elektrolytreiche Gewässer mit Härtegraden im mittelharten bis harten Bereich besiedeln. Die Siedlungsgewässer sind extrem arm an Ammonium und weisen aufgrund nur sehr geringer Mengen an im Wasser gelöster organischer Substanz eine klare Wasserfarbe auf und sind außerdem reich an freier im Wasser gelöster Kohlensäure (CO$_2$) sowie an Eisen-und Manganverbindungen.

Sind *Pilularia globulifera* und *Eleocharis multicaulis* in W Europa Arten von Weichwasser-Standorten und werden als sog. 'Weichwasser-Arten' bewertet, so müssen sie in den Siedlungsgewässern des östlichen Mitteleuropas als Arten von Hartwasser-Standorten und somit als 'Hartwasser-Arten' angesprochen werden. Dieses unterschiedliche Verhalten dürfte zukünftig für die Erarbeitung der Zeigerwertanalyse der aquatischen Makrophyten Mitteleuropas von besonderer Bedeutung sein. Während sich der Zeigerwert für den absoluten Ionengehalt, d.h. für den Mineralstoffgehalt, sehr wesentlich vom Arealzentrum zur östlichen Arealgrenze hin verändert, bleibt derjenige des relativen Ionengehaltes und insbesondere der des Calcium-Kohlensäure-Systems und der Aziditätsverhältnisse fast unverändert. Diese Tatsache zeigt, daß der Elektrolytgehalt der Standorte des azidophilen oceanischen Florenelementes für dessen Verbreitung nur von untergeordneter Bedeutung ist.

Pilularia globulifera, *Eleocharis multicaulis* und *Juncus bulbosus* besitzen gegenüber zahlreichen hydrochemischen Kenngrößen, insbes. dem Elektrolytgehalt, eine sehr breite Valenz, ein euryökes Verhalten. Die Arten besitzen jedoch gegenüber dem Gehalt an freier Kohlensäure (CO$_2$), Ammonium, dem relativen

Tab. 4. Absoluter und relativer Ionengehalt verschiedener Siedlungsgewässer von *Eleocharis multicaulis* S.M. in West- und Mitteleuropa.

Standorte	Hydrochemische Kenngrößen										Gesamt-ionen
	Na$^+$	K$^+$	Ca^{++}	Mg^{++}	Fe$^{2+/3+}$	SO$_4^-$	Cl$^-$	HCO$_3$	NO$_3$	PO$_4^{3-}$	
Beispiele:											
1: mg/l			2,0				39,0				175,0
2: mg/l			1,3				17,4				88,0
3: mg/l			5,1				13,0				108,0
4: mg/l	7,6	1,8	4,0	12,2	0,8	17,5	10,9	59,8			114,7
mval-%	22,6	2,9	12,1	60,7	1,7	22,1	18,7	59,2			
5: mg/l	8,9	3,4	24,6	2,4	2,9	65,5	11,2	2,2	3,2	0,058	131,5
mval-%	19,7	4,4	60,8	10,0	5,3	69,1	15,7	1,8	2,6	0,93	
6: mg/l	10,9	3,6	52,6	6,2	0,24	172,1	18,2	13,1	5,8	0,086	282,8
mval-%	12,8	2,7	71,6	12,8	0,21	78,6	11,1	4,7	2,1	0,21	
7: mg/l	9,4	3,2	29,1	5,6	2,1	65,5	22,1	6,5	4,2	1,2	151,3
mval-%	16,5	3,2	58,5	18,6	3,2	58,3	26,5	4,7	2,6	1,7	

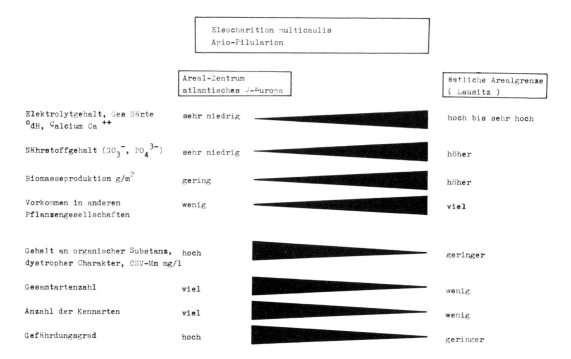

Abb. 1. Übersicht über Veränderungen ökologischer und soziologischer Kenngrößen von Siedlungsgewässern des atlantischen Floren-elementes der *Juncetalia bulbosi* vom Arealzentrum zur östlichen Arealgrenze.

Ionengehalt (SO_4-mval-%) und den biologischen Kriterien der Wassergüte, wie O_2, BSB_5 und organische Belastung (Perman-ganat-Verbrauch mg/l $KMnO_4$) eine sehr enge Valenz, ein stenökes Verhalten. Das chorologische Phänomen des Vor-kommens atlantischer Arten an der östlichen Arealgrenze läßt sich neben mikroklimatischen Ähnlichkeiten vor allem durch die edaphischen Gemeinsamkeiten in der ökologischen Beschaffen-heit aller Siedlungsgewässer West- und Mitteleuropas erklären. Diese edaphischen Voraussetzungen müssen im Gegensatz zu den Verhältnissen im atlantischen Westeuropa an den Stand-orten Mitteleuropas durch die Tätigkeit des Menschen immer von neuem geschaffen werden. Die Azidverhältnisse und der Gehalt an freier im Wasser gelöster Kohlensäure (CO_2) sind die wichtigsten Standortsfaktoren sämtlicher Siedlungsgewässer der *Juncetalia bulbosi*. Die Azidität legt die im Wasserkörper vorherrschende Form der Kohlensäure fest und bestimmt somit das Calcium-Kohlensäure-System der Siedlungsgewässer. Die Vertreter der azidophilen ozeanischen Ufervegetation der *Junce-talia bulbosi* kennzeichnen mineralogen-azidotrophe saure, oli-gohumose, biologisch unverschmutzte, kalkarme aber CO_2-reiche Gewässer mit einem hohen Sauerstoffgehalt und von einer biologisch sehr guten Wasserqualität und einem sandig-tonig bis -kiesig fraktionierten Gewässersediment arm an orga-nischer Substanz. Wie Abbildung 1 zeigt, erfolgt vom Areal-zentrum zur östlichen Arealgrenze hin eine auffällige Zunahme des Elektrolytgehaltes, der Gesamthärte, des Gehaltes an Calci-um und Sulfat, der ökologisch wichtigen Nährstoffe, wie Nitrat und Phosphat und der im Wasser gelösten freien Kohlensäure (CO_2), sowie der Biomasseproduktion der einzelnen Arten. Dagegen ist eine deutliche Abnahme des Gehaltes an im Wasser

gelöster organischer Substanz, als PV-Wert berechnet, und so-mit des dystrophen Charakters der Siedlungsgewässer, der Anzahl der Charakterarten und auch des gegenwärtigen Gefährdungs-grades zu verzeichnen. Aufgrund dieser Zusammenhänge er-klärt sich auch die Tatsache, daß die atlantischen Arten und ihre Gesellschaften an der östlichen Arealgrenze eine wesentlich opti-malere Entfaltung erreichen als im eigentlichen Verbreitungs-zentrum.

Ceratophyllum submersum-reiche Gesellschaften

An der nordöstlichen Arealgrenze des eutrophen temperat-ozeanisch verbreiteten *Ceratophyllum sub-mersum* ist im jungpleistozänen Mittelmecklen-burg während der letzten Jahre eine auffällige Zu-nahme der Vorkommen als Folge anthropogener Vorgänge, insbes. der Intensivierung der Landwirt-schaft, festzustellen. *Ceratophyllum submersum* galt bisher in Mecklenburg als zerstreut bis selten; seit einem Jahrzehnt wurden über 100 neue Stand-orte registriert. Wollert & Bolbrinker (1980) stellten an 254 untersuchten Tümpeln, Seen, Teichen und Gräben die Art in 92 Gewässern allein in den Kreis-gebieten von Teterow und Malchin fest.

Die Verbreitung beschränkt sich auf kleinere, flache, sommerwarme, nährstoffreiche Gewässer mit ganzjährig alkalischer Wasserreaktion, die sich gegenüber der Mehrzahl der Siedlungsgewässer von *C. demersum* durch eine ausgesprochen klare Wasserfarbe aufgrund nur geringer an im Wasser gelöster organischer Substanz (PV < 30 mg/l KMnO$_4$) unterscheiden. Bevorzugt besiedelt werden Sölle, wasserführende Ackerhohlformen auch in Nähe von Ortschaften, Dorfteiche, die von einer starken Eutrophierung bisher verschont geblieben sind, sowie flache Litoralbereiche größerer alkalischer Klarwasserseen. Zahlreiche gegenwärtig mit *C. submersum* bewachsene Seen und Teiche waren früher Badegewässer. Eine Auswertung sämtlicher untersuchter Siedlungsgewässer ergab eine durchschnittliche Größe von 0,1 bis 0,28 ha und eine durchschnittliche Wassertiefe von 0,74 m. Diese geomorphologischen Verhältnisse verursachen eine rasche Erwärmung der Wasserkörper, wie sie für die Siedlungsgewässer im südlichen Mitteleuropa und Südosteuropa charakteristisch ist. Die Ausbreitung in Mecklenburg wird vermutlich durch die zunehmende Anwendung von mineralischem Dünger begünstigt, der in nicht ganz unerheblichem Maße in die Gewässer eingespült wird. Ganz sicher führt auch die langjährige Badetätigkeit in den Kleingewässern zu einer auffälligen Erhöhung des Mineral- und Nährstoffgehaltes, insbesondere an Chlorid und Nitrat.

Soziologisches Verhalten

Ceratophyllum submersum entfaltet in Mecklenburg (Tab. 5) dichte, artenarme Dominanzbestände, die oft das gesamte Volumen des Gewässers einnehmen, wie z.B. im Dorfteich bei Demzin (Tab. 7, Beispiel 2), die wir als *Ceratophylletum submersi* Den Hartog 64 betrachten wollen. *Ceratophyllum submersum* ist an diesen Standorten alleinige bestandsbildende Art. Verschiedentlich sind als Vertreter der *Lemnetea* noch *Spirodela polyrrhiza*, *Lemna minor*, *L. trisulca* und vereinzelt auch *L. gibba* sowie als Vertreter der *Potametea-Potamogeton crispus*, *P. natans*, *Polygonum amphibium* und an besonders nährstoffreichen Standorten bereits *Zannichellia palustris* ausgebildet. Je nach dem Nährstoffgehalt und der Wassertiefe lassen sich neben einer typischen Ausbildung (Tab. 5, Aufn. 1 bis 4), durch optimales Auftreten von *Ceratophyllum* gekennzeichnet, eine Subass. v. *Spirodela polyrrhiza* (Aufn. 5 bis 7), in der verschiedene *Lemna*-Arten eine die *Ceratophyllum submersum*-Bestände überlagernde Wasserlinsendecke bilden, und eine Subass. v. *Potamogeton crispus* (Aufn. 8–10) unterscheiden. Im flachen Litoralbereich alkalischer Klarwasserseen mit sandigem Bodensubstrat tritt *Ceratophyllum submersum* außerdem in Kontakt mit Elementen der *Potametea*, *Potamogeton lucens*, *P. crispus*, *Myriophyllum spicatum*, *Elodea canadensis* und *Zannichellia palustris*. Verschiedentlich bilden sich nach Wollert & Bolbrinker (1981) im Sommer dichte Watten von *Cladophora*- und *Spirogyra*-Arten aus.

In Südeuropa tritt *Ceratophyllum submersum* recht häufig in Altwässern und Altarmen verschiedener Flüsse innerhalb von Gesellschaften der *Lemnetea*, wie *Wolffietum arrhizae*, *Lemno-Utricularietum* und *Salvinio-Spirodeletum*, aber auch in Gesellschaften der *Potametea*, so im *Potameto-Ceratophylletum* und dem *Najadi-Ceratophylletum* sensu Pop (1968) auf.

Tab. 5. Ceratophyllum submersum-reiche Ausbildungen im jungpleistozänen Mittelmecklenburg.

No. der Aufnahme	1	2	3	4	5	6	7	8	9	10
Größe der Aufnahmefläche m²	400	200	200	60	20	60	20	20	20	24
Gesamtdeckung in %	100	100	100	100	100	100	100	100	100	100
Gesamtartenzahl	2	2	8	5	5	5	6	10	9	5
Wassertiefe in cm	85	160	90	75	95	85	110	80	75	60
C-Ass.:										
Ceratophyllum submersum	4.5	5.5	5.5	5.5	5.5	5.5	5.5	5.5	5.5	4.5
Lemnetea-Arten:										
Lemna minor	.	.	1.1	1.3	4.5	3.4	2.3	1.3	1.3	+.3
Spirodela polyrrhiza	.	.	.	+.3	2.3	4.5	1.1	1.3	1.1	1.3
Lemna trisulca	.	.	+.3	.	.	2.3	3.4	1.3	1.3	.
Lemna gibba	.	.	.	.	+.1	2.3	.	.	.	.
Potametea-Arten:										
Ranunculus trichophyllus	.	+.3	+.1	.	+.1	.	.	.	+.1	.
Potamogeton natans	.	.	+.1	+.1	.	.	.	2.1	1.1	.
Potamogeton crispus	.	.	+.3	.	.	.	.	2.3	1.1	2.3
Polygonum amphibium	.	.	+.1	.	.	.	.	+.1	.	.
Hydrocharis morsus-ranae	.	.	.	.	.	.	+.1	+.3	.	.
Ceratophyllum demersum	.	.	.	.	.	.	.	1.1	1.3	.
Zannichellia palustris	.	.	.	.	.	.	.	.	1.3	1.1
Phragmitetea-Arten:										
Typha angustifolia	+.1	.	+.1	.	.	.	+.1	.	.	.
Carex elata	.	.	.	+.3	.	.	.	+.1	.	.

Tab. 6. Absoluter und relativer Ionengehalt verschiedener Siedlungsgewässer von *Ceratophyllum submersum* L. in Mittel- und Südosteuropa.

Standorte	Hydrochemische Kenngrößen									Gesamt-ionen
	Na^+	K^+	Ca^{++}	Mg^{++}	SO_4	Cl	HCO_3	NO_3	PO_4^3	
Beispiele:										
1: mg/l	6,2	1,8	52,3	7,2	52,3	27,7	89,8	0,27	2,4	242,4
mval-%	7,6	1,3	74,4	16,7	30,9	22,2	41,8	2,6	2,6	
2: mg/l	24,3	6,8	66,2	7,8	36,4	45,4	181,3	2,2	4,2	348,1
mval-%	20,9	3,3	63,4	12,3	14,6	24,6	57,0	0,68	2,5	
3: mg/l	9,8	4,2	26,4	4,1	14,1	13,3	91,6	0,86	0,19	164,5
mval-%	19,5	4,9	60,3	15,4	13,4	17,2	68,5	0,63	0,26	
4: mg/l	31,5	7,6	15,7	17,3	23,2	7,8	195,0	1,2	1,8	300,9
mval-%	25,2	4,9	19,8	40,8	12,1	5,5	80,5	0,05	0,14	
5: mg/l	33,1	12,5	15,4	17,4	20,9	7,3	198,2	1,9	1,2	307,9
mval-%	36,4	8,1	19,4	36,1	11,0	5,2	82,1	0,8	0,95	

Innerhalb des *Potameto–Ceratophylletum* W Rumäniens erreicht *Ceratophyllum submersum* das Optimum der Entfaltung und ist mit hoher Abundanz und Stetigkeit vertreten (V 2–4); *Potamogeton fluitans* und *Lemna minor* sind die wichtigsten Begleiter. Diese Ausbildung läßt sich am besten mit den *Ceratophyllum submersum*-Beständen Mecklenburgs vergleichen. Aus dem Donau-Überschwemmungsraum in Ungarn gibt Karpati (1963) *Ceratophyllum submersum* innerhalb des *Salvinio–Spirodeletum* gemeinsam mit *C. demersum* an. Auch in den Altwässern der oberen Weichsel, am Zusammenfluß mit dem San, tritt *Ceratophyllum submersum* gemeinsam mit *C. demersum* in den Gesellschaften *Hydrochari–Stratiotetum*, *Potametum lucentis* und *Nymphoidetum peltatae* auf (Piorecki, 1975). Aus dem Gebiet der Westfälischen Bucht beschreibt Pott (1980) eine Variante von *Ceratophyllum submersum* innerhalb der Subass. v. *Ceratophyllum demersum* des *Myriophyllo–Nupharetum* mit *Potamogeton trichoides* als weiterer häufiger Art.

Ökologisches Verhalten

Am Beispiel 5 verschiedener Standorte wird die ökologisch-hydrochemische Beschaffenheit *Ceratophyllum submersum*-reicher Gewässer untersucht (Tab. 6). Grundsätzlich handelt es sich bei allen um ganzjährig alkalische nährstoff- und bikarbonatreiche Gewässer, die sich jedoch aufgrund geologischer Verschiedenheiten ihrer Standorte im Gehalt an Natrium, Magnesium, Calcium, Bikarbonat und Chlorid unterscheiden. Am Beispiel des Dorfteiches von Demzin (Beispiel 2) zeigt uns das Kreisdiagramm ein elektrolytreiches Gewässer, reich an Hydrogenkarbonat und an Nitrat. Bemerkenswert sind weiterhin die hohen Chloridwerte, die auf eine intensive anthropogene Beeinflussung hinweisen. Nach dem relativen Ionendiagramm sind Calcium und Hydrogenkarbonat die vorherrschenden Ionen, gefolgt von Natrium unter den Kationen und Chlorid unter den Anionen. Die Standorte Südosteuropas (Beispiele 4 u. 5) zeichnen sich ebenfalls durch elektrolytreiche Verhältnisse aus, allerdings ist der Anteil des Chlorides auffällig geringer. Aufgrund der besonderen geologischen Verhältnisse sind Natrium, Magnesium und Hydrogenkarbonat die vorherrschenden Ionen, während der Anteil des Calcium auffällig zurückgegangen ist.

Schlußbetrachtung: Durch biotische Einflüsse kommt es an der nordöstlichen Arealgrenze zu einer spontanen Massenentfaltung des kalziphilen und alkaliphilen *Ceratophyllum submersum* und somit zu einer Häufung von Vorkommen und einer gewissen Arealerweiterung. Überwiegen an diesen Vorpostenstandorten artenarme Reinbestände eines eigenen *Ceratophylletum submersi*, so tritt die Art im südlicheren Arealgebiet innerhalb von *Lemnetea*- und *Potametea*-Gesellschaften und dort oft gemeinsam mit *Ceratophyllum demersum* auf. Durch die Tätigkeit des Menschen werden ein höherer Elektrolytgehalt und eine Anreicherung der Gewässer mit Nitraten und Chloriden verursacht, ohne jedoch den Wasserkörper mit im Wasser gelöster organischer Substanz zu belasten, wodurch die ursprünglich klare Wasserfarbe der Siedlungsgewässer erhalten bleibt. Es bleibt jedoch abzuwarten, welche Veränderungen sich mit fortschreitender Sukzession bzw. bei einer ständigen jährlichen Akkumulation der abgestorbenen Biomasse in diesen kleinen Gewässern zukünftig in der ökochemischen Beschaffenheit abspielen werden. Ein Verdrängen der Bestände von *Ceratophyllum submersum* durch solche von *C. demersum* wäre durchaus denkbar.

Wir bestätigen die bereits vor zwei Jahrzehnten von Hejný (1960) getroffene Feststellung, daß es sich bei *Ceratophyllum submersum* um eine periodisch und ephemer auftretende Art mit starker Affinität zu biotisch veränderten Gewässern handelt, ähnlich wie die Arten *Potamogeton trichoides*, *P. pusillus* und *Zannichellia palustris*. Zum Unterschied von *Ceratophyllum demersum* zieht die Art offenbar freie, nicht verlandende Gewässer vor.

Aldrovanda vesiculosa-reiche Gesellschaften

Am Beispiel 8 verschiedener Siedlungsgewässer *Aldrovanda vesiculosa*-reicher Gesellschaften Mittel- und Südosteuropas wird ein auffälliger Wechsel im diagnostischen Wert in der Nord–Süd-Verteilung der disjunkten Areale nachgewiesen (Tab. 8). Die tropisch-submeridional-ozeanisch verbreitete *Aldrovanda* tritt in Abhängigkeit ihres jeweiligen Areals in Gesellschaften der *Utricularietea intermedio–minoris*, der *Lemnetea*, der *Potametea* und vereinzelt sogar der *Charetea* auf.

Soziologisches Verhalten

Aldrovanda vesiculosa tritt im nordöstlichen Mitteleuropa, etwa in der Höhe Berlin – Torun – Bialystok – Grodno – Wilnus, von Pommern bis Litauen, in Siedlungsgewässern der *Utricularietea intermedio-minoris* auf (Tab. 7, Aufn. 1 bis 5). Die Art besiedelt Schlenken und mit Wasser gefüllte Torfstiche in nährstoffarmen kalkarmen bzw. kalkreichen Zwischenmooren und Sümpfen. Je nach dem Kalkgehalt, d.h. nach dem vorherrschenden Calcium-Kohlensäure-System, bildet *Aldrovanda* zusammen mit Torfmoosen oder aber mit Braunmoosen und den *Utricularia*-Arten *U. intermedia, U. minor, U. ochroleuca* und *U. australis* sowie mit *Sparganium minimum* eine charakteristische Vegetation innerhalb von Ausbildungen des *Sphagno-Utricularietum*, des *Scorpidio-Utricularietum* bzw. des *Drepanoclado-Utricularietum*. Außerdem siedelt *Aldrovanda* in bryo-

Tab. 7. *Aldrovanda vesiculosa*-reiche Gesellschaften in Mittel- und Südosteuropa. S. Appendix 1.

	1	2	3	4	5	6	7	8	9	10
No. der Aufnahme	1	2	3	4	5	6	7	8	9	10
No. des Ionendiagramms	1		2	3	4		5	6	7	8
Größe der Aufnahmefläche m²	8	12	12	10	12	12	20	20	20	12
Gesamtdeckung in %	95	80	100	80	75	65	90	85	85	70
Gesamtartenzahl	7	12	9	14	10	9	12	12	9	6
Wassertiefe in cm	18	40	15	30	12	45	35	25	18	40
Aldrovanda vesiculosa	4.5	3.4	4.5	3.4	2.3	3.4	4.5	3.4	4.5	3.4
Utricularietea-Arten:										
Utricularia intermedia	1.3	2.3	1.3	2.3	3.4	.	.	.	.	.
Utricularia minor	2.1	1.1	1.3	1.1	2.3	.	.	.	.	.
Sparganium minimum	+.1	+.1	1.3	1.1	1.1	.	.	.	.	.
Utricularia australis	+.3	1.3	.	.	.	.	.	.	.	.
DV-Sphagno–Utricularion:										
Sphagnum obesum	1.3	3.4	.	.	.	.	.	.	.	.
Sphagnum auriculatum	1.3	2.3	.	.	.	.	.	.	.	.
Sphagnum subsecundum	.	2.3	.	.	.	.	.	.	.	.
DV-Scorpidio–Utricularion:										
Scorpidium scorpioides	.	.	+.3	2.3	3.4	.	.	.	.	.
Drepanocladus intermedius	.	.	2.3	1.3	3.4	.	.	.	.	.
Drepanocladus revolvens	.	.	3.4	3.4	.	.	.	.	.	.
Lemnetea-Arten:										
Hydrocharis morsus-ranae	.	1.1	+.1	+.1	+.1	1.1	+.3	+.1	+.3	.
Riccia fluitans	.	+.1	.	+.1	1.1	+.1	+.1	.	+.1	.
Spirodela polyrrhiza	.	.	.	.	+.3	+.3	1.1	1.1	2.3	1.3
Lemna trisulca	.	.	.	.	1.3	+.3	2.3	1.3	2.3	.
Lemna minor	.	.	.	.	.	.	2.3	+.3	1.3	1.3
Salvinia natans	.	.	.	.	.	.	1.3	+.3	+.3	.
Stratiotes aloides	.	.	.	.	.	.	1.3	2.3	.	.
Potametea-Arten:										
Potamogeton natans	.	.	.	+.1	.	+.1	+.1	+.1	+.1	1.1
Utricularia vulgaris	.	.	.	.	.	+.1	1.1	1.1	.	.
Nuphar lutea	.	+.1	.	.	.	1.1	.	.	.	.
Myriophyllum spicatum	.	.	.	.	.	+.1	+.1	.	.	.
Nymphaea alba	.	.	.	.	.	.	.	+.1	.	+.1
Hydrilla verticillata	.	.	.	1.3	.	.	.	.	.	.
Myriophyllum verticillatum	.	.	.	1.3	.	.	.	.	.	.
Polygonum amphibium	.	.	.	.	.	.	.	+.1	.	.
Weitere Arten:										
Typha angustifolia	.	+.1	+.1	+.1	.	.	+.1	+.1	+.1	.
Chara tomentosa	.	.	.	2.3	.	.	.	.	.	.
Typha laxmannii	.	.	.	.	.	.	.	.	.	3.4

phytenfreien Beständen des *Utricularietum intermedio-minoris* und des *Sparganietum minimae* oder tritt als selbständige *Utricularia minor–Aldrovanda vesiculosa*-Ges. auf (Pietsch, 1975, 1979).

Die Ausbildungen des *Sphagno–Utricularion* kennzeichnen ausgesprochen kalkarme, saure bis schwach saure mäßig nährstoffarme Siedlungsgewässer. Die *Scorpidium-* und *Drepanocladus*-reichen Ausbildungen des *Scorpidio–Utricularion* charakterisieren dagegen kalkreichere neutrale bis alkalische nährstoffarme Standorte. Je nach der Wassertiefe, dem HCO_3- und dem Nährstoffgehalt lassen sich außerdem verschiedene Varianten unterscheiden, die charakteristische Kleinstmosaike mit einem spezifischen Indikationswert für die jeweiligen Standortsverhältnisse darstellen.

Im nordöstlichen Mitteleuropa, in der Suwalki-Seenlandschaft in NE Polen, lassen sich sogar Vorkommen mit *Hydrilla verticillata, Myriophyllum verticillatum, Chara tomentosa* und *Scorpidium scorpioides* feststellen, wie z.B. am Kruglak-See (Tab. 8, Beispiel 3, Tab. 7, Aufn. 4). Es handelt sich hier um einen mäßig kalkreichen, mäßig nährstoffreichen alkalischen Klarwassersee, dessen östlicher Uferbereich von einem braunmoosreichen Zwischenmoor-Komplex umgeben ist. In den vegetationsfreien Lücken dieser *Carex diandra-* und *C. lasiocarpa*-reichen Bestände siedelt *Aldrovanda* innerhalb von Ausbildungen eines *Scorpidio–Utricularietum*.

Den eigentlichen *Lemna*-reichen Siedlungsgewässern fehlt im nördlichen Mitteleuropa *Aldrovanda vesiculosa* aufgrund des höheren Elektrolytgehaltes und Trophiegrades der Wasserkörper. Nur vereinzelt tritt die Art zusammen mit *Riccia fluitans* in elektrolyt- und nährstoffärmeren Gewässern auf. Diese Ausbildungen lassen sich nach Schwabe-Braun & Tüxen (1981) als *Riccietum fluitantis* Subass. v. *Aldrovanda vesiculosa* auffassen.

Im südlichen Mitteleuropa, vor allem in Südosteuropa, so im pannonischen Raum und im Bereich des Donaudeltas, und schließlich sogar in Kaukasien und im Wolgadelta, tritt *Aldrovanda* ausschließlich in Ausbildungen der *Lemnetea-* und *Potametea*-Gesellschaften in ganzjährig neutral bis schwach alkalischen Gewässern auf, denen jegliche Elemente der *Utricularietea intermedio-minoris* fehlen. Die Art bildet zusammen mit *Spirodela polyrrhiza, Lemna minor, L. trisulca, Hydrocharis morsus-ranae, Salvinia natans* und vereinzelt auch *Stratiotes aloides* eine Vegetation, die vom Baláta-tó in Ungarn erstmalig treffend als *Spirodelo–Aldrovandetum* durch Borhidi & Járai-Komlódi (1959) beschrieben wurde. Verschiedentlich bildet *Aldrovanda* zusammen mit *Typha angustifolia* und *T. laxmannii* charakteristische Bestände, wie z.B. in der Nähe von Krasnodar im nordwestlichen Kaukasus-Vorland (Aufn. 10, Tab. 8, Beispiel 8).

Aldrovanda tritt im nördlichen und nordöstlichen Mitteleuropa als lokale Ordnungs- und Klassen-Kennart der *Utricularietea intermedio-minoris* bzw. als C-Ass. der *Utricularia minor–Aldrovanda vesiculosa*-Ges. auf (Aufn. 1 bis 5). Im südlichen Mitteleuropa und in Südosteuropa ist *Aldrovanda* lokale Kennart der Ordnung *Lemnetalia* und C-Ass. des *Spirodela–Aldrovandetum* (Aufn. 6 bis 9).

Ökologisches Verhalten

Alle *Aldrovanda*-Standorte sind dystrophe Gewässer, reich an Huminsäuren, an im Wasser gelöster organischer Substanz, die PV-Werte bewegen sich zwischen 60 bis 125 mg/l $KMnO_4$, mit einer gewissen Armut an N- und P-Verbindungen, insbesondere an NH_4-N im Wasserkörper ($< 0,10$ mg/l NH_4), eisenarm aber reich an im Wasser gelöster freier Kohlensäure (CO_2). Je nach den Standortsverhältnissen werden elektrolytarme und kalkarme aber ebenso elektrolyt- und kalkreichere Standorte besiedelt. *Aldrovanda vesiculosa* zeigt gegenüber dem Calcium-

Tab. 8. Absoluter und relativer Ionengehalt verschiedener Siedlungsgewässer von *Aldrovanda vesiculosa* L. in Mittel- und Südosteuropa sowie in Kaukasien.

Standorte	Hydrochemische Kenngrößen										Gesamt-ionen
	Na^+	K^+	Ca^{++}	Mg^{++}	$Fe^{2+/3+}$	SO_4^-	Cl^-	HCO_3	NO_3	PO_4^{3-}	
Beispiele:											
1: mg/l	16,7	5,0	39,1	3,8	0,11	72,3	18,2	61,0	5,8	0,094	232,2
mval-%	23,3	4,1	62,4	9,9	0,17	48,0	16,4	31,9	2,9	0,092	
2: mg/l	11,4	5,8	23,8	3,2	0,02	36,1	18,4	45,8	4,2	0,088	149,9
mval-%	23,2	6,9	55,9	12,3	0,004	35,2	24,3	35,2	3,2	0,13	
3: mg/l	19,1	4,8	47,2	6,7	0,14	17,1	10,3	143,9	5,6	0,048	260,3
mval-%	21,6	3,2	60,9	14,3	0,012	9,2	21,4	61,5	2,4	0,04	
4: mg/l	8,9	3,5	71,4	6,8	0,48	8,6	32,3	200,7	5,4	0,096	342,7
mval-%	8,4	1,9	77,2	12,1	0,36	3,9	19,5	71,1	1,9	0,065	
5: mg/l	6,7	2,3	8,2	2,6	0,16	6,8	7,6	30,5	3,8	0,24	70,5
mval-%	29,7	6,1	41,9	21,8	0,58	14,5	21,9	51,0	6,3	0,76	
6: mg/l	4,8	2,3	12,2	7,9	9,3	6,8	21,4	39,1	4,2	0,096	110,4
mval-%	11,3	5,4	32,5	35,1	17,9	9,0	38,4	40,7	4,3	1,91	
7: mg/l	3,9	2,4	19,6	8,1	4,2	8,9	30,4	47,5	3,9	0,086	133,7
mval-%	8,1	2,9	43,1	34,9	10,9	8,6	45,5	39,9	3,4	0,12	
8: mg/l	4,5	2,7	19,6	8,1	4,2	8,9	30,4	47,5	3,9	0,086	131,9
mval-%	9,6	3,1	45,7	30,8	8,9	9,3	43,7	39,1	3,2	0,16	

Kohlensäure-System ein ausgesprochen indifferentes stenökes Verhalten. Die Art besiedelt sowohl schwach saure Calcium-Sulfatgewässer (Tab. 8, Beispiel 1) als auch schwach alkalische Calcium-Hydrogenkarbonat-Gewässer (Beispiele 2 bis 8), wie es in den relativen Ionendiagrammen zum Ausdruck kommt. Trotz der aufgezeigten Verschiedenheiten im diagnostischen Wert der Art und der hydrochemischen Beschaffenheit ihrer jeweiligen Siedlungsgewässer, bestehen doch bei allen in Tab. 8 zusammengestellten Beispielen grundsätzliche Gemeinsamkeiten im Verhalten gegenüber bestimmten ökologischen Standortsfaktoren. *Aldrovanda* besitzt ein deutliches euryökes Verhalten gegenüber einem erhöhten Gehalt an im Wasser gelöster organischer Substanz, an freier Kohlensäure (CO_2), Chlorid und an Silikat sowie gegenüber geringen bis mittleren Calcium- und Mineralstoffkonzentrationen. Außerdem weisen die Siedlungsgewässer des öfteren eine Armut an Sauerstoff auf. Die geringe Valenz gegenüber dem Calcium- und insbesondere dem Elektrolytgehalt kommt in den etwa fast gleichgroßen Kreisflächen der Kreisdiagramme aller 8 Beispiele in Tab. 8 deutlich zum Ausdruck. *Aldrovanda vesiculosa* zeigt somit ein völlig anderes ökologisches Verhalten als wir es bei den Vertretern des azidophilen ozeanischen Florenelementes der *Juncetalia bulbosi* kennengelernt hatten.

Die *Aldrovanda*-Siedlungen sind ein klassisches Beispiel für dynamische Veränderungen freischwebender Wasserpflanzen-Gesellschaften dystropher Gewässer. Die dynamische Balance wird in erster Linie durch die Niederschlags- und Temperaturverhältnisse reguliert, die dazu führen, daß eine Gesellschaft ganz die andere verdrängen und ersetzen kann. So konnte von uns festgestellt werden, daß in trockenen, regenarmen Jahren der Anteil von *Aldrovanda* bis auf wenige Exemplare zurückgeht, während dafür die *Lemna*-reichen Gesellschaften, wie *Lemno-Salvinietum* oder *Spirodelo-Lemnetum* am Beispiel des Baláta-Sees (No. 5) bzw. *Stratiotetum aloidis* am Beispiel des Grünen Sees (No. 6) optimal entwickelt sind. In regenreichen Sommern ist dagegen ein gehäuftes Auftreten von *Aldrovanda* festzustellen, während die *Lemnetea*-Arten nur eine spärliche Vegetation bilden (Tab. 8, Beispiele 5 u. 6). Dieser Wechsel im Auftreten beider Gesellschaften läßt sich aus den Schwankungen im Elektrolytgehalt, dem Calciumgehalt, dem Trophiegrad sowie dem Gehalt an im Wasser gelöster organischer Substanz (als PV-Wert berechnet) anschaulich erklären. In trockenen Sommern kommt es zu einem Anstieg der Menge an anorganischer mineralischer Substanz als Folge der Erhöhung der Mineralisierung im Wasserkörper. Der pH-Wert steigt aus dem schwach sauren Bereich zum Neutralpunkt, der Wasserspiegel wird niedriger. Die Crustaceen als wichtigste Nährtiere wandern in den tieferen Teil des Gewässers und nehmen so an der Oberfläche in der Anzahl ab. Die wenigen vorhandenen *Aldrovanda*-Pflanzen werden durch die *Lemnetea*-Arten überwachsen. In regnerischen Sommern steigt der Wasserspiegel und das *Spirodelo-Aldrovandetum* besiedelt das gesamte Gewässer in großflächiger Ausbildung. Es erfolgt eine starke Einbringung von organischer Substanz aus der Umgebung, die zu einem Anstieg der Konzentration der Huminsäuren führt; der Wasserkörper wird verdünnt, der Elektrolytgehalt und der Gesamthärtegrad nehmen ab. Der pH-Wert wird niedriger und sinkt in den schwach sauren bis sauren Bereich ab. Die Zunahme an ökologischen Nährstoffen erhöht die Anzahl an Zooplanktonorganismen. *Aldrovanda* bevorzugt somit die elektrolyt- und calciumärmeren aber

an Huminsäuren reicheren Standortsverhältnisse während regenreicher Sommer.

Allerdings läßt sich *Aldrovanda vesiculosa* auch während des Sommers vereinzelt in *Lemna*- und *Stratiotes*-reichen Gewässern antreffen, jedoch zu einem Zeitpunkt, an dem es aufgrund beachtlich hoher Phytomasseproduktion der submersen Makrophytenbestände zu einem regelrechten Schwund an Nähr- und Mineralstoffen, insbesondere auch an Sulfat ($< 0,10$ mg/l SO_4) im Oberflächenwasser, gekommen ist. Die Art vermag Standorte bei Verarmung des Oberflächenwassers an ernährungsökologisch wichtigen Stoffen infolge ihrer Fähigkeit zur karnivorischen Lebensweise als ökologische Nische zu besiedeln. Diese Zusammenhänge erklären auch die Tatsache, daß die Größe der Kreisflächen der absoluten Ionendiagramme der *Lemna*-reichen Siedlungsgewässer Süd- und Südosteuropas sogar etwas kleiner sind, als diejenigen des nördlichen Mitteleuropas, denen die *Lemna*-Arten grundsätzlich fehlen.

Literatur

Barber, E., 1893. Beiträge zur Flora des Elstergebietes in der preußischen Oberlausitz. Abh. Naturf. Ges. Görlitz 20: 147–166.

Berta, J., 1961. Beitrag zur Ökologie und Verbreitung von Aldrovanda vesiculosa L. Biologia 16: 561–573.

Borhidi, A. & Jarai-Komlodi, M., 1959. Die Vegetation des Naturschutzgebietes des Baláta-Sees. Acta Bot. Sci. Hung. 5: 259–320.

Dierßen, K., 1973. Die Vegetation des Gildehauser Venns. Beih. Ber. Naturhist. Ges. Hannover 8: 6–120.

Dierßen, K., 1975. Littorelletea uniflorae. In: R. Tüxen (ed.), Prodromus der europäischen Pflanzengesellschaften, Lfg. 2. Vaduz.

Drude, O., 1902. Der hercynische Florenbezirk. Die Vegetation der Erde, 5. Leipzig.

Gräbner, P., 1925. Die Heide Norddeutschlands und die sich anschließenden Formationen in biologischer Betrachtung. Die Vegetation der Erde, 2. Aufl. Leipzig.

Hejný, S., 1960. Ökologische Charakteristik der Wasser- und Sumpfpflanzen in den Slowakischen Tiefebenen (Donau- und Theißgebiet). Bratislawa.

Jäger, E., 1964. Zur Deutung des Arealbildes von Wolffia arrhiza (L.) Wimm. und einiger anderer ornithochorer Wasserpflanzen. Ber. Deutsch. Bot. Ges. 77: 101–111.

Jäger, E., 1968. Die pflanzengeographische Ozeanitätsgliederung der Holarktis und die Ozeanitätsbindung der Pflanzenareale. Feddes Repert. 79: 157–335.

Kárpáti, V., 1963. Die zönologischen und ökologischen Verhältnisse der Wasservegetation des Donau-Überschwemmungsraumes in Ungarn. Acta Bot. Acad. Sci. Hung. 9: 323–385.

Maucha, R., 1932. Hydrochemische Methoden in der Limnologie. In: Die Binnengewässer, 12.

Meusel, H., Jäger, E. & Weinert, E., 1965. Vergleichende Chorologie der zentraleuropäischen Flora. 2 Bde. Jena.

Pietsch, W., 1968. Die Verlandungsvegetation des Sorgenteiches bei Ruhland in der Oberlausitzer Niederung und ihre pflanzengeographische Bedeutung. Ber. Arbeitsgem. sächs. Botaniker, Dresden, NF 8: 55–91.

Pietsch, W., 1973. Vegetationsentwicklung und Gewässergenese in den Tagebauseen des Lausitzer Braunkohlen-Reviers. Arch. Naturschutz u. Landschaftsforsch. 13: 187–217.

Pietsch, W., 1975. Zur Soziologie und Ökologie der Kleinwasserschlauch-Gesellschaften Brandenburgs. Gleditschia 3: 147–162.

Pietsch, W., 1977. Zur Soziologie und Ökologie von Aldrovanda vesiculosa L. in Mittel- und Südost-Europa. Studia Phytologica in honorem Jubilantis A.O. Horvat. Pecs. 107–111.

Pietsch, W., 1978. Zur Soziologie, Ökologie und Bioindikation der Eleocharis multicaulis-Bestände der Lausitz. Gleditschia 6: 209–264.

Pietsch, W., 1979. Zur Bioindikation einiger Vertreter des atlantischen Florenelementes in der Altmark und der Lausitz. Doc. phytosociol. 4: 827–840.

Piorecki, J., 1975. Trapa natans L. w Kotlinie Sandomierskiej (ekologia, rozmieszczenie i ochrona). Rocz. Przemyski 15–16: 374–400.

Pop, I., 1968. Flora si Vegetatia Cimpiei Crisurilor. Acad. Rep. Romania, Bucuresti.

Pott, R., 1980. Die Wasser- und Sumpfvegetation eutropher Gewässer in der Westfälischen Bucht. Abh. Landesmus. Naturk. Münster/Westf. 42: 1–156.

Schoof-van Pelt, M. M., 1973. Littorelletea, a study of the vegetation of some amphiphytic communities of western Europe. Diss. Nijmegen, 216 pp.

Schwabe-Braun, A. & Tüxen, R., 1981. Lemnetea minoris. Prodromus der europäischen Pflanzengesellschaften 4. Vaduz.

Wollert, E. & Bolbrinker, P., 1980. Zur Verbreitung sowie zum ökologischen und soziologischen Verhalten von Ceratophyllum submersum L. in Mittelmecklenburg. Arch. Freunde Naturgesch. Mecklenburg 20: 35–46.

Accepted 9. 3. 1984.

Appendix 1

Fundortsnachweis Tabelle 7: Aldrovanda vesiculosa-*reiche Gesellschaften*

1. Aufn.: W Ufer des Plagesee bei Brodowin, Kr. Eberswalde; 16.8.1980.
2. Aufn.: Schlenken im Zwischenmoorkomplex am Pehlitzwerder am Parsteiner See bei Brodowin, Kr. Eberswalde; 16.8.1980.
3. Aufn.: Schlenken im Zwischenmoorkomplex am Jez. Plaska bei der Ortschaft Plaska, östlich Augustow am Augustowski-Kanal, NE Polen; 25.9.1980.
4. Aufn.: Jez. Kruglak im Naturschutzgebiet Perkuć, bei Paniowo, östlich Augustow, am Augustowski-Kanal; NE Polen; 24.9.1980.
5. Aufn.: Naturschutzgebiet Swierczów, Schlenken im Zwischenmoorkomplex; südlich Wlodawa, Ost-Polen; 11.9.1980.
6. Aufn.: Naturschutzgebiet Swierczów, Restsee mit Stratiotes-Bestand; südlich Wlodawa; Ost-Polen; 11.9.1980.
7. Aufn.: Baláta-See nördlich Szomogyszob, Südungarn; 27.6.1974.

8. Aufn.: Grüner See (Zöld-tó) südwestlich Vojko in der Slowakei, nordwestlich Krolowsky; 30.8.1973.
9. Aufn.: Sumpfgebiet von Caraorman im Donaudelta, Rumänien; 8.8.1973.
10. Aufn.: Sumpfgewässer bei Petrovskaja und Troickaja, nordwestl. Krasnodar, im Kaukasusvorland; 21.9.1971.

Appendix 2

Nachweis der Siedlungsgewässer der Beispiele der Tabellen 3, 4, 6 u. 8

Tab. 3: Siedlungsgewässer von *Pilularia globulifera* L.
1: Britannien, Frankreich, Pool between Loudeac and Merdrignac, Côtes-du-Nord, Schoof-van Pelt (1973). Tab. 14, Nr. 24.
2: Niederlande, Overijssel, Pool in Lonnekerveld, Enschede, Schoof-van Pelt (1973), Tab. 14, Nr. 18.
3: Abflußgraben im Jeggauer Moor bei Trippigleben, Kr. Gardelegen, Bez. Magdeburg, Altmark, D.D.R.; 14.9.1981.
4a: D.D.R., Restgewässer des Tonabbaus bei Bröthen, Kr. Hoyerswerda, Bez. Cottbus, Lausitzer Niederung; 19.7.1981.
4b: D.D.R., Blauer See, Restgewässer des Tonabbaus bei Bernsdorf, Kr. Hoyerswerda, Bez. Cottbus, Lausitzer Niederung; 19.7.1981.

Tab. 4: Siedlungsgewässer von *Eleocharis multicaulis* S.M.
1: Irland, Pool along the road from Annagar to Loughanure, The Rosses, Donegal, Schoof-van Pelt (1973), Tab. 13, Nr. 12.
2: Frankreich, Britannien, Pool east of Paimpont, Vilaine, Schoof-van Pelt (1973), Tab. 10, Nr. 7.
3: Niederlande, Nord-Brabant, Schoof-van Pelt (1973), Tab. 9, Nr. 27.
4: B.R.D., NW Deutschland, Gildehauser Venn, Dierßen (1973).
5: D.D.R., Jävenitzer Moor, Kr. Gardelegen, Bez. Magdeburg, Altmark; 14.9.1981.
6: D.D.R., Restgewässer des Torfabbaus im NSG Bröthen-Zeißholzer Moor, Kr. Hoyerswerda, Bez. Cottbus, Lausitzer Niederung; 4.8.1981.
7: Polen, Heidegewässer im küstennahen Dünenbereich der Ostsee, östl. Bialogóra, in Nordpolen, Woj. Gdansk; 28.9.1976 (Pietsch, 1978).

Tab. 6: Siedlungsgewässer von *Ceratophyllum submersum* L.
1: B.R.D., Westfälische Bucht; Angaben nach Pott (1980).
2: D.D.R., Dorfteich von Demzin, Kr. Malchin, Bez. Neubrandenburg; Mecklenburg; 13.9.1981.
3: Polen, Altwasser bei Chwalowice am San, in SE Polen bei Sandomircz, Piorecky (1975).
4: Ungarn, Kleiner Donauarm bei Baja, Ferenc-Kanal, Südungarn, Karpati (1963).
5: Rumänien, Altwasser und Kanal am Pescaria-Radvani, nördlich Salonta, Pop (1968); 18.8.1973.

Tab. 8: Siedlungsgewässer von *Aldrovanda vesiculosa* L.
1: D.D.R., Plagesee bei Chorin, Kr. Eberswalde, nördlich Berlin; 16.8.1980.

2: Polen, Zwischenmoor bei Plaska, östlich Augustow am Augustowski-Kanal, Woj. Suwalki; 25.9.1980.

3: Polen, Kruglak-See im NSG Perkuć, östlich Paniowo, NE Polen, Woj. Suwalki; 24.9.1980.

4: Polen, Restsee im NSG Swierzczów, im Leszna-Wlodawaer-Seengebiet, östl. Lublin; 11.9.1980.

5: Ungarn, Baláta-See, nördlich Szomogyzsob, Südungarn; 27.6.1974.

6: Grüner See (Zöld-tó), südwestlich Vojko, in der Südostslowakei; Berta (1970); 30.8.1973.

7: Rumänien, Sumpfgebiet von Caraorman, im Donaudelta; 8.8.1973.

8: Sumpfgewässer bei Petrovskaja und Troickaja, nordwestlich Krasnodar, Kaukasus-Vorland; 21.9.1971.

Chorological phenomena of the *Molinietalia* communities in Czechoslovakia

E. Balátová-Tuláčková*

Botanical Institute, Czechoslovak Academy of Sciences, Stará 18, 662 61 Brno, Czechoslovakia

Keywords: Czechoslovakia, *Molinietalia* communities, Phytogeography

Abstract

Forty *Molinietalia* communities were analyzed from the phytogeographical point of view in order to establish their distribution tendencies in Czechoslovakia in relation to the floristic regions of Hercynicum, Pannonicum and Carpaticum. Only few moist meadow communities of the *Molinietalia* order are found in the whole area under study; most of them have their distribution center in one of the floristic regions mentioned above. Eleven *Calthion* and two *Molinion* associations are confined to the Hercynicum, eight of them strictly. It is only in the Moravian and Slovak Pannonicum that all associations of the alliances *Cnidion venosi* and *Veronico longifoliae–Lysimachion* occur; they appear there together with two *Molinion* communities, the distribution of which is, however, a little wider. Three *Calthenion* communities show a still wider distribution, extending into the adjoining areas. One *Calthenion* community occurs in the Bohemian Pannonicum and its surroundings, two *Calthenion* associations have their distribution center in the Carpaticum, but they are not strictly confined to it.

Introduction

This article calls attention to the distribution tendency of the *Molinietalia* communities occurring in the floristic regions of Czechoslovakia, with special respect to Bohemia and Moravia. No consideration is given to the communities with few relevés or to those with a very wide distribution.

Phytogeographical division of Czechoslovakia

According to Dostál (1960) the following phytogeographical regions can be distinguished in Czechoslovakia:

1. *Hercynicum,* the region of the central European forest flora, influenced by the Subatlantic climate, with predominantly silicate substratum and acid soils. It can be divided into three subregions, i.e. the Euhercynicum, the Subhercynicum and the Sudeticum, the last often being regarded as an independent region.

The *Euhercynicum,* where the influence of the Subatlantic climate manifests itself most significantly, includes the border mountains of Bohemia, ranging from the Novohradské hory Mountains to the Krušné hory Mountains in the north, and also the Slavkovský les Mountains, the central part of the Brdy Highlands, the Jihlavské vrchy and the Žďárské vrchy Highlands.

The *Subhercynicum* is a transition between the Sudeticum and/or the Pannonicum and the Euhercynicum. It consists of the Lužické hory Mountains, of their neighbour regions, and also of most of the mountainous regions bordering the Euher-

* The author wants to express her thanks to Doc. Dr Miroslav Smejkal, CSc., for critical revision of the text.

Vegetatio 59, 111–117 (1985).

cynicum and the Sudeticum (partly), that is to say, of a considerable part of western Bohemia, southern Bohemia (almost up to Prague) and western Moravia.

The *Sudeticum* includes territories in which the montane flora and the submontane flora have developed under the influence of Pleistocene glaciers. The influence of the Subatlantic climate is insignificant. The Sudeticum includes the northern border territories from the Jizerské hory Mountains up to the Hrubý Jeseník Mountains as well as the Nízký Jeseník and the Oderské vrchy Highlands.

2. *Pannonicum.* The Pannonicum includes the lowland regions, in Bohemia up to an altitude of 300 to 400 m, in south Moravia and in Slovakia up to 400 to 500 m or even 600 m. It is rich in xerothermic species and is influenced, more or less, by a subcontinental climate and its substratum is mostly rich in nutrients. In the planar belt neutral to alkaline soils (including salt soils) predominate, with some areas of moving sand dunes.

3. *Carpaticum.* This region is limited to the Carpathians and their borders. The flora is influenced by Siberian, Alpine and Balcanic elements and rich in endemic species. The Carpaticum includes the Moravian Karst, northeastern and eastern Moravia and most of Slovakia.

The *Molinietalia* associations occurring in Czechoslovakia and their distribution

About forty associations of the order *Molinietalia* have been described from Czechoslovakia. They are classified into five alliances (cf. Balátová-Tuláčkova, 1981a, b): *Cnidion venosi* Bal.-Tul. 65 (6 associations), *Molinion caeruleae* W. Koch 26 (about 7 ass.), *Calthion* Tx. 37 em. Bal.-Tul. 78 (22 ass., 15 of them being classified into the suballiance *Calthenion* (Tx. 37) Bal.-Tul. 78 and 7 into the suballiance *Filipendulenion* (Lohm. in Oberd. et al. 67) Bal.-Tul. 78), *Veronico longifoliae–Lysimachion* (Pass. 77) Bal.-Tul. 81 (3 ass.) and *Alopecurion pratensis* Passarge 64 (3 ass.).

In Czechoslovakia only few *Molinietalia* associations are distributed throughout the territory from the west to the east. This applies to, e.g. the *Scirpetum sylvatici* (Ralski 31) Schwick. 44, the *Filipendulo-Geranietum palustris* W. Koch 26, the *Lysimachio vulgari–Filipenduletum* Bal.-Tul. 78, the

Molinietum caeruleae W. Koch 26 sensu auct., the *Stellario–Deschampsietum cespitosae* Freitag 57 and the *Alopecuretum pratensis* (Regel 25) Steffen 31.

Most of the other associations have their distribution center in one of the three main regions (Table 1). This is apparently connected, above all, with the following factors:
(1) the distribution area of the characteristic species linked with the history of floristic evolution;
(2) climatic conditions (temperature, precipitation, air humidity, duration of sunshine and of drought, and frost periods);
(3) water conditions on the site (water retention, inundation – also in relation to soil acidity);
(4) geological substratum, nutrient richness of the soil.

On the basis of their distribution, two to three groups of *Molinietalia* associations can be established in Czechoslovakia.

Hercynicum

The Hercynic region includes those *Molinietalia* communities whose center of distribution lies in regions of suboceanic (seldom of oceanic) character. Relations to the boreal floristic region are very limited; they exist, for instance, in the case of the *Scirpo–Juncetum filiformis* Oberd. 57. The investigated communities belong to the *Calthion*, with the exception of the *Junco–Molinietum caeruleae* Preisg. 51 and the *Succiso–Festucetum commutatae* Bal.-Tul. (59) 65. Of the *Calthenion* associations it is, first of all, the *Crepido–Juncetum acutiflori* Oberd. 57, vicariant of the *Juncetum acutiflori* Br.-Bl. 15, which intrudes into the Bohemian Euhercynicum and Subhercynicum from the west. It occurs there on the edge of its continuous area and therefore lacks species of oceanic distribution, with the exception of *Juncus acutiflorus,* the association characteristic species (cf. also Celiński et al., 1978). The *Valeriano officinali–Filipenduletum* Siss. in Westh. et al. 46 *(Filipendulenion)* shows a similar feature in its distribution.

The two associations with *Juncus filiformis,* viz. the *Scirpo–Juncetum filiformis* occurring mainly in the colline belt and the montane association *Junco filiformi–Polygonetum* Bal.-Tul. 81, are also limited to the Euhercynicum and the Subhercynicum,

Table 1. Distribution of the *Molinietalia* communities in Czechoslovakia.

Community[a] (alliance)	Hercynicum			Pannonicum	Carpaticum
	Eu-	Sub-	Sudet.		
Crepido–Juncetum acutiflori	+	+			
Scirpo–Juncetum filiformis	+	+			
Junco filiformi–Polygonetum	+	+			
Polygono–Cirsietum palustris	+	+	(+)		
Angelico–Cirsietum palustris	+	+	(+)		
Polygono–Cirsietum heterophylli	+	+	+		
Deschampsio–Cirsietum heterophylli	+				+
Polygono–Trollietum	+	+	+		+
Valeriano officinali–Filipenduletum	+	+	(+)		
Chaerophyllo hirsuti–Filipenduletum	+	+	+		+
Cirsio heterophylli–Filipenduletum	+	+	+		
Junco–Molinietum	+	+		(+)	
Succiso–Festucetum commutatae	+	+	+		
Cnidion venosi				+	
Veronico longifoliae–Lysimachion				+	
Silaetum pratensis		+		+	
Serratulo–Festucetum commutatae		+		+	
Scirpo–Cirsietum cani		+		+	+
Caricetum cespitosae		+		+	+
Cirsietum rivularis caricetosum cespitosae		+		+	+
Cirsietum oleracei caricetosum cespitosae		+		+	
Cirsietum rivularis (= salisburgensis)	(+)	+	+	+	+
Trollio–Cirsietum salisburgensis			+		+

[a] Rare and widespread associations are omitted.

but extending into western Moravia. They have no connection with the Atlantic region; *Juncus filiformis* is a boreal species (cf. also Oberdorfer, 1957).

The other communities of this group extend from the Euhercynicum and/ or the Subhercynicum into the neighbouring (sub)regions. These are the *Junco–Molinietum caeruleae* (cf. also the map of its distribution published in Foucault & Géhu, 1978), the *Succiso–Festucetum commutatae*, the *Chaerophyllo hirsuti–Filipenduletum* Niem., Heinr. et Hilb. 73 and the two associations of *Cirsium palustre*, viz. the submontane-montane *Polygono–Cirsietum palustris* Bal.-Tul. 74 and the *Angelico–Cirsietum palustris* Bal.-Tul. 73, whose distribution optimum lies in the colline belt.

The *Junco–Molinietum caeruleae* also occurs, though rarely, in the Slovak Záhorie Lowlands (Pannonicum), the *Succiso–Festucetum commutatae* extends from the subregions of the Euhercynicum and the Subhercynicum into the neighbouring (sub)region of the Sudeticum and the Carpaticum. On the other hand, the areas of *Chaeropyllo hirsuti–Filipenduletum, Polygono–Cirsietum palustris* and *Angelico–Cirsietum palustris* spread into the subregion of Sudeticum. A similar tendency is also shown by the relatively rare association *Polygono–Trollietum altissimi* (Hundt 64) Bal.-Tul. 81.

The *Polygono–Cirsietum heterophylli* Bal.-Tul. 75 and the *Deschampsio–Cirsietum heterophylli* Bal.-Tul. 83, the montane or regional vicariant of the former, deserve special attention. The distribution of both associations is connected with the area of *Cirsium heterophyllum,* an Euro-Siberian montane species showing a boreal type of distribution; its southern branch crosses into the territory of Czechoslovakia (cf. Hundt, 1964; Balátová-Tuláčková, 1975). The center of the *Cirsium heterophyllum* associations lies here in the submontane-montane belt of the Hercynicum mountains forming a natural boundary of the country including the western part of the Sudeticum subregion. (Only the *Deschampsio–Cirsietum heterophylli* has, so far, been known from the Carpaticum; it has been documented by one unpublished relevé from the mon-

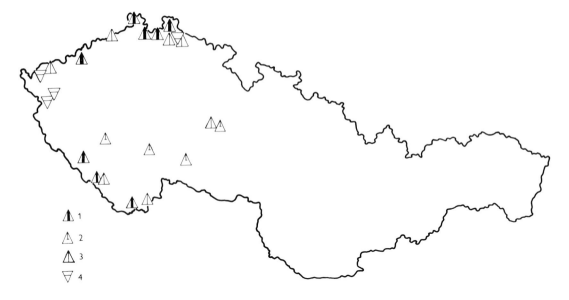

Fig. 1. Distribution of some associations occurring in the Hercynicum only. (1) *Crepido–Juncetum acutiflori,* (2) *Junco filiformi–Polygonetum,* (3) *Scirpo–Juncetum filiformis,* (4) *Valeriano–Filipenduletum.*

tane belt of the High Tatra Mountains.) The same applies also to the *Cirsio heterophylli–Filipenduletum* Neuh. et Neuh.-Nov. in Bal.-Tul. 78.

Pannonicum

Many moist meadow communities of four alliances of the *Molinietalia* (the *Alopecurion* is not taken into consideration) have their distribution center in the Pannonicum, two of them, viz. the *Cnidion venosi* and the *Veronico longifoliae–Lysimachion,* being restricted only to this region. Their area fades away towards the west in most cases in southern Moravia and/or in the northeastern part of Austria (Niederösterreich). The following associations are involved: *Lathyro palustri–Gratioletum* Bal.-Tul. 66, *Gratiolo–Caricetum praecocisuzae* Bal.-Tul. 66, *Cnidio–Violetum pumilae* Korneck 62, *Juncetum atrati* Vich. in Bal.-Tul. 69; *Lysimachio–Filipenduletum picbaueri* Bal.-Tul. 81, *Stachyo–Thalictretum flavi* Bal.-Tul. 81 and *Veronico longifoliae–Euphorbietum lucidae* Bal.-Tul. et Kneževič 75. The other associations, viz. the *Cnidio–Violetum elatioris* K. Walther in Tx. 54 and the *Serratulo–Plantaginetum altissimae* Ilijanić 67, are known from Slovakia only, as far as Czechoslovakia is concerned.

All the communities of the *Cnidion venosi* and

the *Veronico longifoliae–Lysimachion* endure both inundations and drying up of the soil (Balátová-Tuláčková, 1969); consequently, they are confined to river valleys in Moravia and Slovakia.

The subhalophilous *Serratulo–Plantaginetum altissimae,* limited in Czechoslovakia to the Podunajská nížina Lowland, extends further southwards; there is one locality in northeastern Austria (Niederösterreich) and a few ones in northeastern Croatia and northern Serbia (Ilijanić, 1968; Balátová-Tuláčková, 1969). This and the other *Cnidion venosi* associations occurring in southwestern Slovakia show phytogeographical affinity to the inundated meadows of the *Trifolio–Hordetalia* Horvatić 63, described from Serbia e.g. by Cincović (1956, 1959) and Jovanović-Djunić (1965) by the presence of *Clematis integrifolia,* while those occurring in eastern Slovakia are related to the flora of the subcontinental southeastern border regions, witness the presence of *Oenanthe silaifolia* M. Bieb. ssp. *hungarica* (E. Simon) Bertová 73.

With regard to the *Molinion* communities, only two of them, viz. the *Silaetum pratensis* Knapp 54 and the *Serratulo–Festucetum commutatae* Bal.-Tul. 66 are most abundant in the Pannonian region. Their area covers the Moravian and Slovak Pannonicum – here known from the Záhorie Lowland only; besides, they occur in the more continental

115

regions of Bohemia. On the other hand, most of the continental *Calthion* communities extend far beyond the Pannonicum. They are as follows: *Scirpo-Cirsietum cani* (Klapp 65) Bal.-Tul. 73, *Caricetum cespitosae* Steffen 31 and the subassociation *Cirsietum rivularis caricetosum cespitosae* Bal.-Tul. in Ambrož et Bal.-Tul. 62. However, a vicariant of the latter subassociation, the *Angelico-Cirsietum oleracei caricetosum cespitosae* Bal.-Tul. 81, seems to be confined to the Bohemian Pannonicum and its border regions. All these communities avoid the Pannonian river valley areas; they are more frequent in their marginal parts. As to the *Scirpo-Cirsietum cani*, its center of distribution lies in the colline belt of the Pannonicum (this also concerns Bohemia), from where the association extends into the adjoining phytogeographical regions along the rivers – in Bohemia and Moravia into the Subhercynicum, in Slovakia into the border areas of the Carpaticum. A similar description applies to the *Cirsietum rivularis caricetosum cespitosae* occurring also in the continental Slezská nížina Lowlands (in Slovakia it is known from the continental Innercarpathian Spišská kotlina Basin only).

The *Caricetum cespitosae*, with lower nutrient requirements, occurs mainly in Bohemia; it is limited to the Pannonicum of western Bohemia and to the warmer regions of the Subhercynicum (cf. Balátová-Tuláčková, 1981; Blažková, 1973). However, it is also found in the Žďárské vrchy Hills. In Slovakia it is known from the Inner-Carpathian Turčianská kotlina Basin (Bosáčková, 1974).

The distribution of the *Scirpo-Cirsietum cani* and of the communities with *Carex cespitosa* in Czechoslovakia is linked with the distribution areas of their characteristic species: *Cirsium canum* represents a subcontinental European species with a Pannonian tendency, while *Carex cespitosa* is one of the continental Eurasiatic species with a boreal tendency.

Carpaticum

There are only two *Molinietalia* communities of the *Calthion* alliance whose center of distribution lies in the Carpathians, i.e. the *Cirsietum rivularis* (= *salisburgensis*) Nowiński 27 and the *Trollio-Cirsietum salisburgensis* (K. Kuhn 37) Oberd. 57. The center of distribution of the first association is bound to the planar to colline belt, that of the latter to the submontane to montane belt. Both associations are widespread in the Carpathians of Slovakia, from where they extend up to northeastern Bohemia (across Moravia in western direction).

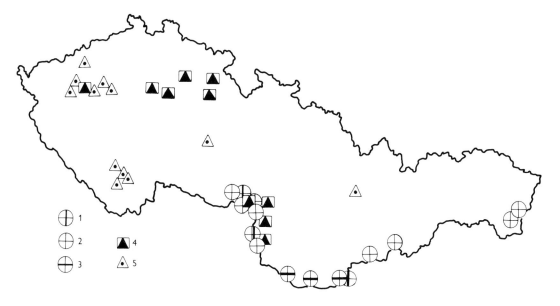

Fig. 2. Distribution of some associations and alliances with a continental distribution tendency. (1) *Veronico longifoliae–Lysimachion,* (2) *Cnidion venosi* (with the exception of the *Serratulo–Plantaginetum altissimae*), (3) *Serratulo–Plantaginetum altissimae,* (4) *Silaetum pratensis,* (5) *Caricetum cespitosae.*

116

Fig. 3. Distribution of the associations with the center of distribution in the Carpaticum (the amount of information available from Slovakia is limited at present). (1) *Cirsietum rivularis* (= *salisburgensis*), (2) *Trollio–Cirsietum salisburgensis.*

They are confined to a substratum more or less rich in calcium.

The *Angelico–Cirsietum oleracei* Tx. 37, a vicariant of the *Cirsietum rivularis,* shows another tendency of distribution. This association occurring in an area of younger moraines in the North German Lowlands appears in Czechoslovakia mainly in northern, central and eastern Bohemia and also in the Silesian Lowlands (the Hlučínská pahorkatina Hilly country), northeastern Moravia and in the colline belt of southern Moravia, southern Slovakia and southeastern Slovakia. Although the *Angelico–Cirsietum oleracei* can be found in all floristic regions of Czechoslovakia, its distribution area and those of the *Cirsietum rivularis* and of both associations with *Cirsium palustre,* viz. the *Angelico–Cirsietum palustris* and the *Polygono–Cirsietum palustris,* are in some regions mutually exclusive. The reason can be found in the different distribution tendency of the above-mentioned *Cirsium* species as well as in the nutrient level of the substratum.

References

Balátová-Tuláčková, E., 1969a. Beitrag zur Kenntnis der tschechoslowakischen Cnidion venosi-Wiesen. Vegetatio 17: 200–207.

Balátová-Tuláčková, E., 1969b. Zur Kenntnis des tschechoslowakischen Serratulo-Plantaginetum altissimae Ilijanić 1967. Acta bot. Croatica 28: 31–38.

Balátová-Tuláčková, E., 1972. Flachmoorwiesen im mittleren und unteren Opava-Tal (Schlesien). Vegetace ČSSR. A 4. 201 pp.

Balátová-Tuláčková, E., 1975a. Das Succiso-Festucetum commutatae und das Serratulo-Festucetum commutatae als Beispiel der Konvergenz in der Pflanzensoziologie. In: H. Dierschke (ed.), Vegetation und Substrat (Rinteln, 31.3.–3.4.1969). Ber. Internat. Symposien IVV (ed. R. Tüxen), pp. 117–130. Cramer, Vaduz.

Balátová-Tuláčková, E., 1975b. Cirsium heterophyllum-Feuchtwiesen und ihre pflanzensoziologische Charakteristik. Folia Geobot. Phytotax. 9: 153–166.

Balátová-Tuláčková, E., 1981a. Phytozönologische und synökologische Charakteristik der Feuchtwiesen Böhmens. Rozpr. ČSAV, Ser. Mat.-Natur. 91(2). 90 pp.

Balátová-Tuláčková, E., 1981b. Beitrag der Syntaxonomie der Wiesen-Hochstaudengesellschaften. In: H. Dierschke (ed.), Syntaxonomie (Rinteln 31.3.–3.4.1980). Ber. Internat. Symposien IVV, pp. 375–384. Cramer, Vaduz.

Blažková, D., 1973. Pflanzensoziologische Studie über die Wiesen der Südböhmischen Becken. Studie ČSAV 10. 170 pp.

Bosáčková, E., 1975. Rastlinné spoločenstvá slatinových lúk na Záhorskej nížine. Čs. Ochr. Prírody 15: 173–247.

Celiński, F., Wika, S. & Cabala, S., 1978. Les prairies marécageuses à Juncus acutiflorus en Silésie (Pologne). In: J.-M. Géhu (ed.), La végétation des prairies inondables. Lille 1976. Coll. Phytosoc. 5: 205–217. Cramer, Vaduz.

Cincović, T., 1956. Tipovi livada u Posavini. Zbor. rad. Poljop. fak. Beograd. 4: 1–26.

Cincović, T., 1959. Livadska vegetacija u rečnim dolinama zapadne Srbije. Ibidem 7: 1–62.

Dostál, J., 1960. The phytogeographical regional distribution of the czechoslovak flora. Sborn. čs. spol. zeměp. 65: 193–202.

Ellenberg, H., 1978. Vegetation Mitteleuropas mit den Alpen. 2nd ed. Ulmer, Stuttgart. 982 pp.

Foucault, B. de & Géhu, J.-M., 1978. Essai synsystematique et chorologique sur les prairies à Molinia coerulea et Juncus acutiflorus de l'Europe occidentale. In: J.-M. Géhu (ed.), Sols turbeux, Lille 1978. Coll. Phytosoc. 7: 135–164d.

Hundt, R., 1964. Die Bergwiesen des Harzes, Thüringer Waldes und Erzgebirgers. Pflanzensoziologie (Jena) 14: 1–264.

Ilijanić, Lj., 1968. Die Ordnung Molinietalia in der Vegetation Nordostkroatiens. Acta bot. Croat. 26/27: 161–180.

Jovanović-Dunjić, R., 1965. Zavisnost močvarnih i livadskih fitocenoza od visine podzemne vode u dolini Velike Morave. Zaštita prirode, 29/30: 25–49.

Jurko, A., 1972. Druhotné spoločenstvá. In: M. Lukniš (ed.), Slovensko. Príroda, pp. 574–617. Obzor, Bratislava.

Kovář, P., 1981. The grassland communities of the Southeastern Basin of the Labe-river I. Syntaxonomy. Folia Geobot. Phytotax. 16: 1–43.

Oberdorfer, E., 1957. Süddeutsche Pflanzengesellschaften. Pflanzensoziologie 10: 1–564.

Passarge, H., 1964. Pflanzengesellschaften des nordostdeutschen Flachlandes I. Ibidem, 13: 1–324.

Ružičková, H., 1978. Trollio-Cirsietum Kuhn 1937 v Liptovskej kotline. Biológia 33: 307–314.

Rybníček, K., Balátová-Tuláčková, E. & Neuhäusl, R., 1982. Přehled rostlinných společenstev rašeliništ a mokřadních luk Československa. Stud. Čs. Akad. Věd. (in print).

Špániková, A., 1971. Fytocenologická štúdia lúk juhozápadnej časti Košickej kotliny. Biol. Pr. 17(2): 1–105.

Tüxen, R., 1937. Die Pflanzengesellschaften Nordwestdeutschlands. Mitt. Florist.-Soziol. Arb.gem. Niedersachsen 3: 1–170.

Westhoff, V. & Held, A. J. den, 1969. Plantengemeenschappen in Nederland. Thieme et Cie, Zutphen. 324 pp.

Accepted 18.1.1984.

The chorologic pattern of European *Nardus*-rich communities

F. Krahulec

Botanical Institute, Czechoslovak Academy of Sciences, 252 43 Průhonice, Czechoslovakia

Keywords: Classification, Clinal distribution, Europe, Insular distribution, *Nardus* communities

Abstract

Two types of *Nardus*-rich communities may be distinguished in Europe according to their geographic pattern:

I. Communities of high mountains showing discontinuous variation. Different florogenesis in island-like high mountains results in a number of alliances characteristic for particular mountain systems: *Trifolion humilis* Quézel 57 (Atlas, N Africa), *Plantaginion thalackeri* Quézel 53 (Sierra Nevada), *Campanulo–Nardion* Rivaz-Martinez 63 (mountains of the central Iberian peninsula), *Nardion* Br.-Bl. 26 (from Pyrenees, Auvergne to W and E Carpathians), *Potentillo ternatae–Nardion* Simon 58 (S Carpathians, Pirin, Rila), *Jasionion orbiculatae* Lakušić 66 (Dinarids), *Trifolion parnassi* Quézel 64 (S Greece), *Nardo-Caricion rigidae* Nordhagen 37 (Iceland, Scotland, Scandinavia, W Sudeten).

II. Communities at lower altitudes with continuous variation, where particular syntaxa are difficult to distinguish. Along the climatic gradient from a subatlantic to a subcontinental climate, a cline in floristic composition of the communities under discussion is found. Towards the eastern borderline, the communities are poorer in characteristic species. Here only one alliance is considered – *Violion caninae* Schwickerath 44. A similar situation occurs along the altitudinal gradient in mountains. Only in mountains with a belt of alpine communities the montane *Nardeta* are saturated with alpine species. This type of communities was described as *Nardo–Agrostion tenuis* by Sillinger (1933).

The chorologic pattern of European *Nardus*-rich communities seems to be of two different types. Communities above the timberline are island-like and particular mountain systems are characterized by a number of different and rather distinct communities (see Fig. 1).

The alliance *Trifolion humilis* was described by Quézel (1957) from the Atlas in N Africa. Various alliances have been described for particular mountain ranges on the Iberian peninsula: the *Plantaginion thalackeri* Quézel 53 for the Sierra Nevada; the *Campanulo–Nardion* Rivaz-Martinez 63 for mountains of the central part of the peninsula are characterized. The mountains of western and central Europe bear rather uniform communities of the

Nardion alliance, which is found from the Pyrenees (e.g. Braun-Blanquet, 1948) throughout the mountains of central France (Auvergne), throughout the Alps and N and central Appenines (e.g. Lüdi, 1943) to the Sudeten mountains and to the western and eastern Carpathians (e.g. Borza, 1934). Communities of the Balkan mountain ranges are more differentiated. The *Potentillo ternatae–Nardion* was described by Simon (1958) from Pirin, but it also occurs in the southern Carpathians and in the Rila. *Nardus*-rich communities of the Dinarids are usually classified into the alliance *Jasionion orbiculatae* Lakušić 66. From southern Greece the *Trifolion parnassi* was described by Quézel (1964) as a vicariant alliance to the *Plantaginion thalackeri* in the

120

Fig. 1. Distribution of *Nardus*-rich alpine communities with island-like pattern (simplified). (1-*Trifolion humilis,* 2-*Plantaginion thalackeri,* 3-*Campanulo–Nardion,* 4-*Nardion,* 5-*Potentillo ternatae–Nardion,* 6-*Jasionion orbiculatae,* 7-*Trifolion parnassi,* 8-*Nardo–Caricion rigidae*).

eastern Mediterranean area. From the mountains of Scandinavia the alliance *Nardo–Caricion rigidae* was described by Nordhagen (1937). Communities of this alliance are also distributed on Iceland (Hadač, 1972), Scotland and to a small extent in central Europe in the Sudeten in close proximity of mires on mountain plains of the Giant Mountains (Krkonoše).

The different florogenesis of particular mountain systems is reflected in distinct communities. They are ranked in different alliances without serious problems. The differences of conceptions of particular authors predominantly concern the ranking of the above-mentioned alliances into higher units. Traditionally, some alliances (*Campanulo–Nardion, Nardion, Potentillo ternatae–Nardion*) are assigned to the order *Nardetalia.* On the other hand, some authors prefer to assign them into the order *Caricetalia curvulae* and thus to the class *Juncetea trifidi* Hadač in Klika et Hadač 44. Some

Fig. 2. Eastern limits of some selected characteristic species of the *Violion caninae* Schwickerath 44. (1-*Genista anglica* L., 2-*Festuca tenuifolia* Sibth., 3-*Polygala serpyllifolia* Hose, 4-*Lathyrus linifolius* (Reichard) Bässler, 5-*Ranunculus nemorosus* DC., 6-*Juncus squarrosus* L.).

central European authors (e.g. Sillinger, 1933) evidenced that these communities are more similar to other grassland communities from above the timberline. This is reflected in the description of a new class for all grass-rich mountain communities, i.e. *Nardo-Calamagrostetea villosae* by Jeník *et al.*

(1980). Similarly communities of the *Jasionion orbiculatae* were assigned to the order *Seslerietalia comosae* which is incorporated into the class *Juncetea trifidi* by Jugoslavian authors (e.g. Blečić & Lakušić, 1976). We may conclude that *Nardus*-rich communities of the alpine parts of the European mountains are rather distinct and classification problems are not connected with the delimitation of particular communities, but with different interpretations on higher syntaxonomical levels.

Another situation is found with the *Nardus*-rich communities of lower altitudes of Europe. From western and central Europe the alliance *Violion caninae* was described by Schwickerath (1944). This alliance is usually characterized as subatlantic (for example by Oberdorfer, 1957, 1978; Moravec in Holub *et al.*, 1967, Preising 1949 sub nominem *Nardo-Galion saxatilis*). However, some questions remain unanswered: (1) Where lies the eastern boundary of this alliance? (2) Are there other alliances of *Nardus*-rich communities in the lower parts of Europe? I don't know the answers to these questions, but, in my opinion, there is a very distinct cline in the floristic composition of these communities. With the decreasing influence of the subatlantic climate the decrease in the number of species characteristic of the *Violion caninae* can be observed. This is illustrated in Figure 2, where the eastern limits of some important taxa for the delimitation of *Violion caninae* are drawn. It can be seen that particular species have different eastern limits. Therefore, it is impossible to use this method for the delimitation of the eastern boundary.

Toward the eastern part of their distribution area the communities contain less and less species typical of the lower syntaxa, while no species restricted to these communities are added. It is the same phenomenon as described by Werger & van Gils (1976): near their geographical boundaries the characteristic species of the communities are lacking. That is true not only for the gradient throughout Europe, but also for the altitudinal gradient. With increasing altitude a gradual decrease in number of characteristic species is seen. This is found in mountain ranges without a belt of alpine communities. In mountains where alpine communities are present, saturation by species of alpine *Nardus* communities takes place. The similar situation as on the eastern boundary of distribution area of the *Violion caninae* is very probably on the western boundary. A gradual transition into subatlantic communities is very probably the main explanation for the difficulties encountered by Barkman (1975).

The syntaxonomical implications are that particular syntaxa of *Nardus*-rich communities at lower altitudes will be difficult to distinguish. In my opinion, the alliance *Violion caninae* is distributed throughout the majority of subatlantic and subcontinental parts of Europe. Towards the boundaries of its distribution this alliance is characterized by order and class species and a great number of species typical for the alliance are lacking. Identical with this alliance is very probably the *Calluno-Festucion capillatae* Horvat 62 prov. because of its similar floristic composition. Another alliance may be distinguished in the montane zone of mountains with alpine communities: the *Nardo-Agrostion tenuis* as described by Sillinger (1933) and which comprises communities containing species of alpine and lowland *Nardus* communities.

References

Barkman, J. J., 1975. Le Violion caninae existe-t-il? Colloq. Phytosoc. 2: 149–156.

Blečić, V. & Lakušić, R., 1976. Prodromus biljnih zajednica Crne Gore. (Prodromus der Pflanzengesellschaften von Montenegro.) Glass. Republ. Zavoda Zašt. Prirode – Prirodnjačkog Muzeja 9: 57–98.

Borza, A., 1934. Studii fitosociologice in Munţii Retezatului. Bul. Grād. Bot. Muz. Bot. Univ. Cluj 14: 1–84.

Braun-Blanquet, J., 1948. La végétation alpine des Pyrénées orientales. Barcelone.

Hadač, E., 1972. Fell-field and heath communities of Reykjanes peninsula, SW Iceland. (Plant communities of Reykjanes peninsula, Part 5.) Folia Geobot. Phytotax. 7: 349–380.

Holub, J., Hejný, S., Moravec, J. & Neuhäusl, R., 1967. Übersicht der höheren Vegetationseinheiten der Tschechoslowakei. Rozpr. Čs. Akad. Věd, Řada Mat. Přírod. Věd 77/3: 1–75.

Horvat, I., 1962. Vegetacija planina zapadne Hrvatske. Prirodoslovna Istraživanja 30. Acta Biologica 2: 1–179.

Jeník, J. Bureš, L. & Burešová, Z., 1980. Syntaxonomic study of vegetation in the Velká Kotlina cirque, the Sudeten mountains. Folia Geobot. Phytotax. 15: 1–28.

Klika, J. & Hadač, E., 1944. Rostlinná společenstva střední Evropy. Příroda 36: 249–259, 281–295.

Lakušić, R. 1966. Vegetacija livada i pašnjaka na planini Bjelasici. (Die Vegetation der Wiesen und Weiden des Bjelasica Gebirges.) Godišnjak Biol. Inst. Sarajevo 19: 25–186.

Lüdi, W., 1943. Über Rasengesellschaften und alpine Zwergstrauchheide in den Gebirgen des Apennin. Berichte Geobot. Forsch.-Inst. Rübel, Zürich 23–68.

Nordhagen, R., 1937. Versuch einer neuen Einteilung der subalpinen-alpinen Vegetation Norwegens. Bergens Mus. Årb., Naturvid. Rekke, Bergen 7: 1–88.

Oberdorfer, E., 1957. Süddeutsche Pflanzengesellschaften. Pflanzensoziologie 10: 1–564.

Oberdorfer, E. & coll., 1978. Süddeutsche Pflanzengesellschaften. II, 2nd ed. Pflanzensoziologie 10: 1–355.

Preising, E., 1949. Nardo-Callunetea. Mitt. Flor.-Soz. Arbeitsgem. N.F. 1: 12–25.

Quézel, P., 1953. Contribution à l'étude phytosociologique et géobotanique de la Sierra Nevada. Mém. Soc. Broter. 9: 5–77.

Quézel, P., 1957. Peuplement végétal des hautes montagnes de l'Afrique du Nord. Paris.

Quézel, P., 1964. Végétation des hautes montagnes de la Grèce méridionale. Vegetatio 12: 289–385.

Rivas-Martinez, S., 1963. Estudio de la vegetación y flora de las Sierra de Guaderrama y Gredos. An. Inst. Bot. Cavanilles 21: 5–325.

Schwickerath, M., 1944. Das Hohe Venn und seine Randgebie-te. Pflanzensoziologie 6: 1–278.

Sillinger, P., 1933. Monografická studie o vegetaci Nízkých Tater. Praha.

Simon, T., 1958. Über die alpinen Pflanzengesellschaften des Pirin-Gebirges. Acta Bot. Acad. Sci. Hung. 4: 159–189.

Werger, M. J. A. & Gils, H. van, 1976. Phytosociological classification in chorological borderline areas. J. Biogeogr. 3: 49–54.

Accepted 7 March 1984.

Communities of *Berteroa incana* in Europe and their geographical differentiation*,**,***

L. Mucina[1] & D. Brandes[2]
[1] *Department of Geobotany, Institute of Experimental Biology and Ecology, Slovak Academy of Sciences, Sienkiewiczova 1, 814 34 Bratislava, Czechoslovakia*
[2] *University Library of the Technical University, Pockelsstrasse 13, D-3300 Braunschweig, F.R.G*

Keywords: *Berteroetum incanae, Dauco–Melilotion*, Europe, Geographical race, *Onopordetalia*, Synchorology, Syntaxonomy

Abstract

Two geographical races have been established within the *Berteroetum incanae* in Europe. The *Galium mollugo* race of the *Berteroetum incanae* is characteristic of the western part of the distribution area of the association whereas the relevés from the eastern part of Europe are classified as the *Acosta rhenana* race of the *Berteroetum incanae*. Adventitious *Berteroetum incanae* from the Netherlands has been shown to be a separate subunit within the *Galium mollugo* race of the *Berteroetum incanae*.

Introduction

Berteroa incana is a biennial rosette herb producing a high number of viable seeds. It is a competitive ruderal (sensu Grime, 1979). Phytosociologically it is limited to the order *Onopordetalia*, and especially the alliance *Dauco–Melilotion*. Communities in which *Berteroa* occurs as a dominant are found in many parts of Europe, though not very frequently. In this paper attention will be focussed on the geographical differentiation of these communities.

* A number of phytosociological relevés has been used by courtesy of Th. Müller, Nürtingen, F. Runge, Münster, V. Westhoff, Nijmegen, H. Lienenbecker, Steinhagen, H. Passarge, Eberswalde, W. Nezadal, Erlangen, W. Kunick, Stuttgart, F. Grüll, Brno, I. Jarolímek and M. Zaliberová, both Bratislava. For valuable comments on the manuscript we are indebted to V. Westhoff, E. van der Maarel, R. Neuhäusl and J. J. Barkman. The English of an earlier version of the manuscript has been kindly corrected by Mr Henry Moreton.
** The nomenclature of species follows Tutin *et al.* (1964–1980) with some changes according to Smejkal (1980).
*** *Acknowledgement*. The work of the senior author was partly supported by the University of Nijmegen Research Fellowship in Biology 1980/1981.

Material and methods

From Austria, Bulgaria, Czechoslovakia, Denmark, Federal Republic of Germany, German Democratic Republic, The Netherlands, Federal Republic of Germany, Poland and Roumania 211 relevés of *Berteroa incana* communities have been collected and combined into local rather homotoneous tables. With the aid of a computer these have been arranged into a synoptic table, in which species occur with their constancy class values. In columns representing local tables with less than 5 relevés the species are indicated by their real presence values. Condensed local tables (constancy values only) have been added to the synoptic table afterwards.

Syntaxonomical handling was accomplished by traditional methods of the Braun-Blanquet approach (Braun-Blanquet, 1964; see also Westhoff & van der Maarel, 1978). The columns in the table have been ordered from west to east in Europe. *Centaureo diffusae–Berteroetum* tables were added at the end of the synoptic table.

The numerical handling of the data includes numerical classification and ordination techniques. Ward's method (WM; Sum-of-squares clustering sensu Orlóci, 1967) and Complete linkage clustering (CLC; see Sneath & Sokal, 1973 for details) have been adopted as numerical classification methods. The original abundance and cover values of the Braun-Blanquet scale have been transformed according to ordinal transformation of van der Maarel (1979). The computation was performed with the CLUSTAN 1C package of Wishart (1978). Euclidean distance (ED) with WM and Similarity ratio (SR) with CLC (see Wishart, l.c. for further information) were used as resemblance functions. Reciprocal averaging (RA; Hill, 1973, program DECORANA, Hill, 1979) is the ordination technique used. Comparisons between classifications have been performed using Goodman-Kruskal coefficient of association (Goodman & Kruskal, 1954) and the program GOODM (Goldstein & Grigal, 1972).

Results

Syntaxonomy

On the basis of the Braun-Blanquet approach two associations were established within the ensemble of *Berteroa incana* communities (Table 1 see Appendix): *Berteroetum incanae* Sissingh et Tideman in Sissingh 50 and *Centaureo diffusae–Berteroetum* Oberdorfer 57.

Stands of the *Berteroetum incanae* are usually dominated by *Berteroa incana*, sometimes also by *Acosta rhenana* (in eastern Europe). They have two or three layers. Besides the dominating species also *Plantago lanceolata, Reseda lutea, Silene alba, Echium vulgare* and *Lolium perenne* form the upper herb sublayer. Some grasses reach as high as 1 m (*Agropyron repens, Arrhenatherum elatius*) and locally form another, third, rather loose sublayer. Some other tall herb species are found either in a vegetative stage (rosette) of the specimens of lower vitality. The lowermost sublayer is composed of dwarf or procumbent herbs as *Medicago lupulina, Alyssum alyssoides, Sedum sexangulare* and *Arenaria serpyllifolia*. In some place also a moss layer is found, usually including *Bryum argenteum, Brachythecium albicans, Hypnum vaucheri* and *Rhacomitrium canescens*.

The *Berteroetum incanae* colonizes well-drained, warm sandy to loamy (loess) soils of moderate nitrogen supply. The community occurs in very special habitats in the western part of its area, where it is restricted to docks and railway stations and more or less dependent on the steady supply of adventitious species. Originally, the community was described from tips near corn-mills in the south and southeast Netherlands. Recently it started spreading here along roads on sandy soils. On sandy soils it also occurs in Lower Saxony, the northern part of Franconia, Brandenburg and Mecklenburg. The habitats of the *Berteroetum incanae* in eastern Europe are more diverse. They are most frequently found on moderately alkalic soils along roads, on abandoned places of various origin, very often on loess slopes and river dykes or in limestone quarries (Mucina, 1981b).

Berteroa incana is considered the only character-species of the association *Berteroetum incanae*, with its optimum fidelity in western Europe. Though not dominant, *Berteroa incana* is quite common in other *Dauco–Melilotion* and *Onopordion* communities in eastern Europe (e.g. Gutte, 1972; Gutte & Hilbig, 1975; Kępczyński, 1975; Czaplewska, 1980; Mucina, 1981a, b).

According to the geographical variation the community can be divided into two subunits. The *Galium mollugo* race is characteristic of the northwestern part of the area (The Netherlands, Denmark, northern and western parts of F.R.G., a part of East Germany). *Holcus lanatus, Cerastium arvense, Carduus nutans, Lamium album, Senecio vulgaris* etc. are differentials for the unit.

The *Acosta rhenana* race, from Czechoslovakia, Poland, Hungary and the Balkans, is differentiated by a large number of species, particularly of (sub)continental distribution (Table 1), many of which are shared with the *Centaureo–Berteroetum*.

Since all differential species except *Cichorium intybus* and *Crepis rhoeadifolia* have their sociological optimum in other syntaxa than in the *Dauco–Melilotion* alliance, the units discussed here are not considered as regional associations, but as geographical races (sensu Oberdorfer, 1957, 1968; Werger & van Gils, 1976; Westhoff & van der Maarel, 1978).

In western Europe the *Berteroetum incanae* occurs in contact with *Arrhenatheretalia* communities, whereas in eastern Europe it is more frequently found in contact with the *Festuco–Brometea* and *Sedo–Scleranthetea* (*Festucetalia valesiacae* and *Sedo–Scleranthetalia*).

As a rule the dominant species of the *Centaureo diffusae-Berteroetum* is *Acosta diffusa* (= *Centaurea diffusa*), *Berteroa incana* being a sub-dominant. The stands of the community are rather large in size (Oberdorfer, 1957), and rather open (50% average cover (Seybold & Müller, 1972).

The community occurs in particular along railways and at railway yards, on well-drained sandy soils with high amounts of ash and dross (Gutte & Hilbig, 1975; Sowa, 1975). As the sites are exposed to direct sun irradiation, the temperature of the substratum can become rather high. This favours not only the dominance of *Acosta diffusa*, but also the occurrence of other adventitious species as *Linaria genistifolia*, *Psyllium scabrum*, *Salsola kali* subsp. *ruthenica* as was observed in West and East Germany (Oberdorfer, 1957; Gutte & Hilbig, 1975). All these species are known to be thermophilous. The community as a whole seems to be adventitious throughout its distribution area.

Synecologically, the community is related to the group of communities dominated by *Salsola kali* ssp. *ruthenica*, *Psyllium scabrum* and *Corispermum* species. According to Oberdorfer (1957) there are syndynamical relations between the *Centaureo diffusae-Berteroetum* and the *Festuco-Sedetalia* and *Festuco-Brometea*. The *Centaureo diffusae-Berteroetum* is documented from West Germany, West Berlin, East Germany and Poland (see Appendix). *Acosta diffusa* is a regional character-species of the association. Differential species against the *Berteroetum incanae* include *Psyllium scabrum*, *Lepidium densiflorum*, *Oenothera parviflora* and *Herniaria glabra*. The segregation is strengthened by the absense of differential species against the *Berteroetum incanae* (*Silene alba*, *Dactylis glomerata*, *Urtica dioica* etc.). This group is largely composed of nitrophilous and mesophilous species, mostly of *Arrhenatheretalia* and *Sisymbrietalia*.

Cluster analyses

The first cluster dichotomy of WM (Fig. 1) splits the Dutch relevés of Sissingh (1950) and the rest of the material. Phytosociologically cluster H represents the *Berteroetum incanae* relevés on which the original description of the syntaxon has been based. The second dichotomy yields two clusters: D and B.

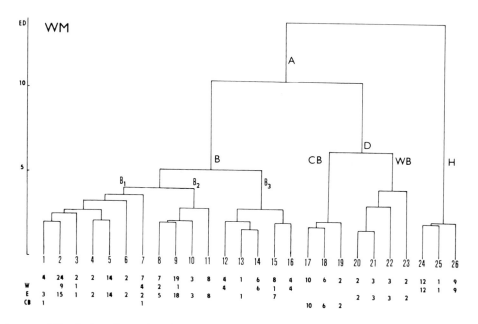

Fig. 1. Dendrogram of Ward's clustering method. Only the upper part of the hierarchy is shown (starting with 26 clusters). The numerals below the dendrogram indicate: total number of relevés, No. of the western race relevés, No. of the eastern race relevés, No. of the *Centaureo-Berteroetum* relevés in a starting cluster resp. B – the mixed cluster of the eastern and western race records; CB – the *Centaureo-Berteroetum* cluster; WB – the cluster of intermediate local tables (Berlin, Bavaria, etc.); H – the cluster of records from The Netherlands.

The former one is composed of two subclusters, viz. CB and WB. The cluster CB is entirely composed of the *Centaureo diffusae–Berteroetum* relevés. It comprises 90% of all *Centaureo–Berteroetum* relevés analyzed. The dendrogram branch 17 represents the *Centaureo–Berteroetum* from a part of Poland (the district of Lublin and the surroundings of Łodż), the others within the cluster CB are composed of relevés coming from Wrocław (Poland), West Berlin and southern F.R.G. The cluster WB comprises two local tables of the *Berteroetum incanae* s.str., namely those from West Berlin (the branches 20, 21; Fig. 1) and Bavaria (the branches 22, 23; Fig. 1). These correspond to the columns 14 and 11 of Table 1 resp. The tables come from the region where the two races of the *Berteroetum in-*

canae co-occur. Nevertheless, according to the overall floristic composition these local tables have been placed into the eastern race of the *Berteroetum incanae*.

The largest cluster B is a mixture of relevés of both *Berteroetum incanae* races. From the phytosociological point of view a somewhat clearer pattern arises on the level of the subclusters B_1, B_2 and B_3 within B. Cluster B_1, although composed mostly of the eastern race relevés, remains heterogeneous. Cluster B_2 is homogeneous. Cluster B_3 is also homogeneous.

The pattern in the WM dendrogram on the 4-cluster level, reasonably corresponds to that of the CLC (Fig. 1 and 2, Table 2).

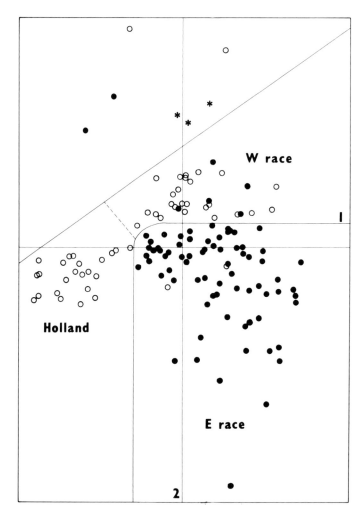

Fig. 2. Reciprocal averaging ordination plane (1st/2nd axes). Full circles – E race; empty circles – W race; asterisks – *Centaureo–Berteroetum*.

Table 2. Goodman-Kruskal comparisons of classifications of the *Berteroa incana* communities

	1	2	3
1 Syntaxonomical classification	x	0.395	0.400
2 Complete linkage clustering		x	0.727
3 Ward's method			x

Ordination

An ordination of the total data set produced a typical horseshoe on the ordination plane of axes 1 and 2. The bulk of relevés have been concentrated around the joining of the two wings of the horseshoe. The wings themselves include outlying *Centaureo-Berteroetum* respectively of *Acosta rhenana*-dominated relevés. After removing the outliers the subsequent ordination revealed a much clearer pattern. Some of the *Centaureo-Berteroetum* relevés are outliers being scattered along the positive side of axis 2. The geographical races of the *Berteroetum incanae* are separated along a diagonal (Fig. 3). The relevés from The Netherlands are separated from the *Acosta rhenana* race along axis 1. The remainder of the western material is more or less separated from the eastern one along axis 2. There is an overlap between the races along axis 2, which corresponds to the existence of mixed clusters, as depicted in Figures 1 and 2.

The data used for the numerical analyses are scattered throughout central Europe approximate-

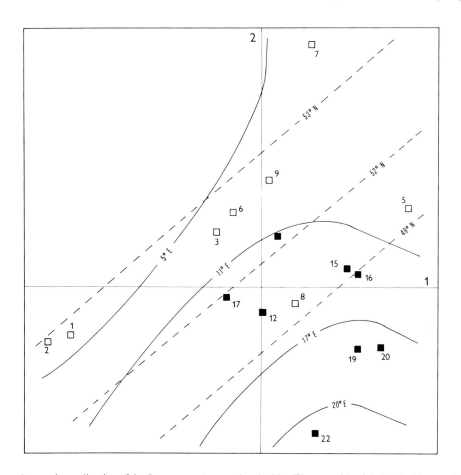

Fig. 3. Reciprocal averaging ordination of the *Berteroetum incanae* local tables. The centroids of the local tables are plotted. The solid and dashed lines represent geographical longitude and latitude resp. Full squares – E race; empty squares – W race; 1, 2 – the Netherlands; 3 – Westfalen; 5 – Karlsruhe; 6 – Bielefeld; 7 – Bremen; 8 – Maindreieck near Ochsenfurt; 9 – Braunschweig; 10 – Hagenow, Schwanenbeck-Alpenberge, Berlin; 12 – Bamberg, Redniztal in Bavaria; 15 – Brno; 16 – Niederösterreich, Burgenland; 17 – Bydgoszcz; 19 – western Slovakia; 20 – Bratislava; 22 – the Východoslovenská Nížina Lowland.

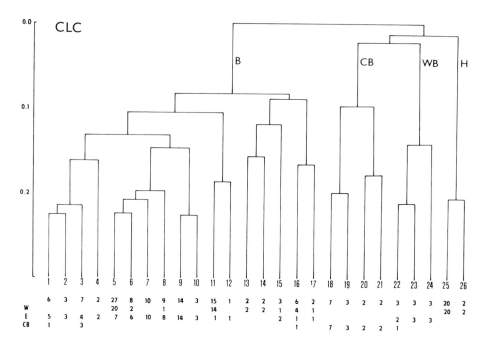

CLC

0.0

0.1

0.2

B   CB   WB   H

	1	2	3	4	5	6	7	8	9	10	11	12	13	14	15	16	17	18	19	20	21	22	23	24	25	26
	6	3	7	2	27	8	10	9	14	3	15	1	2	2	3	6	2	7	3	2	2	3	3	3	20	2
W					20	2		1			14		2	2	1	4	1								20	2
E	5	3	4	2	7	6	10	8	14	3	1	1		2	1	1						2	3	3		
CB	1		3											1				7	3	2	2	1				

Fig. 4. Dendrogram of complete linkage clustering. For further explanation see Figure 1.

Discussion

ly within the range 53 °N to 48 °N and 5 °E to 22 °E. It might seem that the area of the *Berteroetum incanae* in Europe is much larger in west–east direction than in the south–north one, but the community is also reported from Bulgaria (Table 1, column 24) and Sweden (Olsson, 1978).

The values for N latitude and E longitude for the local tables of the *Berteroetum incanae* s. str. show a diagonal isoline pattern (Fig. 4). The tables from Bavaria, Brandenburg, the surroundings of Braunschweig and Berlin (Table 1, columns 8-14) are considered 'transitional' as they are found where the races meet (Fig. 4). The distribution of the *Berteroetum incanae* in Europe is not continuous although the two races are not separated by a gap. There is a transitional zone between 10° to 13 °E and 49° to 52 °N where representatives of the races co-occur.

If we compare the RA-diagram with the WM dendrogram we find that the first WM dichotomy corresponds to the segregation of the clusters along axis 1, and the second one to the segregation along axis 2 (cf. Figs. 1 & 3).

There are two general opinions concerning the width of the *Berteroetum incanae* as a syntaxon. A 'narrow' concept is promoted by Westhoff & den Held (1975), Passarge (1964), Pop & Hodişan (1970) and to some extent also Soó (1973) and Mucina (1981b). In this conception the *Melandrio-Berteroetum* (a synonym of the *Berteroetum incanae*), *Verbasco–Berteroetum* Passarge 1957, *Centaureo diffusae–Berteroetum*, and some times also the *Rorippo pyrenaicae–Berteroetum* Pop et Hodişan form a group of vicarious associations (Passarge, 1964) or better: an association group (sensu Westhoff & van der Maarel, 1978). Seybold & Müller (1972) representing a 'broad' concept include not only the *Verbasco–Berteroetum* and *Centaureo–Berteroetum*, but also the *Verbasco–Chondrilletum* Tillich 1969 into the *Berteroetum incanae* s. l. Our current opinion is intermediate to the extent that we regard the *Centaureo–Berteroetum* as a separate association. This distinction is kept also by German and Czechoslovakian authors (Oberdorfer *et al.*, 1967; Oberdorfer & Müller, 1979; Hejný *et al.*, 1979). Most of the *Centaureo–Berteroetum* relevés come from Poland. This might be a reason why traditionally the relevés of the *Berteroetum incanae* s.str. are classified under the *Centaureo–Berteroetum* (Kępczyński, 1975; Czaplewska, 1980). The *Verbasco–Berteroetum* and *Rorippo–Berteroetum* could not be recognized in the synoptic table as separate associations, thus they were included into the list of synonyms of the *Berteroetum incanae* s.str. The *Melandrio–Berteroetum* and *Verbasco–Berteroetum* associations are included into the western race of the *Berteroetum incanae*, whereas the *Rorippo–Berteroetum* association is placed within the eastern race of the *Berteroetum incanae*. The *Cynoglosso–Berteroetum* (Olsson, 1978) from southern Sweden could not be included in the synthesis because a different sampling technique was used. Because of the lack of relevés from Hungary and the Balkan Peninsula we could not confirm the *Centaureo micranthae–Berteroetum* (Ubrizsy, 1955; see also Soó, 1971; Mititelu & Barabaş, 1972) as a regional association. Ubrizsy in Soó (1971) mentioned *Cephalaria transsilvanica* as a character-species of the latter unit. The one relevé from Bulgaria (Table 1, column 24) contains a number of submediterranean species. A reconsideration of the syntaxonomy of the *Berteroa incana*-dominated communities in Bulgaria may be appropriate if more material is available (see also Mucina, 1979).

Since the first descriptions (Westhoff *et al.*, 1946; Sissingh, 1950; Lebrun *et al.*, 1949; Oberdorfer, 1957), as early as 1966, (Ubrizsy, 1955; Passarge, 1964; Soó, 1964; Gutte, 1966) the *Berteroa incana* communities have been assigned to the *Onopordion acanthii*. Later, after Görs (1966) had described the *Dauco–Melilotion*, German and Czechoslovakian authors considered them as members of this newly described alliance (Seybold & Müller, 1972; Oberdorfer *et al.*, 1967; Oberdorfer & Müller, 1979; Hejný *et al.*, 1979; Gutte, 1972; Gutte & Hilbig, 1975; Mucina, 1981b). Only Westhoff & den Held (1975), and Polish, Roumanian and Hungarian authors still hold to the idea of the *Onopordion acanthii* as being the higher syntaxon of the *Berteroetum incanae* (cf. Pop & Hodişan, 1970; Fijałkowski, 1971; Mititelu & Barabaş, 1971, 1972; Rostański & Gutte, 1971, Kępczyński, 1975; Czaplewska, 1980). Until now the *Dauco–Melilotion* has not frequently been used in Poland, Hungary and Roumania. The description of a separate alliance *Berteroion incanae* by Radke (1979) is not supported by data.

The lower-ranked syntaxa described within the *Berteroetum incanae* s.str., namely the *Berteroetum incanae medicaginetosum* and *typicum* of Sissingh (1950), the *Berteroetum incanae*, typical variant, and the variant with *Salvia nemorosa* of Mucina (1981b) and the *Berteroetum incanae rorippetosum pyrenaicae* of Pop & Hodişan (1970), seem to be of very local importance. Like the race of *Salsola kali* of the *Centaureo–Berteroetum* (Gutte, 1972) they do not represent separate units in the synoptic tables.

Comparing both races of the *Berteroetum incanae* in terms of species diversity one recognizes a more general pattern, with a higher species diversity and higher number of plant associations per alliance towards southeastern Europe. This has also been noted within the *Malvion neglectae* and *Onopordion acanthii* (cf. Mucina, 1979, 1981a). This holds also for natural vegetation types, e.g. when comparing (sub)alpine chalk grasslands or beech woods of central Europe to those of the Balkan (see Neuhäusl in Dierschke, 1981). The causes of this phenomenon have not been appropriately evaluated yet. One could look for causes in dramatical florogenetical processes during Pleistocene glaciation which, gave origin to various floral refugia in eastern Europe, and to diversification centra of species. Diverse flora migration routes in the Holocene and a lower intensity of human impact upon the landscape in eastern Europe as well as a diversity of stands may also be considered.

A question may arise while inspecting the dendrograms why the first dichotomy does not correspond to the division into the *Berteroetum incanae* and *Centaureo–Berteroetum*. The *Centaureo–Berteroetum* appeared closer to the eastern race of the *Berteroetum incanae* and seemed to be rather a subunit of the eastern race than a separate association. Nevertheless, the *Centaureo–Berteroetum* is considered an association due to the high fidelity value of *Acosta diffusa*. The species, however, is weighted by SR the same way as any other discriminant character. Although there are many good discriminant characters (differential species) between the *Berteroetum incanae* and *Centaureo–Berteroetum*, the eastern race of the *Berteroetum incanae* and *Centaureo–Berteroetum*, have a lot of species in common, since the distribution areas of both phytocoena are largely the same.

The comparisons between the results of the numerical and syntaxonomical classifications, respectively (Table 2), yield rather low resemblance values. The differences might be explained by (1) differing classification criteria adopted in both approaches, the weighting of certain species (see the preceding paragraph), and (2) by the fact that the numerical classifications are based on relevés whereas the syntaxonomical treatment is based upon local tables.

References

Brandes, D., 1977. Die Onopordion-Gesellschaften der Umgebung Braunschweigs. Mitt. Florist.-Soziol. Arbeitsgem., N.F. 19/20: 103–113.

Braun-Blanquet, J., 1964. Pflanzensoziologie. Grundzüge der Vegetationskunde. 3. Aufl. Springer, Wien, New York. 865 pp.

Czaplewska, J., 1980. Zbiorowiska roślin ruderalnych na terenie Aleksandrowa Kujawskiego, Ciechocinka, Nieszawy i Wrocławka. Stud. Soc. Sci. Torunensis, Sect. Bot. 11: 1–76.

Dierschke, H., 1981. Zur syntaxonomischen Bewertung schwach gekennzeichneter Pflanzengesellschaften. In: H. Dierschke (ed.). Syntaxonomie, Ber. Int. Symp. Int. Ver. Vegetkde., pp. 109–122. J. Cramer, Vaduz.

Fijałkowski, D., 1971. Zbiorowiska synantropijne wyrobisk krędowych w Chelmie i Rejowcu na Lubelszczyznie. Mater. Zakł. Fitosoc. Stosow. Uniw. Warszaw. 27: 273–289.

Fijałkowski, D., 1978. Synantropy roślinne Lubelszczyzny. Lubel. Tow. Nauk., Prace Wydz. Biol. 5: 1–260.

Goldstein, R. & Grigal, D. F., 1972. Computer programs for the ordination and classification of ecosystems. Ecol. Sci. Div. Publ., Oak Ridge National Laboratory, 417 pp.

Goodman, L. A. & Kruskal, W. H., 1954. Measures of association for cross classifications. J. Am. Stat. Ass. 49: 732–764.

Grüll, F., 1982. Málo známá pionýrská společenstva rostlin na obnažených půdách staveništ' města Brna. Preslia 54: 149–166.

Gutte, P., 1966. Die Verbreitung einiger Ruderalpflanzengesellschaften in weiterer Umgebung von Leipzig. Wiss. Z. Univ. Halle, Ser. Math.-Nat. 15: 937–1010.

Gutte, P., 1972. Ruderalpflanzengesellschaften West- und Mittelsachsens. Feddes Repert. 83: 11–122.

Gutte, P. & Hilbig, W., 1975. Übersicht über die Pflanzengesellschaften des südlichen Teiles der DDR. XI. Ruderalgesellschaften. Hercynia N.F. 12: 1–30.

Hejný, S., Kopecký, K., Jehlík, V. & Krippelová, T., 1979. Přehled ruderálních rostlinných společenstev Československa. Rozpr. Čs. Akad. Věd., Ser. Math. Nat. 89/2: 1–100.

Hill, M. O., 1973. Reciprocal averaging: An eigenvector method of ordination. J. Ecol. 61: 237–249.

Hill, M. O., 1979. DECORANA. A Fortran program for detrended correspondence analysis and reciprocal averaging. Ecology and Systematics, Cornell University, Ithaca. iii + 52 pp.

Hülbusch, K. H. & Kuhbier, H., 1979. Zur Soziologie von Senecio inaequidens DC. Abh. Naturw. Verein Bremen 39: 47–54.

Kępczyński, K., 1975. Zbiorowiska roślin synantropijnych na terenie miasta Bydgoszczy. Acta Univ. Nicolae Copernici, Ser. Biol. 17: 3–87.

Lebrun, J., Noirfalise, A., Heinemann, P. & Vanden Berghen, C., 1949. Les associations végétales de Belgique. Bull. Soc. Roy. Bot. Belg. 82: 105–199.

Lienenbecker, H., 1968. Die Graukressen-Gesellschaft (Berteroetum incanae) im östlichen Westfalen. Natur u. Heimat 28: 126–127.

Maarel, E. van der, 1979. Transformation of cover-abundance values in phytosociology and its effects on community similarity. Vegetatio 39: 97–114.

Mititelu, D. & Barabaş, N., 1971. Vegetaţia Văii Trotuşului (sectorul Urecheşti – Tg. Trotuş). Stud. Comunic. Bacău: 791–820.

Mititelu, D. & Barabaş, N., 1972. Vegetaţia ruderală şi segetală din interiorul şi imprejurimile Municipiului Bacău. Stud. Comunic. Bacău: 127–148.

Mucina, L., 1979. Synantropná vegetácia stredného Považia I. – Spoločenstvá radu Onopordetalia. Thesis, Comenius University, Bratislava.

Mucina, L., 1981a. Die Ruderalvegetation des nördlichen Teils der Donau-Tiefebene. 1. Onopordion acanthii-Verband. Folia Geobot. Phytotax. 16: 225–263.

Mucina, L., 1981b. Die Ruderalvegetation des nördlichen Teils der Donau-Tiefebene. 2. Gesellschaften des Dauco–Melilotion-Verbandes auf ruderalen Standorten. Folia Geobot. Phytotax. 16: 347–389.

Oberdorfer, E., 1957. Süddeutsche Pflanzengesellschaften. Pflanzensoziologie 10: 1–564.

Oberdorfer, E., 1968. Assoziation, Gebietsassoziation, Geographische Rasse. In: R. Tüxen (ed.). Pflanzensoziologische Systematik. Ber. Int. Symp. Int. Ver. Vegetkde. Stolzenau/Weser 1964, pp. 124–141. Junk, Den Haag.

Oberdorfer, E. et al., 1967. Systematische Übersicht der westdeutschen Phanerogamen und Gefässkryptogamen-Gesellschaften. Schriftenr. Vegetkde. 2: 7–62.

Oberdorfer, E. & Müller, Th., 1979. Pflanzensoziologische Exkursionsflora. 4. Aufl. Eugen Ulmer, Stuttgart. 997 pp.

Olsson, H., 1978. Vegetation of artificial habitats in northern Malmö and environs. Vegetatio 36: 65–82.

Orlóci, L., 1967. An agglomerative method for classification of plant communities. J. Ecol. 55: 193–205.

Passarge, H., 1959. Pflanzengesellschaften zwischen Trebel, Grenzbach und Peene (O-Mecklenburg). Feddes Repert. 138: 1–56.

Passarge, H., 1964. Über Pflanzengesellschaften des Hagenower Landes. Arch. Nat. Meckl. 10: 31–51.

Pop, I. & Hodişan, I., 1970. Studiu fitocenologic asupra unei asociaţii nitrofile nouă pentru România. Stud. Univ. Babeş-Bolyai, Ser. Biol. 1: 5–8.

Radke, G. J., 1980. System of ecologic and geographic landscape classification, shown in the central Europe region. Acta Bot. Acad. Sci. Hung. 26: 169–180.

Rostański, K. & Gutte, P., 1971. Roślinność ruderalna miasta Wrocławia. Mater. Zakł. Fitosoc. Stos. Uniw. Warszaw. 27: 167–215.

Seybold, S. & Müller, Th., 1972. Beitrag zur Kenntnis der Schwarznessel (Ballota nigra agg.) und ihre Vergesellschaftung. Veröff. Landesstelle Naturschutz Baden-Württemberg 40: 51–126.

Sissingh, G., 1950. Onkruid-associaties in Nederland. Thesis Wageningen. s'Gravenhage. 224 pp.

Smejkal, M., 1980. Komentovaný katalóg moravské flóry. Universita J. E. Purkyně, Brno. 301 pp.

Sneath, P. H. A. & Sokal, R. R., 1973. Numerical Taxonomy. W. H. Freeman & Co., San Francisco. 573 pp.

Soó, R., 1964. A magyar flóra és vegetáció rendszertani-növényföldrajzi kézikönyve. I. Akadémiai kiadó, Budapest. 589 pp.

Soó, R., 1971. Aufzählung der Assoziationen der ungarischen Vegetation nach den neueren zönosystematisch-nomenklatorischen Ergebnissen. Acta Bot. Acad. Sci. Hung. 17: 127–179.

Sowa, R., 1971. Flora i zbiorowiska ruderalne na obszarze województwa łódzkiego ze szczególnym uwzględnieniem miast i miasteczek. Universitet Łódzki, Łódź. 282 pp.

Tutin, T. G. et al. (eds.), 1964–1980. Flora Europaea. Vols. 1–5. University Press, Cambridge. 464, 455, 370, 505 & 452 pp.

Ubrizsy, G., 1955. Magyarország ruderalis gyomnövénytársulásai II. Ökológiai és szukcessziö-tanulmányok. Növénytermelés 4: 109–126.

Ullmann, I., 1977. Die Vegetation des südlichen Maindreiecks. Hoppea 36: 5–190.

Werger, M. J. A. & Gils, H. van, 1976. Phytosociological classification in chorological borderline areas. J. Biogeogr. 3: 49–54.

Westhoff, V., Dijk, J. W. & Passchier, H., 1946. Overzicht der Plantengemeenschappen in Nederland. 2nd ed. Uitg. K.N.N.V. & N.J.N., Amsterdam.

Westhoff, V. & Held, A. J. den, 1975. Planten-Gemeenschappen in Nederland. 2nd ed. W. J. Thieme & Cie, Zutphen. 324 pp.

Westhoff, V. & Maarel, E. van der, 1978. The Braun-Blanquet approach. In: R. Whittaker (ed.). Classification of Plant Communities, 2nd ed., pp. 287–399. Junk, The Hague.

Wishart, D., 1978. Clustan 1C. User Manual. St. Andrews Computing Centre, London. 175 pp.

Accepted 19.1.1984.

Appendix

Columns in the synoptic table

Table 1. *Berteroetum incanae* and *Centaureo diffusae-Berteroetum* in Europe.

Centaureo-Berteroetum (D):

Psyllium scabrum
Lepidium densiflorum
Ceratodon purpureus
Oenothera parviflora
Herniaria glabra

Berteroetum incanae (D):

Silene alba
Dactylis glomerata
Urtica dioica
Arrhenatherum elatius
Acetosa vulgaris
Knautia arvensis
Descurainia sophia
Cardaria draba
Galium verum
Geranium pusillum
Atriplex patula
Poa annua
Equisetum arvense
Cyanus segetum

Raphanus raphanistrum
Sonchus arvensis
Stellaria media
Medicago sativa
Hypericum perforatum
Tithymalus cyparissias
Acetosella vulgaris
Trifolium arvense
Chrysaspis campestre

Dauco-Melilotion:

Anchusa officinalis
Daucus carota
Echium vulgare
Linaria vulgaris
Melilotus officinalis
Melilotus alba
Oenothera biennis
Reseda lutea
Potentilla intermedia
Picris hieracioides
Acetosa thyrsiflora
Tragopogon dubius
Verbascum densiflorum
Medicago varia
Verbascum phlomoides

Crepis polymorpha 1 (5,25); Crepis tectorum I (17), III (28); Erigeron annuus I (20), 1 (26); Isatis ti nctoria 2 (5), 1 (25); Oenothera depressa I (22), II (30); Odontites vulgaris II (22); Verbascum nigrum I (6,10); Crepis setosa I (20); Oenothera chicago ensis I (13); Oenothera hoelscheri 1 (27);

Onopordion acanthii:

Artemisia absinthium
Reseda luteola

Hyoscyamus niger I (2,19); Lappula squarrosa I (19),2 (23); Verbascum blattaria I (19),1 (25); Lavatera thuringiaca l (22); Marrubium peregrinum I (16), 2 (21); Marrubium vulgare I (10); Nepeta cataria I (2); Stachys germanica 1 (23); Xeranthemum annuum 1 (21); Cota austriaca 4 (21);

Artemisietea vulgaris:

Artemisia vulgaris
Medicago lupulina
Ballota nigra
Tanacetum vulgare
Cirsium vulgare
Arctium minus
Solidago gigantea
Rumex obtusifolius
Rubus caesius

Carduus crispus II (9), I (19); Lapsana communis I (19), 1 (20); Anthriscus sylvestris I (19); Arctium lappa I (17); Armoracia rusticana I (13); Calyste gia aepium I (19); Chaerophyllum temulum I (10); Chelidonium majus I (16); Conium maculatum I (1); Dips acus laciniatus I (19); Echinops sphaerocephalus I (19); Geranium pyrenaicum I (20); Helianthus tuberosus I (24), II(11); Leonurus cardiaca I (30); Tor ilis japonica 1 (4); Silene dioica I (3); Solanum dulcamara I (14); Malva sylvestris 2 (7), 1 (24);

Chenopodietea:

Conyza canadensis
Sisymbrium loeselii
Anisantha sterilis
Sisymbrium altissimum
Sonchus oleraceus
Sinapis arvensis
Hordeum murinum
Setaria viridis
Malva neglecta
Diplotaxis tenuifolia
Echinochloa crus-galli
Digitaria sanguinalis
Atriplex tatarica

Amaranthus albus I (22,28); Ambrosia artemisiifolia I (1,2); Atriplex acuminata 1 (8,25); Bunias orient alis I (7), 5 (23); Erucastrum gallicum 3 (4), 1 (25); Erysimum cheiranthoides I (6), II (17); Tithymalus helioscopius I (12,19); Tithymalus peplus I (12,19); Galeopsis tetrahit II (6), I (18); Geranium colum binum I (8), I (16); Lepidium virginicum I (1), III (12) Matricaria recutita I (1,30); Persicaria macu lata I (1,17); Solanum nigrum II (1), III (28); Urtica urens I (1,2); Xanthium strumarium 1 (17,30); Aethusa cynapium I (1); Amaranthus powellii (2); Atripl ex oblongifolia 1 (26); Chenopodium strictum I (22); Diplotaxis viminea I (12); Eragrostis minor I (22), 1 (21); Fumaria vaillantii I (19); Galinsoga urtici folium I (12); Iva xanthiifolia I (28); Setaria glauca III (22); Portulaca oleracea 1 (21);

Secalietea:

Apera spica-venti
Anthemis arvensis
Papaver rhoeas
Viola arvensis

Allium vineale I (1,2); Avena fatua 1 (7), I (19); Nigella arvensis ; (19,22); Papaver argemone 1 (7), (12); Vicia villosa I (13,20); Chamaepitys trifida I (19); Anagallis arvensis I (12); Aphanes arvensis I (12); Camelina microcarpa I (19); Camelina sativa I (28); Consolida regalis I (19); Kickxia elatine I (28); Kickxia spuria I (19); Lathyrus tuberosus I (19); Linaria arvensis I (19); Neslia paniculata 2 (23) Nonnea pulla I (19); Silene noctiflora I (28); Stachys annua I (19);

Agropyretea repentis:

```
Poa compressa          .  .  .  .  2  .  2  1  I  .  .  I III  .  .  .  I  I  .  2  .  1  1 III  3 II II
Carex hirta            .  I  .  .  .  .  .  1  .  .  .  .  .  .  .  .  I  .  I  .  .  .  .  .  .  II  .
Tithymalus esula       .  .  .  .  .  .  .  .  .  .  .  I  I  .  .  .  I  .  I  .  .  .  .  .  I  .  .
Falcaria vulgaris      .  .  .  .  .  .  .  1  .  .  .  .  .  .  .  .  I  .  II .  .  .  .  .  .  .  .
Cota tinctoria         .  .  .  .  .  .  .  1  .  .  .  .  .  .  .  .  I  .  .  .  .  .  .  .  .  .  .
```

Plantaginetea majoris:

```
Polygonum aviculare agg. IV IV  .  .  .  1  1  . III  .  .  I  4  . III  V  .  . III  .  .  .  .  . II IV
Plantago major          III II  I  1  . I  .  .  . I  .  .  2  . II  I  I  . I  .  .  . II  2  . I
Rumex crispus            I  .  .  .  1  .  1  .  .  .  .  I  I  .  .  I  .  .  .  .  .  .  .  .  .  .  .
Scorzoneroides autumnalis III I  .  .  .  .  .  I  I  .  .  .  .  .  .  .  .  .  .  .  .  . II  .  I  I
Agrostis stolonifera     I  .  1  .  .  .  I  . II  .  .  .  .  .  .  .  .  .  .  .  .  2  .  .  .
Potentilla reptans       .  .  .  .  .  .  1  .  .  .  .  .  .  .  .  .  I  .  .  .  .  .  .  .  .
Polygonum arenastrum     .  .  .  .  .  .  .  .  .  .  .  .  I  4  .  .  .  .  .  .  .  .  .  .  .
```

Glechoma hederacea I (1,22); Inula britannica I (28); Matricaria suaveolens I (17); Mentha longifolia 1 (23); Podospermum laciniatum I (19); Centaurium pulchellum 1 (21); Argentina anserina II (30); Potentilla supina I (19); Prunella vulgaris III (30); Ranunculus repens I (1), 1 (4); Rumex triangulivalvis I (2);

Convolvulo-Chenopodiea:

```
Elytrigia repens        V  V  .  1  . III  2  3 II II  V  . IV  .  .  I III III IV II  1 III  .  .  1 III  2  . II
Convolvulus arvensis    IV III .  1  4 III  .  2  I III  V II III  .  .  I III II IV II  . I  .  .  3  I  .  .
Chenopodium album        V  V  .  .  .  2  .  I  . .  .  I  . I  II III  IV  I  I  .  I II  .  .  .  I  . III  .
Capsella bursa-pastoris III III .  .  .  1  1  . I  . III  . III IV  I  I  . I  2  .  .  .  .  .  .  .  . III
Matricaria perforata     .  .  .  1  2 II  1  2 II  . IV  I  4  .  . II III  .  I  3  .  . III  .
Cirsium arvense          .  .  .  1  2  .  .  . I  . I  .  I III  2  . IV  . I  I  .  2  1 III  .  .
Fallopia convolvulus    II III .  .  .  .  .  .  I  .  .  3  .  I  I  I  .  I  .  . II  .  .  .
```

Molinio-Arrhenatheretea:

```
Achillea millefolium     V  V  1  1  3  V  .  1 IV  V IV III III IV  4  I  V IV IV IV  . IV  2  .  2  2 II  3 II  V
Plantago lanceolata     III III 2  1  2  V  .  2 II II IV  V III  .  . IV II III III IV  3  V  2  .  1  1 III  1  .  I
Trifolium repens         I  I  1  .  . III  1  I  I  .  .  .  .  .  .  .  I  I  .  . II  .  .  1  1  .  1  I
Taraxacum sect. Vulgaria II II  2  1  .  .  . I  . IV  I  II III  1  3 II  V II  I  .  . III  . III  2 II  .
Lolium perenne          IV III 2  .  . II  1  . II  . III IV IV  V  2 II  I II III III II  .  .  .  .  3  I  I
Poa pratensis           III  V  .  . III  .  . IV  V II III  . IV  4  . IV IV  I  I  .  2  .  .  .  2  I II
Pastinaca sativa         .  .  .  2  I  .  2  I  .  .  .  .  .  .  .  I  . II  .  . I  2  . I  .  .
Agrostis tenuis         IV  V  .  . III  .  I  I  I IV  .  .  .  .  .  .  I  .  .  .  . III  1 I  .  I
Festuca rubra           II III 1  .  .  .  .  1  .  .  .  .  .  .  .  I  .  .  .  .  .  .  1  1  .  I
Hypochoeris radicata     .  .  .  .  .  .  .  .  .  .  .  .  .  .  .  I  .  .  .  .  .  .  .  .  .
```

Anthoxanthum odoratum I (1), II (6); Campanula rotundifolia I (1), II (2); Cerastium holosteoides 1 (23), II (28); Festuca pratensis I (6), 1 (27); Galium album I (16,19); Poa palustris I (19,22); Tragopogon pratensis I (9); Agrostis gigantea 1 (26); Bellis perennis I (30); Geranium molle I (1); Lathyrus pratensis 1 (23); Leucanthemum vulgare agg. 1 (23); Phleum pratense I (19); Pimpinella major I (16); Poa trivialis I (1); Rorippa pyrenaica 2 (23); Thymus serpyllum I (22); Veronica chamaedrys I (30); Vicia sepium I (22); Vicia cracca I (16,20);

Festuco-Brometea:

```
Eryngium campestre       .  .  .  .  .  .  .  .  .  .  .  .  .  .  .  . II  .  1  I  .  .  .  .
Artemisia campestris     .  .  .  .  2  .  1  . I  I II  . III IV  .  .  I II II  I  . I  .  1  3  .  . V  .
Poa angustifolia         .  .  .  .  4  .  .  1  I  . I  . I II  . II  .  I  I  . II  .  . 4  1  .  .
Medicago falcata         I  V  .  .  .  .  .  .  .  .  .  I  . II  I  I  . II  .  .  .  1  .  .  .  I
Pilosella sp. div.       I  .  .  .  .  .  .  .  .  .  .  .  .  .  .  .  .  .  .  .  .  .  I  1 II
Festuca ovina            .  .  .  .  I  .  I  I II II II  .  .  I  .  .  .  .  .  .  .  .  .  .  .
Salvia verticillata      I  .  .  .  .  .  .  .  .  .  .  .  .  .  .  .  .  I  .  .  .  .  .  .
Salvia nemorosa          I  .  .  .  .  .  .  .  .  .  .  .  .  .  . II  .  .  .  I  .  .  .
Erysimum diffusum        .  .  .  .  .  .  .  .  .  .  .  .  .  .  .  .  . I  I  I  .  .  .  .
```

Tithymalus waldsteinii I (20,28); Hieracium umbellatum I (6,28); Linaria genistifolia I (20); Potentilla arenaria I (28,29); Scabiosa ochroleuca I (19), 1 (27); Taraxacum sect. Erythrosperma I (1,2); Verbascum lychnitis 1 (5,25); Achillea collina I (19), 2 (21); Asparagus officinalis I (9); Asperula cynanchica I (19); Carex humilis I (19); Colymbada scabiosa I (16); Tithymalus seguierianus I (19); Festuca pseudovina I (22); Festuca tenuifolia I (1); Festuca trachyphylla I (19); Leontodon incanus I (19); Linum austriacum I (19); Onobrychis arenaria I (19); Ononis repens I (8); Plantago media 1 (23); Ranunculus bulbosus I (2); Sanguisorba minor I (19); Senecio jacobea I (10); Silene otites I (22); Tragopogon orientalis I (16); Verbascum austriacum I (16); Pseudolysimachion spicatum I (18); Viola saxatilis 2 (23); Medicago minima 1 (21); Leopoldia comosa 1 (21); Artemisia austriaca 1 (21);

Sedo-Scleranthetea:

```
Erodium cicutarium       I  I  .  .  .  .  .  1  . III  I  .  .  I  I  I  3  .  .  . I  1  .  I
Sedum acre              I  I  .  .  1  .  1  . I  . II  . I  .  . 3  .  .  .  1  .  .
Bromus mollis           IV III .  . IV  1  . I  I III IV IV  . IV  .  . I  I  3  I  .  .  1  I  .  I II
Anisantha tectorum       I  .  .  .  1  .  .  . II IV  . II  . I III II II  .  1  2 1  4  1 II  2 IV
Arenaria serpyllifolia   .  .  1  .  .  I  1  1  . II  .  .  .  .  . II  2  .  2  1  .  .  .  . II
Potentilla argentea      I IV  .  .  .  .  .  . II III  .  .  . I III  1 II  .  .  .  .  2  I
Senecio viscosus         .  .  .  3  .  1  .  I  . II  .  .  .  .  .  .  3  .  .  . I  .
Vicia hirsuta            .  I  .  1  .  .  .  .  .  .  .  .  .  .  .  .  .  .  .  .  .  .
Anthemis ruthenica       I  .  .  .  .  .  .  .  .  .  .  .  .  .  .  .  1  .  .  .
Bromus squarrosus        .  .  .  2  .  .  .  .  .  .  .  . I  2  .  . 2  .
Scleranthus annuus       .  .  .  2  .  .  .  1  .  .  .  .  . I  .  .  1
```

Acinos arvensis I (19, 29); Alyssum alyssoides I (19,29); Arabidopsis thaliana I (12), 1 (26); Cardaminopsis arenosa I (14), 2 (23); Helichrysum arenarium I (13,17); Scleranthus annuus I (12), 1 (23); Armeria elongata I (12); Cerastium semidecandrum I (1); Corynephorus canescens I (17); Draba nemoralis 1 (23); Poa bulbosa 2 (21); Phleum bertoloni I (2); Hylotelephium maximum I (3); Sedum sexangulare I (12); Valerianella dentata 2 (23); Veronica arvensis 2 (23), 1 (21); Vulpia bromoides I (28); Vulpia myuros 1 (24), 2 (21);

Other species (Trifolio-Geranietea, Thlaspietea rotundifolii, Rhamno-Prunetea, Querco-Fagetea and others):

Epilobium angustifolium 1 (7), I (9); Linaria repens I (5), 1 (25); Rosa canina I (14, 19); Sambucus nigra I (1,14); Agrimonia eupatoria I (21); Ailanthus altissima juv. I (20); Arctium sp. 1 (27); Aster sp. I (28); Avena sativa I (19); Bromus sp. I (19); Clematis vitalba I (20); Clinopodium vulgare I (16); Crepis sp. I (19); Fragaria viridis I (19); Helianthus annuus I (28); Helianthus sp. I (13); Hieracium caespitosum I (28); Holcus mollis I (1); Inula conyza I (16); Juglans regia I (19); Lathyrus sylvestris I (7); Mentha sp. I (28); Oenothera ammophila I (28); Origanum vulgare I (20); Oxybaphus nictaginea I (2); Reseda gracilis 1 (5); Robinia pseudacacia I (20); Rubus fruticosus agg. I (20); Panicum miliaceum 1 (7); Pinus sylvestris I (9); Potentilla erecta II (30); Prunus avium I (19); Prunus sp. I (23); Syringa vulgaris I (16); Taraxacum obliquum I (28); Trifolium medium I (15); Veronica sp. I (20); Vicia tenuifolia I (20); Achillea coarctata I (24); Anchusa barreleri 1 (24); Silene longiflora 1 (24); Abietinella abietina I (19); Tortula muralis

Bryophyta:

```
Musci indet.             .  .  .  1  . III  . . III  .  .  V  .  .  .  .  .  .  .  .  .  .  .
Bryum argenteum         III IV  .  .  .  .  .  .  .  .  .  .  .  . I  2  .  .  .  .  .  1  . II
```

Brachythecium albicans I (2), 2 (27); Brachythecium sp. I (1,22); Pohlia sp. II (1), III (2); Aloina rigida III (30); Barbula convoluta I (30); Bryum caespiticium 2 (27); Campylium chrysophyllum I (30); Hypnum vaucheri I (19); Rhacomitrium canescens I (1); III (30); Bryum badium I (20).

Berteroetum incanae, Galium mollugo race:

1. Sissingh (1950), The Netherlands, 15
2. Sissingh (1950), The Netherlands, 7
3. Runge (unpubl.), Westfalen (F.R.G.), 3
4. Westhoff (unpubl.), Molsbergen (Denmark), 1
5. Th. Müller (unpubl.), Ulm, Soflingen, Krs. Ludwigsburg (F.R.G.), 4

6. Lienenbecker (1968, unpubl.), vicinity of Bielefeld (F.R.G.), 9
7. Hülbusch & Kuhbier (1979), Bremen (F.R.G.), 2
8. Ullmann (1977), Maindreieck near Ochsenfurt (F.R.G.), 3
9. Brandes (1977, unpubl.), vicinity of Braunschweig (F.R.G.), 11
10. Passarge (1959), Brandenburg, Ost-Mecklenburg (G.D.R.), 11

Berteroetum incanae, Acosta rhenana race:
11. Nezadal (unpubl.), Rednitztal near Nürnberg (F.R.G.), 5
12. Brandes (unpubl.), Bamberg (F.R.G.), 6
13. Passarge (1964, unpubl.), Hagenow, Schwanenbeck-Alpenberge, Berling (G.D.R.), 7
14. Kunick (unpubl.), West Berlin (F.R.G.), 5
15. Grüll (1982), Brno (Czechoslovakia), 4
16. Forstner (unpubl.), Niederösterreich, Burgenland (Austria), 7
17. Kępczyński (1975), Bydgoszcz (Poland), 15
18. Czaplewska (1980), Aleksandrowie Kuj., Ciechocinek, Nieszawa, Wrocławek (Poland), 8
19. Mucina (1981b, unpubl.), the western part of Slovakia (Czechoslovakia), 21
20. Jarolímek (unpubl.), Bratislava (Czechoslovakia), 20
21. Mucina (unpubl.), the southern part of the Podunajská Nížina Lowland (Czechoslovakia), 4
22. Mucina & Zaliberová (unpubl.), the Východoslovenská Nížina Lowland (Czechoslovakia), 10
23. Pop & Hodişan 1970, Valea Someşului Rece (Roumania), 2
24. Mucina (unpubl.), Melnik (Bulgaria), 1

Centaureo diffusae–Berteroetum:
25. Oberdorfer (1957), Karlsruhe, Mannheim (F.R.G.), 2
26. Th. Müller (unpubl.), Krs. Ludwigsburg (F.R.G.), 2
27. Kunick (unpubl.), West Berlin (F.R.G.), 2
28. Gutte (1966), vicinity of Leipzig (G.D.R.), 10
29. Rostański & Gutte (1971), Wrocław (Poland), 4
30. Sowa (1971), Łódz, Tomaszów Mazowiecki (Poland), 5
31. Fijałkowski (1978), woj. Lubelskie (Poland), 5

Syntaxonomische Wertung chorologischer Phänomene*

H. Passarge
13 Eberswalde, Schneiderstr. 13, D.D.R.

Keywords: Areal, Edaphic-ecologic, Geographic subdivision, *Papaveretum argemones,* Syntaxonomy

Abstract

On the basis of superregional constant-species combination and edaphic-ecological subdivisions according to trophic status and moisture regime the following synchorological phenomena are discussed; distribution of associations, areal boundaries, relation to vicariant neighbour units and syngeographical division, with the *Papaveretum argemones* as a model. With the regional ass. *Chamomillo–Papaverenetum* and *Myosotido–Papaverenetum* as examples a new rank for vicariant associations is proposed.

Vorbemerkung

Überall modifizieren ökologische Faktoren die Zusammensetzung von Flora und Vegetation. Vorkommen von Pflanzen und Pflanzengesellschaften werden dabei teils kleinstandörtlich teils arealmäßig begrenzt. Obwohl derartige Grenzen aus dem komplexen Zusammenwirken obwaltender Ökobedingungen resultieren, hat es sich bei ihrer systematischen Bewertung – ähnlich wie in der Taxonomie – als sinnvoll erwiesen, die überwiegend regional-geographischen, meist klimatisch (oder historisch) begründeten Unterschiede neben den lokal-ökologischen (±edaphischen) syntaxonomisch gesondert zu berücksichtigen. Zumindest bei niederen bzw. Hilfsrangstufen führt dies zu einer Zweigleisigkeit im System mit erhöhter Aussagekraft, falls die registrierten Vegetationsverschiedenheiten richtig gedeutet werden. Von regionaler Vikarianz bei Vegetationseinheiten sollte man nur sprechen, wenn sich die betreffenden Ausbildungen sowohl geographisch ± ausschließen (bzw. nur partiell überlappen) als auch durch ausfallende bzw. neu hinzukommende arealgeographisch spezi-

fische Zeigerpflanzen (Florenelemente) auszeichnen. Vikariierende Formen niederen Ranges sind somit stets an syngeographische Differenzialarten (Schwickerath, 1944) gegenüber der Ausgangseinheit gebunden.

Am Beispiel des *Papaveretum argemones* (Libb. 32) Krusem. et Vlieg. 39 sollen einige synchorologische Aspekte aufgezeigt und ihre syntaxonomische Wertung erörtert werden.

Großräumige Konstanz der Artenverbindung und Verbreitungsgebiet

Geographische Vikarianz ist nur prüfbar an Hand einer floristisch-coenologisch fest umrissenen Ausgangseinheit als Ausdruck relativ konstanter ökologischer Bedingungen. Das *Papaveretum* kann als *Aphanion*-Gesellschaft definiert werden mit *Papaver argemone, P. dubium* (regional *Vicia villosa, V. dasycarpa, V. tetrasperma*) und zahlreichen Frühlingsephemeren: *Veronica triphyllos* (regional *Myosotis stricta, M. discolor*), *Holosteum umbellatum* (regional: *Arenaria serpyllifolia, Veronica hederifolia, Arabidopsis thaliana, Erophila verna, Gagea pratensis*). Zu den bezeich-

* Die Sippennomenklatur richtet sich nach Oberdorfer (1979).

138

nenden Begleitern gehören einzelne Arten der *Scle-ranthus*-Gruppe; vorherrschend ist meist *Apera spica-venti*. Die Ass. gedeiht auf etwa karbonatfreien humos-sandigen, im Verbreitungszentrum auch sandig-lehmigen Böden mittelmäßiger Trophie (R 4–6, N 5–6 nach Ellenberg, 1974), die meist nur frühjahrsfrisch (F 4–5) sind, relativ schnell abtrocknen und sich gut erwärmen (T 5–7).

Diese Verhältnisse herrschen vornehmlich im Wintergetreide des Standortbereiches der Roggen-Kartoffelwirtschaft. Bevorzugte Wuchsorte sind Talsandgebiete, sandig-kiesige Ablagerungen in Becken, Fluß- und Stromtälern, sonstige binnenländische Sandgebiete und diesen entsprechende durchlässige Silikatverwitterungsböden in relativ sommerwarmer Lage. Der Verbreitungsschwerpunkt der Assoziation liegt im temperaten Mitteleuropa, wobei das *Papaveretum* im küstenfernen nördlichen Tiefland zwischen Rhein und Weichsel offenbar die größte Siedlungsdichte erreicht. Die lehmigen und mergeligen Böden der Löß- und Kalkhügelländer sowie alle submontan-montanen Gebirgslagen aussparend, wurde die Assoziation bisher von S-Skandinavien (Merker, 1959) bis Oberitalien (Pignatti, 1957) bestätigt und reicht von W-Europa (Büker, 1942; Lacourt, 1977) bis nach O-Polen und zur Slowakei. Bisher fehlende Nachweise aus einigen Regionen (England, Böhmen, Österreich, Jütland) dürften auf derzeitige Kenntnislücken zurückzuführen sein. Nach Rivas-Martinez, Izco & Costa (1971) kommt die Assoziation selbst noch in N-Spanien (Leon) vor *(Bunico–Vicietum tetraspermi)*. Zwar verbergen sich diese Belege nicht selten unter sehr verschiedenen Namen, doch sprechen Arten wie *Papaver argemone*, *Veronica triphyllos*, *Holosteum umbellatum*, *Vicia tetrasperma* neben *Aphanes arvensis*, *Scleranthus annuus* usw. auch in N-Spanien für die Zugehörigkeit zum *Papaveretum argemones*.

Kleinstandörtliche Untergliederung

Geringfügige edaphisch-ökologische Verschiedenheiten verändern zwar nicht die bezeichnende Artenverbindung der Assoziation wohl aber können sie diese partiell bereichern – vom zentralen Typus aus gesehen – indem einige gesellschaftsfremde Arten als Trennarten bestimmter Sonderformen eindringen. Die Differenzialarten greifen von räumlich und ökologisch benachbarten Vegetationseinheiten über und dokumentieren zu

diesen vermittelnde Übergangsausbildungen, die syntaxonomisch heute als Subassoziation (evt. Subass.-Gruppe), Variante, Subvariante ausgewiesen werden. Ökologisch handelt es sich um kleinstandörtliche Modifikationen im Trophie-, Wasser-, seltener auch Wärmehaushalt des Bodens, wobei letztere z.T. schon zu den klimaabhängigen, ± regional-geographischen Abwandlungen überleiten. Für die Ranghöhe dieser Untereinheiten sollte der Differenzierungsgrad (relative Zahl der Trennarten), aber auch die regionale Gültigkeit ausschlaggebend sein. So zeichnet sich die Subassoziation gegenüber der Variante normalerweise durch mehr bzw. konstantere Trennarten aus, und ihr Geltungsbereich erstreckt sich über das gesamte Areal der Regional-Assoziation. Ein weiteres Merkmal (objektiver Syntaxonomie) ergibt sich aus dem Umstand, daß die Differenzialarten der Subassoziationen stets nur aus gleichwertigen benachbarten Struktureinheiten kommen, die fast nie zum selben Verband gehören wie die Vegetationseinheit, in die sie partiell differenzierend eindringen. Varianten-Trennarten können sich demgegenüber haüfiger aus anderen als strukturgleichen Kontakteinheiten rekrutieren. Zum Normalfall gehört schließlich noch die Dreigliederung, die neben dem zentralen Typus auf allen Rangebenen gegenüber erstem artenreichere Ausbildungen mit einem Mehr oder Weniger (reich–arm, feucht–trocken usw.) der antagonistischen Fremdeinflüsse registriert. Als wichtige Forderung bleibt für die Assoziation als Grundeinheit im System eine einheitliche Untergliederung, deren Rangstufen nur jeweils einen variablen Faktoreneinfluß zum Ausdruck bringen. Neben Trophie-Subassoziationen (reich–arm) kann es somit keine Wasserhaushalt-Subassoziation, sondern nur Feuchte-Varianten bzw. umgekehrt geben.

Während sich die Originalbeschreibungen von Libbert (1932), Kruseman & Vlieger (1939) auf das Herausarbeiten des neuen Assoziationstypus beschränkten, macht Sissingh (1946, 1950) erste Vorschläge zur Untergliederung. Vom *Papaveretum typicum* Siss. 46 grenzt er das *Papaveretum juncetosum* Siss. 46 ab mit den Differenzialarten: *Juncus bufonius, Gnaphalium uliginosum, Galium aparine, Cerastium holosteoides* und *Ranunculus repens.* Letzteres siedelt auf wechselfeuchten Standorten und ist merklich artenreicher als der Typus. Vanden Berghen (1951) bestätigt diese Untergliederung für Flandern und fügt ein anspruchsvolleres *Papaveretum alopecuretosum* Vanden Berghen 51 mit den Trennarten *Alopecurus myosurioides* und *Ranunculus arvensis* hinzu.

Aus heutiger Sicht ist die *Juncus*-Subassoziation von Sissingh wohl nur als Variante bzw. Subvariante zu werten, denn sie vermittelt zu den nicht strukturgleichen *Isoëto-Nanojuncetea*.

Dagegen weist die *Alopecurus*-Subassoziation zur anspruchsvolleren Halmfrucht-Assoziation *Alopecuro-Matricarietum chamomillae* (Wasscher 41) Meisel 67. Von Tüxen (1937, 1950) wurde das *Papaveretum* zunächst in die weit gefaßte *Alchemilla-Matricaria*-Ass. Tx. 37 mit einbezogen, bis Oberdorfer (1957) bzw. Passarge (1957a, b) die Eigenständigkeit des *Papaveretum argemones* erkannten und gegenüber einem enger gefaßten *Aphano-Matricarietum* Tx. 37 em. Pass. 57 herausstellten. Die Auswertung meines ostelbischen Materials ergab seinerzeit neben der typischen Subassoziation ein *Papaveretum scleranthetosum* Pass. 57 mit den säurefesten Trennarten *Scleranthus annuus, Spergula arvensis, Rumex acetosella, Raphanus raphanistrum, (Setaria viridis, Erodium cicutarium)* und ein reicheres *Papaveretum delphinietosum* Pass. 57 mit den Trennarten *Consolida regalis, Galium aparine, Veronica persica, Euphorbia helioscopia (Sinapis arvensis, Thlaspi arvense, Papaver rhoeas, Valerianella dentata)* – jeweils mit typischen, *Mentha arvensis*- und *Conyza*-Varianten. Zu einem ähnlichen Gliederungsvorschlag kam schon Trentepohl (1956). Sein reichhaltiges Material vom *Raphanetum trifolietosum* bei Darmstadt – nachträglich von Oberdorfer (1957) als zum *Papaveretum* gehörig erkannt – ergab dort *Rumex acetosella*-, typische und *Delphinium*-Varianten neben typischen und *Juncus*-Subvarianten. Auch an sehr viel Material (jeweils über 300 Aufnahmen aus verschiedenen Gebieten) bestätigten Passarge (1964) für das ostelbische Tiefland bzw. Meisel (1967) für W- und NW-Deutschland die großräumige Geltung dieser Untereinheiten.

(Zur Umbenennung der anspruchsvolleren Ausbildung in *Euphorbia exigua*-Subass. (in der Tabelle) bzw. *Atriplex patula*-Subass. im Text (!) bei Vorhandensein von *Delphinium* als Trennart besteht kein Anlaß.) Dabei weisen die angeführten Subass.-Trennarten zu den ökologisch benachbarten *Arnoseridion*- bzw. *Triticion*-Ass. Für das *Papaveretum gageetosum pratensis* Wojcik 65 gilt dies allerdings nicht, denn die Trennarten *Gagea pratensis, Allium vineale, Ornithogalum* sind Elemente der strukturfremden Frühlings-Ephemerenfluren des *Gageo-Allion* Pass. 64. Auch die gegenüber dem dortigen *Papaveretum typicum* zu konstatierende relative Artenarmut der *Gagea*-Subass. mit im Mittel 15,5 Arten gegenüber durchschnittlich 22 Arten beim Typus spricht gegen eine Einstufung als Subassoziation. Bei teilweise regional variablen Trennarten gliedert sich das *Papaveretum argemones* somit einheitlich in Trophie-Subass. und Wasserhaushalt-Varianten/-Subvarianten.

Zur Abgrenzung von Nachbareinheiten und zum Arealgrenzverhalten

Nur im Zentrum seines Areals, etwa dem binnenländischen Tiefland zwischen Weser und Oder-Warthe, ist das *Papaveretum* fast alleinige *Aphanion*-Ass. (Tab. 1) und siedelt hier ziemlich bodenvag auf reinen Sanden, lehmigen Sanden,

Tabelle 1. Anteil des *Papaveretum argemones* (b) an *Aphanion*-Aufnahmen (a) in verschiedenen Arealbereichen (1–3) und Gebieten (jeweils von N nach S geordnet)

Gebiet	(Autor, Jahr)	Aufnahmezahl	
		a : b	b (%)
1. Westlicher Bereich			
NW-Deutschland	(Hofmeister, 1970)	96 : 9	9
Niederlande	(Kruseman & Vlieger, 1939)	38 : 9	24
	(Sissingh, 1950)	40 : 20	50
Belgien	(Vanden Berghen, 1951)	22 : 14	64
NW-Schweiz	(Brun-Hool, 1963)	91 : 24	26
N-Italien	(Pignatti, 1957)	45 : 30	67
2. Zentraler Bereich			
NW-Mecklenburg	(Passarge, 1962)	25 : 11	44
NO-Mecklenburg	(Passarge, 1959a)	19 : 16	84
M-Mecklenburg	(Passarge, 1963b)	22 : 19	86
S-Mecklenburg	(Krausch & Zabel, 1965)	50 : 50	100
N-Havelland	(Passarge, 1957)	14 : 14	100
Spreeland	(Klemm, 1970)	61 : 61	100
O-Spreewald	(Passarge, 1959b)	25 : 25	100
Mittelelbe	(Jage, 1972)	487 : 221	45
Franken	(Nezadal, 1975)	303 : 153	51
O-Bayern	(Rodi, 1966)	46 : 21	46
Württemberg	(Rodi, 1961)	33 : 7	21
3. Östlicher Bereich			
N-Polen	(Passarge, 1963a)	30 : 12	40
M-Polen	(Wojcik, 1965)	109 : 63	58
SO-Polen	(Sicinski et al., 1978)	433 : 156	38
N-Slowakei	(Passarge & Jurko, 1975)	42 : 3	7

Lehmen und ebenso auf Anmoorböden. In dem weit größeren, dieses Kerngebiet umgebenden Arealbereich steht die Assoziation im Kontakt mit verwandten *Aphanion*-Einheiten. Trotz vermittelnder Ausbildungen (Übergänge) fällt es i.d.R. jedoch nicht schwer, das *Papaveretum* gegenüber diesen ± sicher abzugrenzen. Das gilt für die westlichen *Legousietum speculi-veneris* (Krusem. et Vlieg. 39) Siss. (46) 50 und *Alopecuro-Matricarietum* (Wasscher 41) Meisel 67 (meist anspruchsvoller) ebenso wie für das *Aphano-Matricarietum* Tx. 37 em. Pass. 57 oder die östlichen *Consolido-Brometum* (Denissow 30) Tx. et Prsg. 50, *Vicietum tetraspermi* Krusem. et Vlieg. 39 em. Kornas 50 und *Herniario-Polycnemetum* Fijalk. 67. Im Überlappungsbereich des *Papaveretum argemones* mit den genannten *Aphanion*-Ass. kann man häufig ein edaphisches Alternieren feststellen. So etwa, wenn das *Papaveretum* nur auf betont durchlässigen Böden gedeiht, während z.B. das *Aphano-Matricarietum* ausgesprochen staufrische Lehme bevorzugt. Im großen gesehen zeigen diese Trophie-verwandten Assoziationen ein vom temperat-mitteleuropäischen *Papaveretum* merklich abweichendes Areal und können mit gewisser Einschränkung als vikariierende Einheiten (mit erheblicher Arealüberlappung) angesehen werden.

Die äußere Verbreitungsgrenze der Assoziation markieren fast immer Ausbildungen mit deutlich verminderter Zahl an diagnostisch wichtigen Arten. So sinkt in Beispielaufnahmen von der jütländischen Halbinsel nach Norden hin nicht nur die Artenzahl im *Papaveretum*, sondern es fallen noch weit vor der absoluten Arealgrenze so wichtige Arten wie *Papaver argemone*, *Erophila verna*, *Arenaria serpyllifolia*, *Odontites verna* aus. Dennoch besteht an der *Papaveretum*-Zugehörigkeit der bisher nördlichsten Belegaufnahmen bei Hanstholm (nördlich des 57° n.Br.) mit *Papaver dubium*, *Arabidopsis*, *Myosotis stricta* kein Zweifel. Erst die *Viola tricolor-Myosotis arvensis*-Ges. an der N-Spitze Jütlands ist frei von *Papaveretum*-Charakter, obgleich die Bestände ähnlich artenreich wie die ca. 350 km weiter südlich anzutreffenden Normalausbildungen des *Papaveretum* sind. Recht augenfällig ist übrigens die sich nach N hin mit abnehmender Sommerwärme verringernde Halmlänge des Winterroggens (Tab. 3).

Syngeographische Gliederung

Alle Vegetationseinheiten – soweit sie nicht unter Extrembedingungen (z.B. *Lolio-Plantaginetum*) leben – zeigen bei Arealabmessungen (N-S bzw. O-W) von über 1000 km, den unterschiedlichen Klimabedingungen entsprechend merkliche Abwandlungen. Sie kommen teils als blosse Schwerpunktverlagerungen bei einzelnen Taxa (Mengen- oder Stetigkeitszu- bzw. -abnahme), teils im Ausfall bzw. neu Hinzutreten gewisser Arten zum Ausdruck. Soweit es sich hierbei nicht nur um lokale, sondern um großräumig gültige Erscheinungen handelt, die sich in bestimmter Einflußrichtung gezielt verändern, darf man sie als syngeographisch bedingt ansprechen. In der Regel gehen mit derartigen regionalen Veränderungen solche in den Klimabedingungen (Temperatur, Luftfeuchtigkeit, Niederschläge) einher.

Die spezifischen Zeigerpflanzen dieser Einflüsse lassen ein gegenüber dem Arealtypenspektrum der Assoziation abweichendes Arealverhalten erkennen. Diese syngeographischen Differenzialarten greifen gewissermaßen als arealfremde Florenelemente partiell in Bereiche der Assoziation ein, diese als spezielle Regionalausbildungen kennzeichnend. Ähnlich den niederen Hilfsrängen im edaphisch-ökologischen Bereich (Subassoziation, Variante) können wir auch bei den syngeographisch-klimatischen Abwandlungen verschiedene Rangstufen unterscheiden. Abermals sollten dabei der Differenzierungsgrad und die regionale Geltung (groß- oder kleinräumig) Maßstab für die Ranghöhe sein. Mein Vorschlag (Passarge, 1964) ist, die Rasse einer geographischen Variante und die Vikariante (Rassengruppe) der Subassoziation analog zu bewerten. Erst bei namhaften Abwandlungen sollte von vikariierender Regional-Assoziation gesprochen werden.

Ähnlich wie bei der edaphischen Untergliederung gibt es unterschiedliche syngeographische Einflußrichtungen. Formal sprechen wir von zonaler Vikarianz, wenn sich zwei Ausbildungen in vergleichbarer Höhenlage regional ausschließen, wohingegen der Begriff der vertikalen Vikarianz den Höhenstufen-Unterschieden (Höhenformen) vorbehalten bleibt (Holub & Jirasek, 1967). Der Montanstufe nächst verwandte Klimabedingungen können sich allerdings auch zonal in Richtung N wiederholen und etwa einheitlich durch boreal-

Tabelle 2. Syngeographische Gliederung und grossräumige Konstanz diagnostisch wichtiger ...

Spalte	a	b	c	d	e	f	g	h	i	k	l	m	n	o	p	q	r	s	t	u
Aufnahmezahl	30	5	18	6	20	9	24	80	9	24	9	7	9	14	148	18	8	63	26	65
Papaver argemone	5	3	3	5	5	5	2	2	5	1	2	3	5	2	2	5	3	4	4	5
Papaver dubium	5	2	.	5	5	4	.	2	5	5	.	.	3	4	2	5	4	2	3	2
Veronica triphyllos	.	1	0	2	4	5	5	5	.	.	2	3	1	4	2	5	3	5	4	5
Vicia villosa	1	1	.	.	.	.	.	0	1	.	1	.	1	4	2	1	1	2	.	.
Apera spica-venti	5	1	5	4	5	5	5	5	5	3	4	5	5	5	4	5	4	4	5	4
Veronica hederifolia	2	3	4	3	5	5	5	5	3	2	2	4	1	4	3	5	4	1	4	3
Arenaria serpyllifolia	3	1	.	4	4	5	3	3	2	3	3	4	2	1	2	1	4	4	3	2
Aphanes arvensis	1	4	4	4	.	5	5	5	2	1	4	3	2	2	2	1	2	.	2	3
Myosotis stricta	.	1	.	.	.	.	.	3	.	.	3	.	2	1	2	.	4	5	4	3
Trifolium arvense	0	1	.	.	1	.	.	.	.	.	3	3	1	4	1	.	2	0	2	1
Lithospermum arvense	.	.	.	d	.	.	.	d	.	1	.	1	2	1	.	4	4	4	4	4
Rhinanthus s. apterus	.	.	.	.	.	.	.	.	.	.	.	.	3	.	.	.	.	0	2	1
Veronica dilleni	.	.	.	.	.	.	.	.	.	.	.	.	.	.	2	2	.	2	1	2
Gagea pratensis	.	.	1	1	.	.	0	.	.	.	.	.	.	.	2	.	.	2	2	2
Camelina pilosa	.	.	.	.	.	.	.	.	.	.	.	.	.	.	.	.	1	0	.	1
Galeopsis tetrahit	.	.	.	.	.	.	1	.	2+	2+	5	5	4	1	1	.	1	.	.	.
Holcus mollis	.	.	.	.	0	.	.	.	.	.	2	3	1	1	.	.	.	.	.	.
Lapsana communis	.	1	.	.	0	.	.	.	.	1	3	.	3	.	1	.	.	.	.	.
Erysimum cheiranthoides	.	.	.	.	0	.	.	.	2	.	.	.	3	.	1	.	.	.	.	.
Galeopsis bifida	.	.	.	.	1	.	.	.	.	.	2	.	3	.	.	.	1	.	.	.
Aethusa cynapium	.	.	.	1	.	.	0	.	.	2	2	.	.	.	.	.	.	.	.	.
Sherardia arvensis	.	.	.	.	.	.	.	.	.	2	2	1	.	.	.	.	.	.	.	.
Stellaria graminea	.	.	.	.	.	.	.	.	.	2	2	.	.	.	.	.	.	.	.	.
Equisetum sylvaticum	.	.	.	.	.	.	.	.	.	.	1	.	2	.	.	.	.	.	.	.
Chamomilla recutita	3	3	5	3	4	3	3	3	3	0	2	.	.	1	2	.	.	.	.	.
Poa annua	4	1	5	5	4	5	5	5	5	5	1	.	.	.	1	.	2	.	.	.
Myosotis discolor	.	.	4	3	3	1	.	0	.	.	.	1	.	.	.	.	.	.	.	.
Legousia speculum-veneris	2	1	.	.	0	2	.	0	.	1	.	.	.	.	.	.	.	.	.	.

Herkunft der Aufnahmen: a. N-Italien (Pignatti, 1957), b. Frankreich (Lacourt, 1977). c. Belgien (Vanden Berghen, 1951), d.-f. Niederlande (Weevers, 1940, Sissingh & Tideman, 1960 d; Sissingh, 1950 e; Kruseman & Vliegers, 1939 f); g.-H. Niederrhein (Meisel, 1973 g; Wedeck, 1971 h); i. NW-Deutschland (Hofmeister, 1970), k. Schweiz (Brun-Hool, 1963); l. Bayern (Vollrath, 1966), m. Württemberg (Rodi., 1961), n. N-Polen (Passarge, 1963). o. SW-Deutschland (Oberdorfer, 1957); p. Mittelelbe (Jage, 1972), q. Spreewald (Passarge, 1959); r. S-Schweden (Merker, 1959); s. Zentral-Polen (Wojcik, 1965); t.-u. SO-Polen (Wiesniewski, 1968 t; Warlochinska, 1974 u).

Chorologische Syntaxa:

1. *Chamomille–Papaverenetum argemones* (Krusem. et Vlieg. 39) reg. ass. nov. (Spalte a-k).
1.1. Normalvikariante (Spalte b-h).
1.2. *Galeopsis*-Vikariante (Spalte i-k, Material nur teilweise dazugehörig).
2. *Myosotido strictae–Papaverenetum* (Libb. 32) reg. ass. nov. (Spalte l-u).
2.1. Normalvikariante (Spalte o-u).
2.2. *Galeopsis*-Vikariante (Spalte l-n).

montane Florenelemente zum Ausdruck gebracht werden. Die in Mitteleuropa registrierbaren syngeographischen Besonderheiten dokumentieren meist Florenelemente unterschiedlicher Kontinentalität (ozeanisch-kontinental) und/oder Borealität (mediterran-boreal/arktisch bzw. planar-montan/alpin). Dem jeweiligen Ausprägungsgrad entsprechend steht hierbei der eine oder andere Regionaleinfluß im Vordergrund und ist dementsprechend syntaxonomisch übergeordnet zu bewerten.

Die bisher beim *Papaveretum* ermittelten syngeographischen Differenzen ergaben eine subatlantische Rasse mit *Myosotis discolor (versicolor), Legousia, Hypochoeris glabra* und eine subkontinentale mit *Myosotis stricta (micrantha), Lithospermum arvense, Erysimum cheiranthoides* (Passarge, 1957b), in der Folge auch *Myosotis versicolor-* bzw. *Lithospermum-Rasse* (Passarge, 1964) genannt. Die stärker eigenständige Form in Oberitalien (*Alchemillo-Matricarietum papaveretosum* Pignatti 57) wurde als submediterrane *Lolium temulentum*-Vikariante dem *Papaveretum argemones* zugerechnet. Zu ihren syngeographischen Trennarten zählen *Lolium temulentum, Oxalis corniculata, Bromus gussonii, Bunias orientalis* mit submediterranmediterraner Hauptverbreitung, dazu *Cerastium glomeratum.* Letztere verbindet sie mit der spanischen *Bunias*-Vikariante mit ihren mediterran-kontinentalen Elementen: *Bunias erucago, Anchusa azurea, Chondrilla juncea, Poa bulbosa, Filago arvensis.* Weitere Untersuchungen müssen erweisen, ob den mediterranen *Papaveretum*-Ausbildungen die noch grössere Selbständigkeit einer vikariierenden Assoziation (*Bunio-Vicietum tetraspermae* Rivas-Martinez et al. 71) zukäme.

Eine erneute Zusammenstellung des inzwischen sehr umfangreichen Materials aus dem temperaten Hauptverbreitungsgebiet läßt weitere Differenzen und Schwerpunktverlagerungen erkennen. Nach zunehmender Kontinentalität geordnet, folgt auf die westeuropäische *Myosotis discolor*-Rasse (Tab. 2, c-e) eine Ausbildung mit *(Matricaria) Chamomilla* (Tab. 2, f-k), eine *Myosotis stricta*-Form (Tab. 2, l-p) und schließlich die subkontinentale *Lithospermum*-Rasse (Tab. 2, q-u). Unabhängig von dieser W–O Gliederung treten noch subboreal-submontane *Galeopsis*-Ausbildungen (Tab. 2, i-n) in Erscheinung.

Zur syntaxonomischen Bewertung

Versuchen wir, diese syngeographischen Abwandlungen ihrer Bedeutung nach hierarchisch zu werten, so scheint der Teilung in eine westliche *Chamomilla*-reiche Rassengruppe (Tab. 2, a-k) und eine östliche *Myosotis stricta*-Rassengruppe Vorrang zuzukommen. Erstere zeichnet sich durch etwa konstante *Chamomilla* und *Poa annua,* dazu *Myosotis discolor* und *Legousia* aus. Hinzu kommen zahlreiche Trennarten der anspruchsvolleren Ausbildung, die sich wie *Alopecurus myosuroides, Ranunculus arvensis, Avena fatua, Euphorbia exigua, Sherardia arvensis, Silene noctiflorum, Kickxia elatine* etwa auf den ozeanischen Klimabereich beschränken. Sporadische Vorkommen von *Lithospermum arvense* bleiben hier eng an die reichere Subassoziation gebunden.

Sehr verbreitet sind weiter die Krumenfeuchtezeiger der *Gna-*

phalium-Gruppe. In diesem Rahmen differenzieren im unteren Bergland (Brun-Hool, 1963; Richard, 1975) *Galeopsis tetrahit, Lapsana, Vicia cracca, Viola tricolor, Aethusa cynapium,* in Küstennähe auch *Galeopsis segetum, Erysimum cheiranthoides* (Hofmeister, 1970) eine *Galeopsis*-Vikariante (Tab. 2, i, k) von der sonst verbreiteten typischen Vikariante (Tab. 2, a-h). Unter den Lokalrassen sind die westeuropäische *Myosotis discolor*-Rasse (Kruseman & Vlieger, 1939; Sissingh, 1950; Vanden Berghen, 1951) und eine zur folgenden Gruppe vermittelnde *Myosotis stricta*-Rasse (Wedeck, 1971, 1972) zu erwähnen. Selbst die submediterrane *Lolium temulentum*-Ausbildung aus der westlichen Poebene (Pignatti, 1957) läßt sich zwanglos noch dieser *Chamomilla*-Rassengruppe anschließen (Tab. 2a). Arealmäßig reicht die westeuropäische Form bis NW-Italien, W-Schweiz und im nördlichen Tiefland bis ins Wesergebiet.

Die östliche *Myosotis stricta*-Rassengruppe (Tab. 2, l-u) zeichnet sich durch *Myosotis stricta, Trifolium arvense, Lithospermum arvense, Rhinanthus serotinus* ssp. *apterus, Gagea pratense, Veronica dilleni, Conyza canadensis, Camelina microcarpa/pilosa, Descurainia sophia, (Vicia villosa) V. dasycarpa* aus, vom weitgehenden Fehlen der für die *Chamomilla*-Rassengruppe bezeichnenden Arten abgesehen. Wichtigste Trennart der anspruchsvolleren Subassoziation ist meist *Consolida regalis* dazu *Sinapis arvensis, Galium aparine, Valerianella dentata* und regional unterschiedlich *Veronica persica, Lamium purpureum, Thlaspi arvense* und *Euphorbia helioscopia.*

Grundfeuchte Böden tragen eine *Mentha arvensis*-Variante, krumenfeuchte *Gnaphalium*-Subvarianten sind im Binnenland recht selten. Betont trockene Standorte werden durch die *Conyza*-Variante auch mit *Setaria* und *Digitaria*-Arten markiert. In diesem Rahmen hebt sich gegenüber der temperat-mitteleuropäischen Normalvikariante eine subboreal-hochkolline *Galeopsis*-Vikariante ab mit *Galeopsis tetrahit, G. bifida, Lapsana communis, Equisetum sylvaticum, Viola tricolor, Stellaria graminea* und *Erysimum cheiranthoides* (Rodi, 1961; Vollrath, 1966; Sychova, 1959; Passarge, 1963). Ein Beispiel dafür, daß lokal auch kleinstandörtliche Besonderheiten wie erhöhter Grundwassereinfluß bei der *Mentha*-Variante mesoklimatisch *Galeopsis, Erysimum* usw. begünstigen können, geben Aufnahmen von Wnuk (1976). In beiden Vikarianten gibt es Rassen, so die jütländische *Viola tricolor*-Rasse (ohne *Centaurea cyanus, Anagallis, Matricaria inodora*), die subozeanische *Chamomilla*-Rasse, die subkontinentalen *Vicia dasycarpa-, Lithospermum*-Rassen (Passarge, 1957, 1964; Krausch & Zabel, 1965; Jage, 1972; Nezadal, 1975) bzw. die pannonische mit *Polycnemum arvense* (Passarge & Jurko, 1975; Fijalkowski, 1978).

Das Areal der mitteleuropäisch gemäßigt-kontinentalen *Myosotis stricta*-Rassengruppe reicht von der O-Grenze der Assoziation im Weichselraum bis zu den Trockengebieten an Oberrhein und Main, dem küstenfernen niedersächsischen Tiefland zwischen Weser und Elbe sowie nach S-Skandinavien.

Für die taxonomische Bewertung der hochrangigen syngeographischen Differenzen scheint zweierlei wichtig: (1) das *Papaveretum argemones* ist bei mittlerer Artenzahl um 25 (20–30) eine gut abgrenzbare Vegetationseinheit mit hoher floristischer Konstanz über weite Räume hinweg; (2) es gliedert sich in zwei sich arealmäßig etwa ausschließende zonal-vikariierende Rassengruppen, deren syngeographische Differenzen zwar floristisch kaum 20% der mittleren Artenzahl ausmachen, coenologisch aber durch andersartige Untergliederung (*Alopecurus*- bzw. *Delphinium*-Subass.) verstärkt werden. Es scheint daher gerechtfertigt, hierbei von zwei vikariierenden Regional-Assoziationen zu sprechen.

Wenig befriedigend ist allerdings die bisherige Nomenklatur, macht sie doch keinen Unterschied zwischen etwa alternierenden Grundeinheiten der Vegetation (z.B. *Papaveretum argemones* und *Aphano–Matricarietum*) und vikariierenden Assoziationen bzw. Regional-Assoziationen innerhalb der gleichen Grundeinheit. Braun-Blanquet bewertete derartige Unterschiede als Subassoziationen (Moravec, 1975). Heute bleibt diese Rangstufe nur den edaphisch-kleinstandörtlichen Variationen vorbehalten. Doch fehlt der Syntaxonomie meines Erachtens ein Pendant zur vikariierenden Subspezies in der Taxonomie, mit dem sich die Regional-Assoziation als eine Art vikariierender 'Unter-Assoziation' auch nomenklatorisch eigenständig kennzeichnen läßt. In Anlehnung an die von Barkman, Moravec & Rauschert (1976) im Code vorgeschlagenen höheren Hilfsrangstufen (z.B. -enion für den Unterverband) möchte ich für die neue Rangstufe der vikariierenden oder Regional-Assoziationen folgerichtig die Endung '-enetum' vorschlagen. Das *Papaveretum argemones* (Libb. 32) Krusem. et Vlieg. 39 gliedert sich ein in:

1. westliches *Chamomillo–Papaverenetum argemones* (Krusem. et Vlieg. 39) reg. ass. nov. mit den Subass. *typicum* Siss. (46) 50, *alopecuretosum* Vanden Berghen 51, ? *scleranthetosum* sowie typischen und *Juncus bufonius*-(Sub-)Varianten im Einflußbereich eines wintermilden, relativ niederschlagreichen Großklimas (Jahresmittel 9–12 °C, Januar ±0 bis +3 °C; 700–1000 mm Jahresniederschlag) und ein

2. östliches *Myosotido strictae–Papaverenetum argemones* (Libb. 32) reg. ass. nov. mit den Subass. *typicum* Pass. 57, *delphinietosum* Pass. 57, *scleranthetosum* Pass. 57 sowie typischen, *Mentha arvensis*- und *Conyza*-Varianten und regional seltenen *Gnaphalium*-Subvarianten unter temperaten bis gemäßigt-kontinentalen Klimabedingungen (Jahresmittel 7–10 °C; Januar +1 bis –2 °C; Jahresniederschläge 500–750 mm).

In beiden Teilarealen scheinen kühl- bis warmgemäßigte Sommertemperaturen mit Juli-Mitteln zwischen 16–20 °C (außerhalb der Mediterran-Region) vorzuherrschen.

Im Verbreitungszentrum der Assoziation (*Papaveretum*-Anteil im *Aphanion* über 80%), dem binnenländischen Tiefland zwischen Weser (extrazonal auch im Mainzer Trockengebiet) und Oder (Warthe), bewegen sich die Temperaturmittel im Jahresdurchschnitt zwischen 8–9,5 °C (Januar +0,5 bis –1,5 °C; Juli 17–18,5 °C) bei Niederschlagsummen zwischen 550–700 mm.

Literatur

Barkman, J. J., Moravec, J. & Rauschert, S., 1976. Code der Pflanzensoziologischen Nomenklatur. Vegetatio 32: 131–185.

Brun-Hool, J., 1963. Ackerunkrautgesellschaften der Nordwestschweiz. Beitr. geobot. Landesaufn. Schweiz 43.

Büker, R., 1942. Halmfrucht-Gesellschaft. 11. Rundbr. Zentralst. Vegetationskart. Wiss. Mitt. Hannover (Polykopie).

Ellenberg, H., 1974. Zeigerwerte der Gefäßpflanzen Mitteleuropas. Scripta Geobot. 9.

Fijalkowski, D., 1978. Synantropy roslinne Lubelszczyzny. Lub. Towarz. nauk. Warszawa-Lodz. 260 pp.

Grosser, K. H., 1967. Studien zur Vegetations- und Landschaftskunde als Grundlage für die Territorialplanung. Abh. Ber. Naturkundemus. Görlitz. 42: 1–95.

Haffner, P., 1960. Pflanzensoziologische und Pflanzengeographische Untersuchungen im Muschelkalkgebiet des Saarlandes. Natursch. u. Landschaftspfl. im Saarland 2: 66–164.

Hanf, F., 1937. Pflanzengesellschaften des Ackerbodens. Pflanzenbau. 13/14: 449–476.

Hilbig, W., 1973. Übersicht über die Pflanzengesellschaften des südlichen Teiles der DDR. VII. Die Unkrautvegetation der Äcker, Gärten und Weinberge. Hercynia N.F. 10: 394–428.

Hofmeister, H., 1970. Pflanzengesellschaften der Weserniederung oberhalb Bremens. Diss. Bot. 10.

Holub, J. & Jirásek, V., 1967. Zur Vereinheitlichung der Terminologie in der Phytogeographie. Folia Geobot. Phytotax. 2: 69–113.

Holzner, W., 1973. Die Ackerunkrautvegetation Niederösterreichs. Mitt. Bot. 5: 1–157.

Jage, H., 1972. Ackerunkrautgesellschaften der Dübener Heide und des Flämings. Hercynia N.F. 9: 317–391.

Klemm, G., 1970. Die Pflanzengesellschaften des nordöstlichen Unterspreewald-Randgebietes. Verh. Bot. Ver. Prov. Brandenburg 107: 3–28.

Kloss, K., 1960. Ackerunkrautgesellschaften der Umgebung von Greifswald (Ostmecklenburg). Mitt. Flor.-soziol. Arbeitsgem. N.F. 8: 148–164.

Krausch, H. D. & Zabel, E., 1965. Die Ackerunkraut-Gesellschaften in der Umgebung von Templin/Uckermark. Wiss. Z. Päd. Hochsch. Potsdam. Math.-Nat. 9: 369–388.

Kruseman, G. & Vlieger, J., 1939. Akkerassociaties in Nederland. Nederl. Kruidk. Arch. 49: 327–398.

Lacourt, J., 1977. Essai de synthèse sur les syntaxons commensaux des cultures d'Europa. Thèse Univ. Paris-S, Orsay. 149 pp.

Langendonck, H., 1935. Étude sur la flore et la végétation des environs de Gand. Bull. Soc. Roy. Bot. Belg. 67(18) 1: 117–180.

144

Libbert, W., 1932. Die Vegetationseinheiten der neumärkischen Staubeckenlandschaft I. Verh. Bot. Ver. Prov. Brandenburg 74: 10–93.

Malato-Beliz, J., Tüxen, J. & Tüxen, R., 1960. Zur Systematik der Unkrautgesellschaften der west- und mitteleuropäischen Wintergetreide-Felder. Mitt. Flor.-soziol. Arbeitsgem. N.F. 8: 145–147.

Meisel, K., 1967. Über die Artenverbindung des Aphanion arvensis J. et R.Tx. 1960 im west- und nordwestdeutschen Flachland. Schriftenr. Vegetationskde. 2: 123–133.

Meisel, K., 1973. Ackerunkrautgesellschaften. In: W. Trautmann (ed.), Vegetationskarte der Bundesrepublik Deutschland 1: 200 000 Bl. Nr. CC 5502 Köln. pp. 4–57. Bonn-Bad Godesberg.

Merker, H., 1959. Bestandesaufnahme der Ackerunkrautvegetation in einigen westschonischen Gemeinden 1958. Bot. Notis. 112: 134–157.

Militzer, M., 1970. Die Ackerunkräuter in der Oberlausitz Teil II. Die Ackerunkrautgesellschaften. Abh. Ber. Naturkundemus. Görlitz. 45 9: 1–43.

Moravec, J., 1975. Die Untereinheiten der Assoziation. Beitr. Naturkdl. Forsch. SW-Deutschl. 34: 225–232.

Nezadal, W., 1975. Ackerunkrautgesellschaften Nordostbayerns. Hoppea. Denkschr. Regensb. Bot. Ges. 34: 17–149.

Oberdorfer, E., 1957. Das Papaveretum argemone, eine für Süddeutschland neue Getreide-Unkrautgesellschaft. Beitr. naturkdl. Forsch. SW-Deutschl. 16: 47–51.

Oberdorfer, E., 1979. Pflanzensoziologische Exkursionsflora, 4. Aufl. Stuttgart. 997 pp.

Passarge, H., 1957a. Vegetationskundliche Untersuchungen in der Wiesenlandschaft des nördlichen Havellandes. Feddes Repert. Beih. 137: 5–55.

Passarge, H., 1957b. Zur geographischen Gliederung der Agrostidion spica-venti-Gesellschaften im nordostdeutschen Flachland. Phyton 7: 22–31.

Passarge, H., 1963. Beobachtungen über Pflanzengesellschaften landwirtschaftlicher Nutzflächen im nördlichen Polen. Feddes. Repert. Beih. 140: 27–69.

Passarge H., 1964. Pflanzengesellschaften des nordostdeutschen Flachlandes. Pflanzensoziol. 13: 1–268.

Passarge, H. & Jurko, A., 1975. Über Ackerunkrautgesellschaften im nordslowakischen Bergland. Folia Geobot. Phytotax. 10: 225–264.

Pignatti, S., 1957. La vegetazione messicola della colture di Frumento, Segale e Avena nella provincia di Pavia. Arch. Bot. Biogeogr. Ital. 5, 33 4. Ser. 2: 3–79.

Raabe, E. W., 1944. Über Pflanzengesellschaften der Umgebung von Wolgast in Pommern. Arb. Zenstralst. Vegetationskart. Arb. Zentralst. Veget. Kart. 14. Rundbr.

Richard, J.-L., 1975. Les groupement végétaux du Clos du Doubs (Jura Suisse). Beitr. Geobot. Landesaufn. Schweiz 57: 1–71.

Rivas-Martinez, S., Iczo, J. & Costa, M., 1971. Sobre la flora y la vegetación del Macizo de Pena Ubina. Trab. Dep. Bot. Veg. 3: 47–123.

Rodi, D., 1961. Die Vegetations- und Standortsgliederung im Einzugsgebiet der Lein (Kreis Schwäbisch-Gmünd). Veröff. Landest. Naturschutz u. Landschaftspfl. Baden-Württ. 27/28: 76–163.

Rodi, D., 1966. Ackerunkrautgesellschaften und Böden des westlichen Tertiär-Hügellandes. Denkschr. Regensbg. Bot. Ges. 26: 161–198.

Rodi, D., 1967. Die Sandmohnflur (Papaveretum argemone (Libb. 32) Krusem. et Vlieg. 39) des Sandäcker der Tertiär-Hügellandes (Oberbayern). Mitt. Flor.-soziol. Arbeitsgem. N.F. 11/12: 203–205.

Schwickerath, M., 1944. Das Hohe Venn und seine Randgebiete. Pflanzensoziol. 6: 1–278.

Sicinski, J., 1974. Segetal communities of the Szczercowska (Widawska) Depression. Acta Agrobot. 27(2): 5–93.

Sicinski, J., Sowa, R., Warcholinska, A. H., Wiesniewski, J. & Wnuk, Z., 1978. Zroznicowanie florystyczno-ekologiczne zbiorowisk segetalnych w srodkowej Polsce. Mater. Kraj. Konfer. 1978. Lodz.

Sissingh, G., 1950. Onkruid-Associaties in Nederland. 's-Gravenhage. 224 pp.

Sissingh, G. & Tideman, P., 1960. De Plantengemeenschappen uit de Omgeving van Didam en Zevenaar. Med. Landbouwhogesch. Wageningen 60(13): 1–30.

Sychowa, M. 1959. Phänologie des Blühens und Fruchtens einiger Ackergesellschaften in Kostrze bei Kraków. Fragm. Flor. Geobot. 5: 245–280.

Tillich, H.-J., 1969. Die Ackerunkrautgesellschaften in der Umgebung von Potsdam. Wiss. Z. Pädag. Hochsch. 13: 273–320.

Trentepohl, H., 1956. Ackerunkrautgesellschaften westlich von Darmstadt. Schr. Naturschutzst. 3: 151–206.

Tüxen, R., 1937. Die Pflanzengesellschaften Nordwestdeutschlands. Mitt. Flor.-soziol. Arbeitsgem. Niedersachsen 3: 1–170.

Tüxen, R., 1950. Grundriß einer Systematik der nitrophilen Unkrautgesellschaften in der Eurosibirischen Region Europas. Mitt. Flor. soziol. Arbeitsgem. N.F. 2: 94–175.

Tüxen, R., 1954. Pflanzengesellschaften und Grundwasserganglinien. Angew. Pflanzensoziol. 8: 64–98.

Vanden Berghen, C., 1951. Aperçu sur la végétation de la région située a l'ouest de Gand. Bull. soc. Roy. Bot. Belg. 83: 283–316.

Vollrath, H., 1966. Über Ackerunkrautgesellschaften in Ostbayern. Denkschr. Regensburg. Bot. Ges. 26: 117–158.

Warcholinska, A. U., 1974. Communities of segetal weeds of the Piotrków Plain. Acta Agrobot. 27(2): 95–193.

Wedeck, H., 1971. Über das Papaveretum argemones (Libb. 32) Krusem. et Vlieg. 39 in der Niederrheinischen Bucht. Decheniana 123: 19–25.

Wedeck, H., 1972. Über Vorkommen und soziologische Bindung von Montia verna Neck. im Raum Zülpich (Spermatophyta, Portulacaceae). Dechaniana 125: 141–145.

Weevers, Th., 1940. De flora van Goeree en Overflakkee dynamisch beschouwd. Nederl. Kruidkd. Arch. 50: 285–354.

Wnuk, Z., 1976. Segetal weed communities of Przedborz-Malogoszcz Chain and adjacent areas. Acta Univ. Lodz. Fol. Bot. 2 14: 85–177.

Wojcik, Z., 1965. Les associations des champs cultivés en Masovie I. Ekol. Polska. A 13: 641–682.

Wollert, H., 1965. Die Unkrautgesellschaften der Oser Mittelmecklenburg. Arch. Nat. Meckl. 11: 85–101.

Accepted 17.2.1984.

Floristic relationship between plant communities of corresponding habitats in southeast Greenland and alpine Scandinavia*

F. J. A. Daniëls**
Department of Vegetation Science and Botanical Ecology, State University Utrecht, Lange Nieuwstraat 106, 3512 PN Utrecht, The Netherlands

Keywords: Chorology, Corresponding habitat, Floristic similarity, Geographical race, Plant community, Regional association, Scandinavia, Southeast Greenland, Syntaxonomical evaluation

Abstract

Barkman's similarity coefficiënts have been calculated for twelve ecologically related communities of southeast Greenland (SEG) and alpine Scandinavia (SCA). Comparisons were made between corresponding saxicolous lichen communities, dwarf shrub communities, snow bed communities and herb and *Salix* shrub communities. The corresponding SEG and SCA communities of extreme habitats have the same faithful taxa or the same dominant taxa, relatively few or no area-differential (ArD) taxa and they are floristically strongly related. They should be classified in one single association; the geographical variation is expressed in terms of geographical races. Corresponding vegetation types of mesic habitats have low floristic similarity coefficiënts, many ArD taxa and the same dominant taxa, or different faithful taxa. The geographical variation should be expressed here on the association level.

Introduction

A syntaxonomical study is presented of the shrub, dwarf shrub and lichen vegetation of southeast Greenland (SEG), which is low arctic and (sub)-oceanic in climate. Floristically this region is narrowly related to alpine Scandinavia (SCA), but has an impoverished vascular flora (Daniëls, 1975, 1980).

Floristic impoverishment usually implies considerable difficulties in a classification of vegetation upon a floristic basis. In order to achieve a robust classification I have made a number of comparisons with previously classified vegetation types of ecologically and floristically related regions, such as alpine Scandinavia and to a lesser degree the alpine belt of the Alps (ALP). In doing so I more and more got the impression, that vegetation types of corresponding, extreme habitats show a convergence in floristics, while those of corresponding, more mesic habitats show a divergence in their floristic composition. As a matter of fact, such phenomena have already been noticed before.

Du Rietz (1924) emphasized the floristic similarity between the ombrotraphent bog vegetation of the Alps and Scandinavia, and the uniformity of many halophytic communities, such as the *Puccinellietum phryganodis* Hadač 46 of the arctic shores (cf. Thannheiser, 1975) is well known.

Apart from these incidental observations I thought it interesting to check this phenomenon of similarity more precisely for the terrestric, low arctic–northern alpine vegetation in general, closely comparing the community floristics. In this way I hope to present more precise evidence on similarity and at the same time to show general trends in the syntaxonomical evaluation of the classificatory problems involved.

* Nomenclature follows Daniëls (1980), Hultén (1968) and Lid (1963).
**I thank Marinus J. A. Werger for valuable comments on the manuscript.

Vegetatio 59, 145–150 (1985).
© Dr W. Junk Publishers, Dordrecht. Printed in the Netherlands.

Table 1. Comparisons between 12 ecologically corresponding plant communities of southeast Greenland (SEG) and alpine Scandinavia (SCA).

Community pairs		Name of the vegetation types with literature source	Vegetation type	Number of records	Number of taxa	Number of taxa in common	% taxa in common	Number of AfD taxa	% AfD taxa	x same faithful taxa / ● different faithful taxa	x same dominant taxa	Similarity coefficient	▲ same; △ different association
A.	SEG	Umbilicarietum cylindricae, Daniëls (1975, tab. 2: 1–7)	achionophytic saxicolous lichen vegetation	7	9	7	78	0	0	x	x	1.9	▲
	SCA	Umbilicarietum proboscideo-hyperboreae, Creveld (1981, tab. IX: 1–17)		17	19	7	37	3	16				
B.	SEG	Drepanocladus exannulatus soc. complex, de Molenaar (1976, tab. VIII: 1–12)	moss mire vegetation	12	6	3	50	0	0	x		1.2	▲
	SCA	Drepanocladetum exannulati, Dahl (1957, tab. 42: 107–542)		10	7	3	43	2	29				
C.	SEG	Cassiopo–Anthelietum juratzkanae, de Molenaar (1976, tab. XII: 1–12)	late snow bed vegetation	12	16	10	62	0	0	x	x	0.9	▲
	SCA	Anthelietum juratzkanae, Gjaerevoll (1956, tab. 33: 1–X, 34: 1–IV)		16	19	10	53	0	0				
D.	SEG	Empetreto–Betuletum nanae typicum, Daniëls (1980, tab. 12: 16–19)	achionophytic dwarf shrub vegetation	4	43	19	44	2	5	x	x	0.9	▲
	SCA	Cetrarietum nivalis typicum, Dahl (1957, tab. 16: 3–19)		12	46	19	41	4	9				
E.	SEG	Mniobryo–Epilobietum hornemannii, de Molenaar (1976, tab. III: 5–12)	spring vegetation	8	12	4	33	0	0	x		0.7	▲
	SCA	Mniobryo–Epilobietum hornemannii, Dahl (1957, tab. 40a: 1–610)		12	9	4	44	0	0				
F.	SEG	Pleuroclado–Polytrichetum norvegicum typicum, de Molenaar (1976, tab. XII: 13–16)	late snow bed vegetation	4	15	7	47	0	0	x		0.7	▲
	SCA	Polytrichum norvegicum soc., Gjaerevoll (1956, tab. 31: 1–VII)		7	19	7	37	0	0				
G.	SEG	Parmelietum omphalodis sphaerophoretosum, Daniëls (1975, tab. 3: 1–7)	chionophytic saxicolous lichen vegetation	7	13	7	54	0	0	●		0.5	△
	SCA	Parmelietum omphalodo-saxatilis, Creveld (1981, tab. XI: 10–36)		26	28	7	25	6	21				
H.	SEG	Sphagno–Salicetum callicarpaeae, Daniëls (1980, tab. 4: 1–6)	dwarf shrub mire vegetation	6	18	12	67	1	6	●		0.5	△
	SCA	Aulacomnio–Sphagnetum warnstorfii, Dahl (1957, tab. 46: 369–605)		7	57	12	21	17	30				
I.	SEG	Phyllodoco–Salicetum callicarpaeae, Daniëls (1980, tab. 8: 1–63)	chionophytic dwarf shrub vegetation	63	28	11	39	2	7	●		0.4	△
	SCA	Phyllodoco–Vaccinietum myrtilli, Nordhagen (1943, tab. 15)		–	37	11	30	10	27				
J.	SEG	Alchemilletum alpinae, de Molenaar (1976, tab. XVI: 1–13)	chionophytic herb vegetation	13	20	7	35	1	5	x		0.4	△
	SCA	Alchemilletum alpinae, Dahl (1957, tab. 35: 362–568)		5	37	7	19	5	14				
K.	SEG	Caricetum bigelowii, de Molenaar (1976, tab XXIX: 1–22)	chionophytic grass vegetation	22	18	7	39	1	6	x		0.3	△
	SCA	Polytricho–Caricetum bigelowii, Dahl (1957, tab. 31: 78–501)		22	44	7	16	8	16				
L.	SEG	Festuco–Salicetum callicarpaeae, Daniëls (1980, tab. 7: 1–25)	shrub vegetation	25	21	4	19	4	19	●		0.1	△
	SCA	Rumiceto–Salicetum lapponae, Dahl (1957, tab. 38: 36–613b)		13	37	4	11	14	38				

Methods

Twelve plant communities of low arctic southeast Greenland have been compared with those of alpine Scandinavia, which occur in ecologically corresponding habitats (Table 1). Each pair shows a corresponding physiognomy. The Greenlandic communities were described by Daniëls (1975, 1980) and de Molenaar (1976); the Scandinavian communities by Creveld (1981), Dahl (1957), Gjaerevoll (1956) and Nordhagen (1943).

Floristic similarities between the corresponding communities have been calculated with Barkman's similarity coefficiënt (Barkman, 1958; and Fig. 1).

Taxa with presence I or less were left out of consideration. The presence classes II, III, IV and V and 1, 2, 3 and 4 were transformed in the presence percentages 30, 50, 70 and 90. Moreover, the faithful and differential taxa of the corresponding vegetation types have been compared. Two types of differential taxa have been distinguished: (a) area-differential (ArD) taxa which differentiate between two vegetation types because they are absent in one of the types apparently because they are absent from the area of one type for historical or macro-climatological reasons; (b) habitat-differential (HabD) taxa, which occur in the general distribution areas of both vegetation types, but which are restricted to one of the vegetation types as a result of edaphic factors.

To determine the status of ArD taxa the following authors were consulted: Dahl (1950), Poelt (1969), Hansen (1978), Lid (1963), Böcher et al. (1978) and Hultén (1968).

Results

Communities with the highest similarity coefficiënts (1.9–0.7) occur in extreme habitats (community pairs A/F, Table 1, Fig. 2). These habitats are extreme in being exceptionally cool, wet, dry, or experiencing severe solifluction. Usually this extremeness is related to the snow cover on the site, being either very long, or very short to even absent in wintertime.

Communities with low floristic similarity coefficiënts (0.5–0.1) occur under more moderate environmental conditions (community pairs G/L, Table 1, Fig. 2).

Communities of extreme habitats have relatively few ArD taxa, namely 16% of the A/F communities of SEG, 50% of the A/F communities of SCA (Table 1).

Communities of mesic habitats have more ArD taxa: 83% of the G/L communities of SEG, 100% of the G/L communities of SCA (Table 1).

Communities of extreme habitats (community pairs A/F) have the same faithful (x), or the same dominant (x) taxa (Table 1).

Communities of mesic habitats (community pairs G/L) have the same dominant (x) taxa, or different faithful taxa (●) (see Table 1).

Syntaxonomical evaluation

The syntaxonomical evaluation of the phenomena depends on the association concept that is held. Here I understand an association as defined by W. & A. Matuszkiewicz (1973).

A. Communities A/F of the more extreme habitats (see Table 1) show high floristic similarity coefficiënts (1.9–0.7), have the same faithful or the same dominant taxa and have no or few ArD taxa.

In the community pairs C, E and F (snow bed and

Community	A	B		A	B		α	β	γ
Species a	V	II		90	30		60	.	30
Species b	III	IV		50	70		.	20	50
Species c	II	.	2	30	.	3	30	.	.
Species d	.	III		.	50		.	50	.
Species e	.	II		.	30		.	30	.
Species f	I	.		.	.		.	.	.
Species g	+	.		.	.		.	.	.
Species h	r	. .		.	.		.	.	.
							90 Σα	100 Σβ	80 Σγ

Fig. 1. Example of the calculation of similarity between communities A and B. 1. Species constancy tables are compared, low constancy species f to h are disregarded; 2. conversion of constancy values in calculation values; 3. α lists the surplus values for all species with a higher presence in community A; and β similar for community B, while γ lists the common values for species common to A and B. The similarity coefficiënt is calculated as

$$S = \frac{\Sigma\gamma}{\sqrt{\Sigma\alpha \times \Sigma\beta}},$$

in this case $S = \dfrac{80}{\sqrt{90 \times 100}} = \dfrac{80}{95} = 0.8$

148

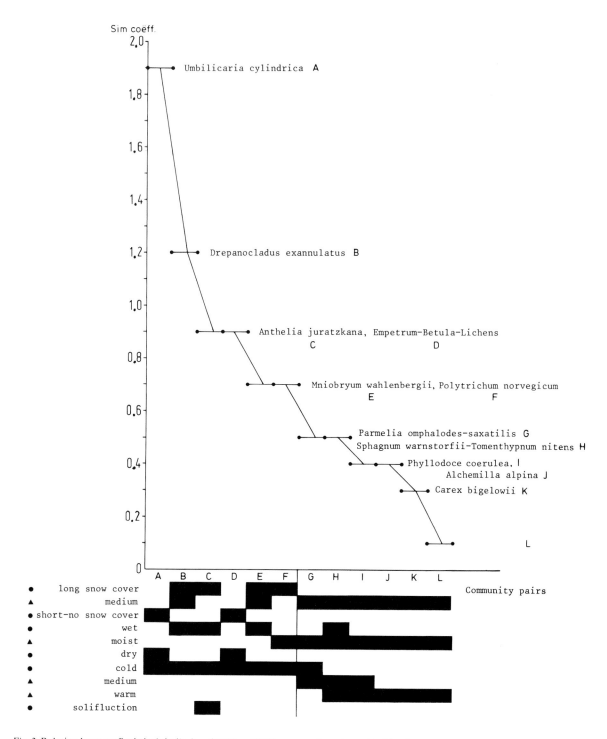

Fig. 2. Relation between floristical similarity of SEG and SCA community pairs and habitat factors. For each pair the prominent species are listed.

spring vegetation) the floristic similarity coefficiënts are relatively high (0.9–0.7). The SEG and SCA communities of this group have the same faithful taxa, resp. *Anthelia juratzkana* (C), *Mniobryum wahlenbergii* (E) and *Polytrichum norvegicum* (F) and there are no ArD taxa.

As a consequence the SEG and SCA communities of each community pair belong to the same association. The floristical differences between the SEG and SCA communities might be expressed in terms of subassociations, which in this case exclude each other also geographically.

In the community pairs A and B (achionophytic saxicolous lichen communities, moss mire communities) the floristic similarity coefficiënts are high (1.9–1.2) and they share the same faithful taxa, *Umbilicaria cylindrica, U. hyperborea, U. torrefacta,* and *U. proboscoidea* (A) or the same dominant taxon, *Drepanocladus exannulatus* (B). Only the SCA communities have ArD taxa: *Umbilicaria polyphylla, Hypogymnia intestiniformis* and *Parmelia stygia* for A and *Salix lapponum* and *Eriophorum angustifolium* for B.

As a consequence the SEG and SCA communities of each community pair belong to the same association. The geographical variation is expressed in geographical races. We distinguish a SEG race (without ArD taxa) and a SCA race (with ArD taxa).

In community pair D (achionophytic lichen-rich dwarf shrub communities) the floristic similarity between the SEG and SCA communities is 0.9. *Betula nana* might be considered as a weakly preferential faithful taxon. Both communities have ArD taxa: *Festuca brachyphylla* and *Salix callicarpaea* for SEG, and *Calluna vulgaris, Vaccinium vitis-idaea, V. myrtillus* and *Cladina alpestris* for SCA. As a consequence we consider both communities to the same association, but as two different geographical races characterized by the ArD taxa mentioned before.

B. Communities G/L of mesic habitats (see Table 1) show low similarity coefficiënts (0.5–0.1), nearly all have ArD taxa and they have the same or different faithful taxa or the same dominant taxa.

In community pair G (chionophytic saxicolous lichen communities) the floristic similarity coefficiënt is rather low (0.5). The SEG and SCA community each have different faithful taxa, resp. *Parmelia omphalodes* for SEG, and *P. saxatilis* and *P. sulcata* for SCA. The SCA community only has ArD taxa: *Umbilicaria polyphylla, U. cinereorufescens, Hypogymnia physodes, Parmelia stygia, P. centrifuga* and *Coelocaulon divergens.*

Consequently both communities belong to different associations. In community pairs J and K (herb and grass communities) the floristic similarity coefficiënts are low (0.3–0.4). The corresponding SEG and SCA communities have the same dominant taxa (faithful taxa are absent), *Alchemilla alpina* (J) and *Carex bigelowii* (K), and both SEG and SCA communities have ArD taxa: *Campanula gieseckiana* for SEG group J and K, *Rumex acetosa, Anthoxanthum odoratum, Solidago virgaurea, Campanula rotundifolia* and *Festuca ovina* for SCA group J, and *Vaccinium vitis-idaea, Festuca ovina, Rubus chamaemorus, Rumex acetosa, Solidago virgaurea, Trientalis europaea, Anthoxanthum odoratum* and *Cladina alpestris* for SCA, group K. I am of the opinion that separate regional associations should be distinguished here, because of the lack of true faithful taxa, the great number of ArD taxa, the additional HabD taxa, and the low floristic similarity coefficiënts.

In community pairs H, I and L (dwarf shrub mire, chionophytic dwarf shrub heath and shrub communities, the floristic similarity between the SEG and SCA communities is low (0.5–0.1), whereas both communities have ArD taxa and different faithful taxa. For SEG, group H these faithful taxa are *Sphagnum warnstorfianum* and *Tomenthypnum nitens,* for SCA, group H this is *Pedicularis oederi;* for SEG, group I these are *Phyllodoce coerulea* and *Diphasium alpinum* and for SCA, group I it is *Vaccinium myrtillus;* for SEG, group L they are *Salix callicarpaea, Festuca rubra* and *Hieracium hyparcticum* and finally for SCA, group L it is *Salix lapponum.* ArD taxa are *Salix callicarpaea* for SEG, group H and *Andromeda polifolia, Salix lanata, S. lapponum, S. myrsinites, S. reticulata, Carex dioica, C. fusca, C. vaginata, Eriophorum angustifolium, E. vaginatum, Festuca ovina, Coeloglossum viride, Pedicularis oederi, Anthoxanthum odoratum, Rubus chamaemorus, Saussurea alpina, Equisetum fluviatile* for SCA, group H; *Salix callicarpaea* and *Coptis trifolia* for SEG, group I and *Cladina sylvatica, Vaccinium myrtillus, V. vitis-idaea, Festuca ovina, Solidago virgaurea, Trientalis europaea, Cornus suecica, Pedicularis lapponica, Anthoxanthum odoratum* and *Rumex acetosa* for SCA, group I; *Hieracium hyparcticum, Salix callicarpaea, Campanula gieseckiana, Coptis trifolia* for SEG, group L, and *Salix lapponum, Vaccinium myrtillus, V. vitis-idaea, Deschampsia cespitosa, Festuca ovina, Geranium silvaticum, Melandrium rubrum, Pedicularis lapponica, Rubus chamaemorus, Rumex acetosa, Solidago virgaurea, Campanula rotundifolia, Anthoxanthum odoratum* and *Trientalis europaea* for SCA, group L. Thus, the differences are of such a magnitude that all communities of the three pairs should be considered as separate regional associations. Community pair L consists of two truly vicarious associations, as *Salix callicarpaea* and *S. lapponum* exclude each other geographically (cf. Hultén, 1968).

Concluding remarks

In stating these conclusions some reservation should be maintained for the following reasons. The number of communities compared and the number of records used is rather small. Furthermore, different investigators have sampled and phytosociologically evaluated the data, while the present selection of the vegetation types to be compared was more or less subjective. Keeping this in mind, I conclude that the SEG and SCA communities of corresponding, but extreme environments show strong floristic relationships, while those of similar but mesic sites are floristically more different in particular in the many ArD taxa. These taxa mainly belong to the relatively rich subarctic, boreal, temperate or montane floras and can survive in low arctic southeast Greenland and at the higher altitudes of alpine Scandinavia in relatively mesic sites only. They strongly contribute to the geographical differences between the related communities, however, so that the geographical differences imply phytosociological differences at the association level.

In extreme environments we find no or few ArD taxa. The phytosociological variation resulting from geographical differences in the corresponding communities is expressed in geographical races, which is below the association level. Thus, the phytosociological similarity between southeast Greenland and alpine Scandinavia is most pronounced in extreme environments, and here the same associations are found.

References

Barkman, J. J., 1958. Phytosociology and Ecology of Cryptogamic Epiphytes. van Gorcum, Assen. 628 pp.

Böcher, T., Fredskild, B., Holmen, K. & Jakobsen, K., 1978. Grønlands Flora. Haase & Søns Forlag, Kobenhavn. 327 pp.

Creveld, M., 1981. Epilithic lichen communities in the alpine zone of Southern Norway. Bibliotheca Lichenologica 17: 1–287.

Dahl, E., 1950. Studies in the Macrolichen Flora of South West Greenland. Medd. om Grønl. 150 (2): 1–176.

Dahl, E., 1957. Rondane, Mountain Vegetation in south Norway and its relation to the environment. Skr. Norske Vidensk. Akad. Oslo I; Mat. Naturv. Klasse 1956 (3): 1–374.

Daniëls, F. J. A., 1975. Vegetation of the Angmagssalik District, Southeast Greenland. III. Epilithic macrolichen communities. Medd. om Grønl. 198 (3): 1–32.

Daniëls, F. J. A., 1980. Vegetation of the Angmagssalik District, Southeast Greenland. IV. Shrub, dwarf shrub and terricolous lichen vegetation. Thesis State University of Utrecht, The Netherlands.

Du Rietz, G., 1924. Studien über die Vegetation der Alpen, mit derjenigen Skandinaviens verglichen. Veröff. Geobot. Inst. Rübel. 1: 31–138.

Gjaerevoll, O., 1956. The Plant Communities of the Scandinavian Alpine Snow-Beds. Det. Kgl. Norske Vidensk. Selsk. Skr. 1956 (1): 1–405.

Hansen, E. S., 1978. Notes on occurrence and distribution of lichens in Southeast Greenland. Medd. om Grønl. 204 (4): 1–77.

Hultén, E., 1968. Flora of Alaska and Neighboring Territorities. Stanford University Press, Stanford, California. 1008 pp.

Lid, J., 1963. Norsk og Svensk Flora. Det Norske Samlaget, Oslo. 88 pp.

Matuszkiewicz, W. & Matuszkiewicz, A., 1973. Prezeglad Fitosocjologiczny Zbiorowisk Lesnych Polski. Cz.1. Lasy bukowe. Phytocoenosis 2: 143–202.

Molenaar, J. de, 1976. Vegetation of the Angmagssalik District, Southeast Greenland. II. Herb and Snow-bed Vegetation. Medd. om Grønl. 198 (2): 1–266.

Nordhagen, R.,1943. Sikilsdalen og Norges fjellbeiter. En Plantensociologisk monographi. Berg. Mus. Skr. 22: 1–607.

Poelt, J., 1969. Bestimmungsschlüssel europäischer Flechten. Cramer, Lehre. 757 pp.

Thannheiser, D., 1975. Beobachtungen zur Küstenvegetation auf dem westlichen kanadischen Arktis-Archipel. Polarforschung 45 (1): 1–16.

Accepted 19.1.1984.

Corresponding *Caricion bicolori-atrofuscae* communities in western Greenland, northern Europe and the central European mountains*

K. Dierßen & Barbara Dierßen**
Botanical Institute, the University, Olshausenstr. 40, 2300 Kiel, F.R.G.

Keywords; Calcareous fen, *Caricion bicolori-atrofuscae, Caricion davallianae,* Geographical race, Glacial refugium, Phytosociology, Synecology

Abstract

The alliance *Caricion bicolori-atrofuscae* comprises plant communities of more or less wet calcareous fens on shallow peat with a high base saturation, occurring in arctic as well as alpine, more seldom subalpine areas. The characteristic species are presented and compared with those from communities of the *Caricion davallianae* growing in lowland and mountain sites of temperate and boreal calcareous fens.

An abbreviated ecological characteristic is given and the contact communities of the different community complexes are pointed out. The regional floristic differences within the communities are shown, and the geographical distribution and regional frequency of the differentiated associations are given. The synsystematic ranks of three central European mountain to alpine associations are discussed in detail (*Caricetum frigidae, Juncetum alpini, Equiseto–Typhetum minimi*). They should be excluded from this alliance.

Generally, the alpine phytocoenoses are more often invaded by species from neighbouring communities, presumably as an effect of competition in sites less favouring fen development in comparison to the oro-arctic and arctic sites, especially those with an oceanic to suboceanic climate.

We conclude, that the observed floristic differences are mainly caused by historical processes and not by the different ecology of the sites in question. This must be taken into consideration when discussing the synsystematic delimitation and rank of the corresponding communities.

Introduction

The vegetation of flushed shallow fens around springs in the alpine belt, of small terraces subjected to solifluction in the surroundings of glaciers and inital peat deposits are of interest both for the ecology and distribution history of plantspecies and their communities. Within the characteristic plant communities of these sites, such species participate as 'key species', which are valuated as 'glacial relicts' in central Europe or as 'overwintering' species during the pleistocene glaciation in boreal and arctic refugial areas. These species are generally characterized by their boreo-arctic–alpine disjunction, some of them also by disjunctions within their boreo-arctic main area, being concentrated near their possible refugial areas. In some cases, their populations may have been isolated since the last glaciation, and have not only built up different ecotypes, but also split into morphological differentiated taxa.

It seems of interest to examine the way of ecological niche formation within those long-isolated populations and to point out the degree of floristic differentiation of the corresponding isolated vegetation types.

* Nomenclature follows: Lid (1974), Nyholm (1954–68).

** *Acknowledgement.* The authors are grateful to the 'Deutsche Forschungsgemeinschaft' for travel grants since 1973 for their investigations in Scandinavia and Greenland.

Caricion bicolori - atrofuscae		Caricion davallianae
CH V	*Juncus triglumis*	*Carex flava* s. str.
	Oncophorus virens	*Carex lepidocarpa*
	Oncophorus wahlenbergii	*Carex hostiana*
	Juncus biglumis	*Eriophorum latifolium*
	Juncus castaneus	*Tofieldia calyculata*
	Meesia uliginosa	*Epipactis palustris*
	Fissidens osmundoides	*Dactylorhiza incarnata*
	Salix myrsinites	*Swertia perennis*
	Carex microglochin and others	*Taraxacum paludosum*
		Primula farinosa and others
DV	*Carex capillaris*	*Carex panicea*
	Tomentypnum nitens	*Trichophorum cespitosum* ssp. *cespitosum*
	Saxifraga aizoides	
	Salix reticulata	

O, but preferential in one of the alliances:

Equisetum variegatum	*Trichophorum alpinum*
Thalictrum alpinum	*Campylium elodes*
Bryum pallens	
Tofieldia pusilla	

Caricetalia davallianae

Campylium stellatum
Riccardia pinguis
Pinguicula vulgaris
Selaginella selaginoides
Juncus alpino-articulatus agg.
Carex dioica
Triglochin palustre
Eleocharis quinqueflora
Odontoschisma elongatum
Fissidens adianthoides and others

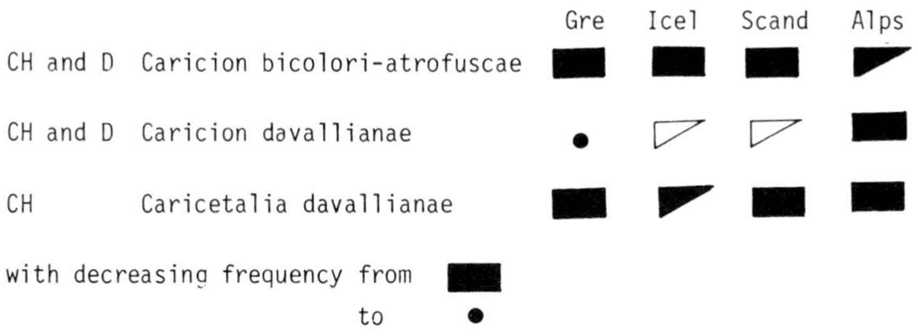

Fig. 1. Floristic similarities and differences between *Caricion bicolori-atrofuscae* and *Caricion davallianae*.

Syntaxonomic delimitation

The floristic differences between the *Caricion bicolori-atrofuscae* and the floristically closely related *Caricion davallianae* as well as the similarities on the order level are given in Figure 1. Some bryophytes, e.g. *Oncophorus* species (to some extent also occurring in calcareous snow fields), *Meesia uliginosa* and *Fissidens osmundoides*, although being most frequent in the *Caricion bicolori-atrofuscae*, also occur in the *Caricion davallianae*. A number of differential species in comparison to the *Caricion davallianae* in Scandinavia and arctic regions overlap with contact communities of the *Carici rupestris–Kobresietea*, for instance *Carex capillaris, Kobresia simpliciuscula, Pedicularis flammea, Thalictrum alpinum, Tofieldia pusilla*. On the other hand, species of the *Carici rupestris–Kobresietea* from less dry sites occasionally grow in associations of the *Caricion bicolori-atrofuscae*, especially Rhododendron lapponicum, Carex scirpoidea, Carex misandra and *Tomentypnum nitens* as a common bryophyte. The delimitation in the field is often complicated. Regionally, the *Kobresietum simpliciusculae* may have an intermediate position between the two classes *Scheuchzerio–Caricetea nigrae* and *Carici rupestris–Kobresietea*.

It may be difficult to establish the boundary towards the *Caricion davallianae* communities in the alpine belt, especially in boreal and arctic areas, where species of the *Caricion davallianae* generally drop out. This seems a problem, too, in the alpine belt of the Alps. In the latter case, the sites with *Caricion davallianae* communities are smaller and more incomplete than in the mountain and subalpine belt. They are more often subjected to activities of erosion and frost action and more or less mixed with *Caricion bicolori-atrofuscae* species, which act as differential species in the alpine belt. As a consequence, the limits between the two alliances are not given by altitudinal belts, but mainly by local edaphic factors.

As a rule, the *Caricion bicolori-atrofuscae* sites are more often open than those of the *Caricion davallianae*. Pioneer species like *Juncus alpino-articulatus* agg., *Eleocharis quinqueflora*, *Triglochin palustre* and *Equisetum variegatum* with a low competitive power therefore are more numerous in, but by no means restricted to the former alliance.

The syntaxonomic delimitation against wetter mesotrophic fen communities is difficult. The *Calliergono–Caricetum saxatilis* for instance has a somewhat intermediate position between the *Caricion nigrae* and the *Caricion bicolori-atrofuscae*. This is also true for the *Carex holostoma* community, whose phytosociological position is uncertain according to Scandinavian data (i.e. Ryvarden, 1969: p. 30). In Greenland, however, it is more closely connected with calcareous substrates and seasonal fluctuations of soil humidity. It is thus enriched with characteristic species of the *Caricion bicolori-atrofuscae*.

Ecological characteristics

According to the relict character of the faithful species, the characteristic sites are extreme with respect to soil and microclimate. Cold water is constantly present in the root horizon and restricts the ion uptake. A high amount of oxygen in the soil water as well as low temperatures during the short vegetation period prevent a remarkable peat accumulation. Solifluction and cryoturbation lead to a disruption of plant roots and soil surface structures.

The cation content of the soil water is comparable with that of *Caricion davallianae* communities; the pH varies between about 5.0 and 8.0; mean pH often lies around 7.0. The cation exchange capacity is normally lower than 50 m.e./100 g organic matter, the mean percentage of base saturation in most cases higher than 75%. The amount of organic matter may change considerably in different sites. In the main root horizon, however, it generally exceeds 50% (Conzelmann, 1979). The habitat differences, i.e. in comparison with the floristically allied syntaxa of the *Caricion davallianae* are primarily microclimatic. They result in the following physical soil conditions: shallow peat deposits with a relatively low content of organic matter, a relatively high percolation and frequent solifluction. In comparison with the communities of the *Carici rupestris–Kobresietea*, the water capacity and water content of the soils are high. With a decreasing content of organic matter, silt and clay by erosion processes, the soil movement may decrease, too, caused by the absence of soil water. The characteristic species of the *Carici rupestris–Kobresietea* require a considerable frost hardiness because of the low amount of snow.

The microhabitats of different communities within the *Caricion bicolori-atrofuscae* mainly differ in soil water content, content of organic matter, texture and degree of frost action. In fairly wet solifluction soils, these parameters can change rapidly within short distances. Especially the wet sites in arctic areas often show a differentiated and diverse micropattern with fairly well-delimited vegetation units.

Contact communities

The connections with possible contact communities are demonstrated in Figure 2. The thickness of the arrows symbolizes the degree of floristic similarity. In most of the contact syntaxa, the vegetation cover is more closed than in the *Caricion bicolori-atrofuscae*, the soil mobility is less pronounced and soil building processes are more advanced except in eroded areas. Hence, the 'infiltration' by

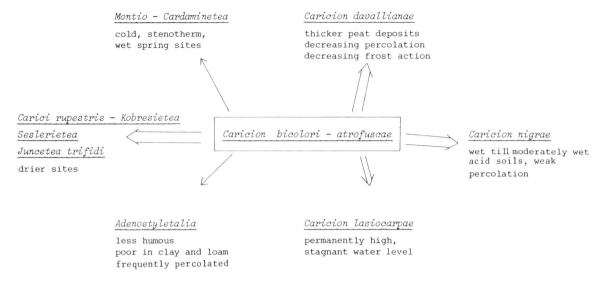

Fig. 2. The *Caricion bicolori-atrofuscae* and its possible contact communities.

species from the contact communities seems an expression of an edaphic stabilization in the direction sketched. Furthermore, the increase of 'foreign influences' seems more pronounced in the Alps and probably must be interpreted as an expression of the diminished cryoturbation as compared to boreal and arctic areas.

On the other hand, the continuous colonization of the sites by the *Caricion bicolori-atrofuscae* communities suggests the presence of a disturbance factor, for instance an increasing solifluction, the dislocation of alluvial soils or a weak sand accumulation in coastal areas, for instance in habitats with the *Carici maritimae–Juncetum baltici*.

Phytogeographical differentiation

As stated above, most of the species and communities in question can be classified as glacial relicts (in the Alps) or as glacial survivors (in boreal and oro-arctic to arctic areas). Of course, this counts also for other species and plant communities with a comparable distribution (i.e. *Salicetea herbaceae, Carici rupestris-Kobresietea*), but not or less so for other alliances within the class *Scheuchzerio-Caricetea nigrae*.

Because of the shallow peat deposits and the high degree of peat mineralization, the direct – palynological – proof of the relict character only seems

possible in exceptional cases. On the other hand, the existence of different ecotypes in isolated populations, which can be verified by ecophysiological tests, makes a long separation plausible. For many species, the distance between the Alps and northern Europe may have been too large to be bridged during the last glacial period. Also on hardly accessible isles like Iceland a substantial enrichment of the flora by species reimmigrating from Greenland, Norway or Scotland seems improbable. To summarize, the syngenesis of *Caricion bicolori-atrofuscae* communities in different areas of northern Europe, Greenland and the Alps probably took place independently. Since potential sites are fairly small and genetical exchange between them difficult, the floristic similarities are surprising.

A synopsis of the regional differential species with a higher constancy is given in Table 1. For the sites in Greenland, especially *Salix arctophila* and *Stellaria longipes* are characteristic; both species have a wide sociological range. Icelandic sites have two species in common with Greenland ones, which are both absent from Fennoscandia: *Salix glauca* ssp. *callicarpaea* and *Lomatogonium rotatum*. Only the latter has its main distribution in one association of the alliance: within the *Carici maritimae-Juncetum baltici*. Its only localities in Fennoscandia are isolated on the Fischer peninsula and Kola on the Barents Sea coast.

In the Scandinavian mountains, *Salix myrsinites*

Table 1. Differential species with a regionally high (>40%) frequency.

			Greenland	Iceland	Scandinavia	Alpes
		Salix arctophila	±	.	.	.
		Stellaria longipes	±	.	.	.
		Salix glauca ssp. callicarpaea	+	+	.	.
CH	V	Primula stricta	(+)	(+)	+	.
CH		Lomatogonium rotatum	+	+	(+)	.
CH	V	Pedicularis flammea	+	.	+	.
CH		Carex parallela	+	.	+	.
CH	V	Salix myrsinites	.	.	±	.
CH	V	Salix arbuscula	.	.	±	.
D	V	Euphrasia frigida	+	+	+	.
	O	Calamagrostis neglecta	+	+	+	.
		Carex bigelowii	+	+	+	.
CH	V	Minuartia stricta	+	.	+	.
CH		Koenigia islandica	+	+	+	.
CH		Juncus balticus[a]	.	+	+	.
	O	Primula farinosa	.	.	.	±
	O	Carex davalliana	.	.	.	±
		Sesleria varia	.	.	.	±
		Salix foetida	.	.	.	±
		Aster bellidiastrum	.	.	.	±
	O	Carex flavella	.	.	.	±

[a] The differences between *Juncus arcticus* and *J. balticus* are fairly weak in areas, where populations of both taxa occur in the same habitat, in the present case in lowland areas of northern Iceland and near the Norwegian Barents Sea coast.

+ = only isolated 'relic' sites.

and *Salix arbuscula* are faithful alliance character species. The *Caricion bicolori-atrofuscae* stands of Greenland, Iceland and Fennoscandia have some species in common, which are absent in central Europe. Generally, the communities in Iceland are relatively poor in character and differential species because of the deviating edaphic conditions and the poor accessibility after the last glaciation.

The alpine communities contain a number of species, which only reach the southernmost mountain areas of northern Europe. Some of them are characteristic of the order *Caricetalia davallianae* of the class *Scheuchzerio–Caricetea nigrae*, others have their main distribution within other classes. No species are restricted to the alliance in the Alps.

Our list of differentiating species is still incomplete. We only used those taxa with higher constancy, which are of some diagnostic value. The number of those accidental species with a low constancy may be high in some communities.

A synopsis of the syntaxa of the alliance is given in Figure 3. Two of them (1-2) are almost entirely restricted to Greenland and other arctic regions.

Three have only been described for the Alps or for central European mountains (12–14). Five of the associations in Greenland and northern Europe have corresponding communities in the Alps (3-5, 8, 9). Some of them have already been described as separate associations. The boreo-arctic communities from comparable sites with identical key species should be separated on the geographical race level only.

The attachment of the central European alpine associations of the *Caricion bicolori-atrofuscae* seems somewhat difficult.

The *Caricetum frigidae* contains hardly any characteristic species of the alliance. Apart from (alliance) differential species *Saxifraga aizoides* (restricted to alpine sites) only *Equisetum variegatum* grows with a fairly high frequency in the *Caricion bicolori-atrofuscae*. Neither species show a high fidelity to this alliance. On the other hand, the incorporation of the *Caricetum frigidae* in the *Caricion davallianae* seems possible and offers a better solution, also from a chorological viewpoint.

This is also true for the *Juncetum alpini,* which

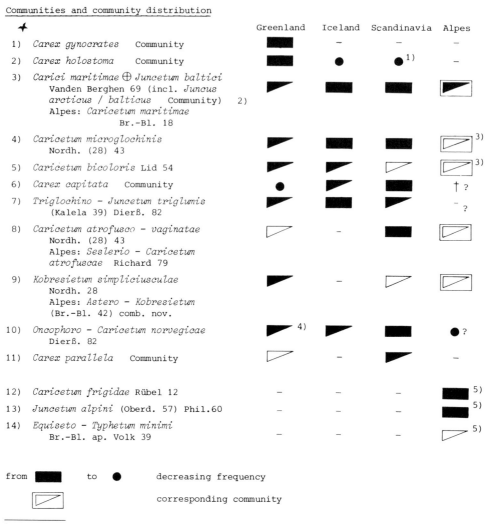

Communities and community distribution

		Greenland	Iceland	Scandinavia	Alpes
1)	*Carex gynocrates* Community	■	–	–	–
2)	*Carex holostoma* Community	■	●	●[1]	–
3)	*Carici maritimae* ⊕ *Juncetum baltici* Vanden Berghen 69 (incl. *Juncus arcticus / balticus* Community) [2] Alpes: *Caricetum maritimae* Br.-Bl. 18	◣	■	■	◣
4)	*Caricetum microglochinis* Nordh. (28) 43	◣	■	■	◹[3]
5)	*Caricetum bicoloris* Lid 54	◣	◣	◹	◹[3]
6)	*Carex capitata* Community	●	◣	■	† ?
7)	*Triglochino – Juncetum triglumis* (Kalela 39) Dierß. 82	◣	■	◣	– ?
8)	*Caricetum atrofusco – vaginatae* Nordh. (28) 43 Alpes: *Seslerio – Caricetum atrofuscae* Richard 79	◸	–	■	◸
9)	*Kobresietum simpliciusculae* Nordh. 28 Alpes: *Astero – Kobresietum* (Br.-Bl. 42) comb. nov.	◣	–	◹	◹
10)	*Oncophoro – Caricetum norvegicae* Dierß. 82	◣[4]	◣	■	●?
11)	*Carex parallela* Community	◸	–	◣	–
12)	*Caricetum frigidae* Rübel 12	–	–	–	■[5]
13)	*Juncetum alpini* (Oberd. 57) Phil.60	–	–	–	■[5]
14)	*Equiseto – Typhetum minimi* Br.-Bl. ap. Volk 39	–	–	–	◹[5]

from ■ to ● decreasing frequency

◹ corresponding community

---

[1] Compare Ryvarden, 1969: p. 30.

[2] Compare footnote Table 1.

[3] Compare Bressoud, 1980.

[4] Mainly *Carex norvegica* ssp. *inserrulata* Kalela.

[5] Should better be adjoined *to* the *Caricion davallianae*.

Fig. 3. Synopsis of the known syntaxa of the alliance and their distribution.

contains more species characteristic of the alliance *Caricion davallianae*. As a rule, this pioneer community of wet, open, unstable soils is only weakly characterized.

The *Equiseto–Typhetum minimi* also offers difficulties. For the characteristic species of both order and class are for the greater part extending absent, the community has a marginal position within the class. The association key species *Typha minima* and *Typha shuttleworthii* have a de-alpine distribution in the colline and montane belt and show no floristic affinities to the *Caricion bicolori-atrofuscae* area in alpine or (oro-)arctic sites from a chorological viewpoint.

The difference between our syntaxonomic opinion and for instance that of Görs (see Oberdorfer, 1977) is based on a deviating judgement of the sociological position of *Equisetum variegatum* and *Juncus alpino-articulatus*. They are both here considered to be character species of the order *Caricetalia davallianae*.

Concluding remarks

In spite of the *Caricion bicolori-atrofuscae* communities being strongly isolated from each other, the development of their phytocoenoses in different regions have a high degree of floristic similarity. The sociological affinities within these communities in Greenland and northern Europe are so high, that a separation seems useful only at the geographical race level. Corresponding syntaxa of central Europe must be separated at the association level, because the alpine communities are impoverished in alliance character and differential species and contain some alpine differential species with a low fidelity within the alliance.

The geographical differentiation within the *Caricion bicolori-atrofuscae* is more pronounced than in other alliances of azonal fen communities, i.e. the *Rhynchosporion albae*, the *Caricion nigrae* and the *Caricion lasiocarpae*, where a separation is generally useful on the geographical race level. On the other hand, the differences are by far less distinct in comparison with the contact communities of the *Carici rupestris-Kobresietea*, which are divided into three orders within the same area (compare Ohba, 1974).

The cause of the floristical differences of the corresponding syntaxa in distinct regions is probably not the duration of the isolation. We suppose that the intrusion of species from contact communities is decisive. This tendency seems more pronounced in the Alps, where the number and size of suitable habitats decrease and the competition from species of neighbouring communities therefore increases. Because of the low intensity of solifluction in the Alps, open pioneer areas are rarer and smaller. In addition, the high evaporation rate does not favour hygrophyte communities, in contrast to boreal and oceanic arctic sites.

An important problem in valuating the synsystematic rank of open, sometimes fragmentarily developed communities of this alliance seems the partly high degree of admixture of species from neighbouring sites. This results in a relatively high heterotoneity of the syntaxa as a whole. A convinced disciple of the Braun-Blanquet method has learned to arrange plant communities synsystematically according to their floristic similarity and to build up a hierarchic system which may be compared – with all necessary restrictions – in the same formal logic way as idiotaxonomical systems. There exists nevertheless at least one fundamental difference: morphologists and systematists, when valuating resemblances of taxa, generally may differentiate between homologous and analogous structures of their objects. This viewpoint has no counterpart in phytosociology. This fact, therefore, points out the limits of objectivity of higher statistic procedures in order to judge floristic similarities. A strong formalism seems unwise, for even the information content of a clearly differentiated table only can be an abstract picture of the reality in nature. It needs a satisfying interpretation with respect to biology, population dynamics, ecology and chorology of the species in question.

In our example, questions of a phytosociological arrangement and differentiation cannot be solved without a satisfying consideration of chorological and historical aspects.

References

Bressoud, B., 1980. La végétation du bas-marais de l'Ar du Tsan (Val de Réchy, 2185 m, Nax, Valais). Bull. Murithienne 97: 3–24.

Conzelmann, A., 1979. Zum Vergleich einiger ökologischer Kenngrößen (Azidität, Austauschkapazität, Aschegehalt) bei torfbildenden Pflanzengesellschaften und Arten. Zulassungsarbeit Biol. Institut II, Freiburg, Polykopie, Freiburg 100 pp.

Dierssen, K., 1977. Zur Soziologie von *Carex maritima* Gunn. Mitt. Flor.-soz. ArbGem. N. F. 19/20: 297–312.

Dierssen, K., 1982. Die wichtigsten Pflanzengesellschaften der Moore NW-Europas. Genf. 382 + 32 pp.

Hulten, E., 1964. The circumpolar plants I. Kungl. Svenska Vetensk. Handl. 8(5), Stockholm 275 pp.

Oberdorfer, E. (ed.), 1977. Süddeutsche Pflanzengesellschaften I, 2nd ed. Jena 309 pp.

Ohba, T., 1974. Vergleichende Untersuchungen über die alpine Vegetation Japans. 1. Carici ruperstris-Kobresietea bellardii. Phytocoenologia 1: 339–401.

Richard, J.-L. & Geissler, P., 1979. A la découverte de la végétation des bords de cours d'eau de l'étage alpin du Valais (Suisse). Phytocoenologia 6: 183–201.

Accepted 19.1.1984.

Vegetation structure and microclimate of three Dutch *Calthion palustris* communities under different climatic conditions*

L. M. Fliervoet & M. J. A. Werger**
Department of Plant Ecology, University of Utrecht, Lange Nieuwstraat 106, 3512 PN Utrecht, The Netherlands

Keywords: Biomass, *Calthion palustris,* Climate, LAI, Microclimate, Vegetation structure

Abstract

Three Dutch *Calthion palustris* communities, situated in different phytogeographic districts which vary in climatic conditions, are compared with respect to vegetation structure and microclimate. The three *Calthion* stands which are similar in soil, management and hydrology, differ slightly in total aboveground biomass in the period just before cutting, but there is a larger difference in the biomass contributed by phanerogams, bryophytes and litter. The structure of the *Calthion* communities varies in vertical distribution of biomass and leaf area (*LAI*), and growth form and leaf size composition. These differences are interpreted in terms of climatic differences such as length of growing season, temperature and wind. Profiles of decreasing light intensity within the vegetation canopy are related to the vertical distribution of biomass, *LAI* and leaf inclination of the various *Calthion* communities. Temperature and saturation deficit of the air on the different sites show profiles of a similar shape which suggests that in such ecologically comparable plant communities, vegetation structure differs under influence of the macroclimate in such a way that the resulting vegetation canopies modify the microclimate within the vegetation to become homologous.

Introduction

Geographical variation in the plant cover of a major phytogeographical unit (e.g. region, district) relates to adaptive responses of plants to geographically changing environmental, mostly climatic, factors. This geographical variation has at least two aspects: 1) A shift in species composition along a climatic gradient may occur, determined by the ecological amplitudes of the species and the degree of climatic change along the gradient. This is the type of chorological variation that traditionally has been studied in vegetation science; 2) Independently, structural and functional features of the vegetation may change. Particularly within the recent IBP and MAB projects this aspect has been studied in a comparison of physiognomically convergent but phylogenetically unrelated ecosystems in different parts of the world under corresponding climatic constraints (e.g. Parsons, 1976; Mooney, 1977; Cowling & Campbell, 1980). These convergence studies showed that the adaptive structural and functional responses to climatic variation can involve single plant organs, the entire plant as an integral unit, as well as the plant community as an integrated system of species populations. The adaptive response of single plant organs have drawn

* Nomenclature of species follows Heukels-van Ooststroom (1977).
** *Acknowledgement.* This study is part of the project on vegetation structure, financially supported by the Foundation for Fundamental Biological Research (BION), of the Dutch Organization for the Advancement of Pure Research (ZWO). Much of the research was carried out at the Division of Geobotany, University of Nijmegen. The authors kindly acknowledge this hospitality. Also the data contributed by B. Aerssens, J. Cortenraad, and S. Troelstra and the typing help of Mrs R. Jaegermann are gratefully acknowledged.

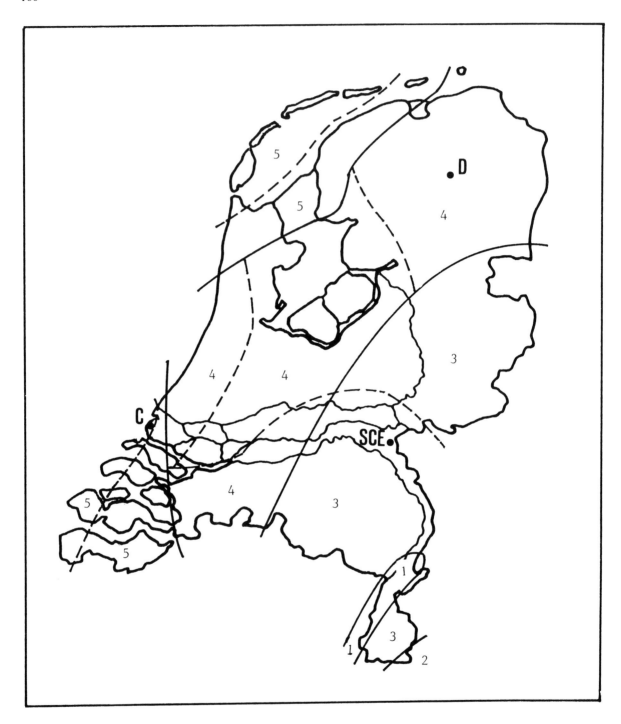

Fig. 1. Climatic districts (and their subdivisions) of the Netherlands according to Barkman (1958): 1. Subcontinental district; 2. Submontane district; 3. Subatlantic district; 4. Atlantic district; 5. Euatlantic district. The map shows the locations of the three study sites: C. Merrevliet, climatologically in the Euatlantic district and phytogeographically in the Haff district; D. Roodzande, climatologically in the Atlantic district and phytogeographically in the Drenthe district; SCE. Bruuk, climatologically in the Subatlantic district and phytogeographically in the Subcentral European district. Climatic differences between the sites are listed in Table 1.

attention since very long (Werger, 1980). In this respect the study of leaf characteristics has been especially rewarding, both in palaeoecological studies (cf. Dolph & Dilcher, 1980), and in analyses of long climatic gradients (e.g. Werger & Ellenbroek, 1978) explaining specific leaf features in terms of environmental effectiveness and cost-profit balances (Orians & Solbrig, 1977; Givnish & Vermeij, 1976; Taylor, 1975; Gates, 1976; Parkhurst & Loucks, 1972). In relation to field situations, these studies, like those from the turn of the century (Warming, 1895; Schimper, 1898; Raunkiaer, 1934), nearly always concerned major climatic gradients, and always considered woody plants, especially their leaves.

In this paper we want to investigate the structural response, both of single organs and of the entire aboveground vegetation structure, in a non-woody vegetation type, to relatively minor variations in macroclimate. We selected three stands of a moist grassland community belonging to the *Calthion palustris* R. Tüxen 37 em. 51 situated on pedologically and hydrologically similar sites in three climatologically different parts of the Netherlands. These *Calthion* stands are extensively used as hay meadows, mown once a year at the end of the growing season, with the hay removed. No fertilizers are used.

Study areas: plant geography, climate and soil

The three *Calthion* stands are all located in nature reserves (Fig. 1). A coastal site (C) is situated in the Merrevliet behind the coastal dunes of Voorne. This area may be considered as a southern extension of the Haff district (van Soest, 1932, who himself included the Dutch area in the Fluviatile district). One site (SCE) lies in the Bruuk near Nijmegen in the Dutch subcentral European phytogeographic district, and the third site (D) at Roodzande, near the river Drentsche Aa, in the Dutch Drenthe phytogeographic district. All phytogeographic districts belong to the Atlantic and central European floristic areas (Mennema et al., 1980). The SCE site forms part of the central European floristic area, but this site possesses still many characteristics of the Atlantic floristic area and can be considered as a transition. The phytogeographic districts do not correspond to the climatic division of the Netherlands, as presented in Figure 1 which shows that the three *Calthion* sites are situated in three distinct climatic districts.

Table 1 shows the differences in macroclimate between the three sites. The C site clearly shows a strongly oceanic climate with the smallest range between maximum and minimum daily temperatures and the lowest number of ice days (i.e. with temperature below zero). The other two sites have more continental features. The differences in annual precipitation are probably irrelevant for this type of vegetation due to soil water conditions. All three *Calthion* sites are on a peaty substrate; the water table remains near the surface throughout the summer, and comes at or above the surface in winter. A comparison of the amount of precipitation in combination with the humidity of the air is probably more relevant. The P/S quotient of Meyer calculated for the months June–August shows a remarkable difference between the D site (0.74) and the other two sites (0.54 for the SCE and 0.59 for C (Barkman & Westhoff, 1969).

For each site the pH (H_2O) value is between 5.5 and 5.9. Soil organic matter content varies from

Table 1. Climatic differences (temperature, wind and precipitation) of the three Dutch phytogeographical districts involved, based on the period 1931–1960 (after Mennema et al., 1980).

Phytogeographic district	Haff (C)	Subcentral European (SCE)	Drenthe (D)
Temperature (°C)			
mean daily minimum (Jan)	0.5 to 1.0	−0.5 to −1.0	−1.5 to −2.0
mean daily minimum (July)	14.0 to 14.5	12.5 to 13.0	11.5 to 12.0
mean daily maximum (Jan)	4.5 to 5.0	4.0 to 4.5	3.5 to 4.0
mean daily maximum (July)	20.0 to 20.5	22.5 to 23.0	21.0 to 21.5
Average annual number of ice days ($t < 0$ °C)	8 to 9	12 to 13	17 to 18
Average wind velocity (m . sec⁻¹)	6.0 to 6.5	3.5 to 4.0	4.5 to 5.0
Average annual precipitation (mm)	700 to 725	725 to 750	750 to 775

Table 2. Calthion palustris communities in three phytogeographic districts in the Netherlands represented by relevés according to Braun-Blanquet.

Phytogeographic district	Drenthe (D)				Subcentral European (SCE)		Haff (C)	
location	Burg-vallen	Roodzande			Bruuk		Merrevliet	
date	26/6	27/6	28/7	28/7	1/7	11/8	15/7	15/7
cover % total	80	95	100	100	100	100	100	100
herbs	70	95	85	100	100	100	90	95
bryophytes	5	30	20	90	80	75	60	60
litter	1	1	1	5	5	5	5	5
number of herbs	13	9	13	14	17	15	11	9
number of grasses	6	5	5	7	11	8	10	6
area (m × m)	1 × 1	1 × 1	1 × 1	2 × 2	1 × 2	1 × 2	1 × 1	1 × 1
Ranunculus acris	2a	2a	+	2b	+	+	+	1
Holcus lanatus	1	2a	1	3	1	1	3	2a
Festuca rubra	2a	2m	2a	1	+	+	2m	1
Cardamine pratensis	1	1		1	r	+	+	+
Anthoxanthum odoratum	1	1	2a	1	1	+	2a	2b
Rumex acetosa	+	2a	1	2b			+	2a
Filipendula ulmaria	+	1	2a	1	+	1		
Valeriana dioica		+		1	1	+		
Rhinanthus serotinus	3	3	2b	1	+			
Caltha palustris	2b	2a	1	2a		+		
Juncus acutiflorus	2a	1	+	1	2a	4		
Crepis paludosa	r		+	+	4	2a		
Trifolium repens	1	1	+	+				
Ranunculus repens	1		1	2b				
Equisetum fluviatile		+	+	1				
Myosotis scorpioides			+	+	+			
Lotus uliginosus				+	+		+	
Angelica sylvestris					2a	4	'	r
Lysimachia vulgaris				+	2a	+		+
Cirsium palustre			r		r	2a		r
Equisetum palustre	+				+	r		
Prunella vulgaris					+	+	+	
Vicia cracca					1	+		
Agrostis canina				+	+	+		
Molinia caerulea					+	+		
Carex acutiformis					+	1		
Ajuga reptans					+		1	
Luzula multiflora					+		+	
Carex disticha					+	+	1	1
Carex nigra		+	2a	+			2a	2a
Lychnis flos-cuculi							2b	2b
Trifolium pratense							2a	3
Plantago lanceolata	1						1	2a
Phragmites australis							1	+
Iris pseudacoris							+	
Juncus subnodulosus							+	+
Salix repens							+	
Orchis majalis	r						+	
Poa pratensis	+			+			+	
Cerastium holosteoides	+							+
Cynosurus cristatus	+							

Table 2. (Continued).

Phytogeographic district	Drenthe (D)				Subcentral European (SCE)		Haff (C)	
location	Burg-vallen	Roodzande			Bruuk		Merrevliet	
date	26/6	27/6	28/7	28/7	1/7	11/8	15/7	15/7
cover % total	80	95	100	100	100	100	100	100
herbs	70	95	85	100	100	100	90	95
bryophytes	5	30	20	90	80	75	60	60
litter	1	1	1	5	5	5	5	5
number of herbs	13	9	13	14	17	15	11	9
number of grasses	6	5	5	7	11	8	10	6
area (m × m)	1 × 1	1 × 1	1 × 1	2 × 2	1 × 2	1 × 2	1 × 1	1 × 1
Carex aquatilis	2a							
Viola palustris		2a						
Orchis maculata					+			
Achillea ptarmica						+		
Carex panicea					+			
Calamagrostis canescens					+			
Taraxacum s. palustria					+			
Succisa pratensis						+		
Rhytidiadelphus squarrosus	2a	3	2b	5	4	3	4	4
Calliergonella cuspidata			+		2b	2b	+	
Pseudoscleropodium purum					1	2a	2a	
Polytrichum juniperinum			1					2a

50% for the D and SCE sites to about 80% for the C site. N and P contents are quite similar taking into account the high percentage of organic matter of the coastal site. Total N values as average percentage of soil dry weight are 2.3%, 2.1% and 1.6%, and exchangeable P values are 9.7, 2.7 and 2.5 mg per 100 g soil dry weight for the C, D and SCE sites, respectively (B. Aerssens, unpubl. data; S. Troelstra, unpubl. data).

The phytogeographic subdivision of the Netherlands (van Soest, 1932) is reflected by the occurrence of species combinations for various parts of the country. Since the *Calthion* is a largely azonal syntaxon, we do not expect phytogeographically differential species amongst its characteristic species. According to Table 2 this is true since among the characteristic and differential species for the three sites, only *Carex aquatilis* is phytogeographically restricted.

Methods

The vegetation was sampled in the main flowering stage when vegetation is maximally developed.

We had to distinguish two stages of the subcentral European *Calthion* during this main flowering period: the first in the beginning of July (SCE-July) with a dominance of *Crepis paludosa* and a second in August (SCE-August), when *Juncus acutiflorus* dominates the vegetation.

The aboveground vegetation was harvested in duplicate in layers of 10 cm in homogeneous plots of 0.5 × 0.5 m.

Total biomass was separated in phanerogams, bryophytes and dead standing crop, and the vascular plants are divided per vegetation layer in leaf, stem and inflorescence fractions. For all fractions dry weights (24 hr at 105 °C) were determined.

Before drying leaf areas were measured for all phanerogams in all plots using a LI-3000-3050A meter. Leaf Area Indices (*LAI*) are calculated for bifacial leaf area. For each stand Leaf Area Ratio (*LAR*) was calculated as the ratio of bifacial leaf area (in cm^2) per aboveground biomass of the phanerogams (in g dry weight).

For leaf size (unifacial) spectra the classification of Raunkiaer (1934) as modified by Taylor (1975) was used (Table 5). Following Gates (1968) and

Table 3. Aboveground biomass, Leaf Area Index, *LAI*, and Leaf Area Ratio, *LAR*, of the *Calthion* communities of the three phytogeographic districts.

Location	Date of harvesting	Aboveground biomass (g . m⁻² dry w)							LAI (m² . m⁻²)	LAR (cm² . g⁻¹ total dry w)
		total	phanero-gams	bryo-phytes	litter	% of dry weight (phanerogams)				
						stem	leaf	inflores-cence		
C	15-7-1981	632 ± 96	431 ± 50	104 ± 18	97 ± 42	51	47	2	8.7 ± 0.1	275 ± 19
D	28-7-1981	502 ± 50	289 ± 41	122 ± 15	91 ± 16	38	60	2	7.4 ± 0.9	325 ± 22
SCE-July	1-7-1981	510 ± 44	456 ± 36	8 ± 5	46 ± 5	46	52	2	7.1 ± 0.6	161 ± 2
SCE-August	11-8-1981	595 ± 53	461 ± 50	59 ± 19	75 ± 5	58	37	5	5.8 ± 0.2	128 ± 12

Taylor (1975) leaf sizes for composite leaves were determined on the separate leaflets.

Growth form spectra were calculated for each stand, using J. J. Barkman's (unpubl.) growth form system, on basis of proportional dry weight contribution of each growth form to the total biomass.

Microclimatic changes within the vegetation canopy were measured in duplicate on similar bright sunny days between 11.00 am and 14.00 pm. We measured decrease in light intensity of PhAR (400–700 nm) using a point light sensor horizontally above the vegetation and a line sensor (both from TFDL, Wageningen) with a light-sensitive surface of 60 × 1 cm placed at various height intervals in the vegetation. Decrease of light intensity was calculated proportionally to the intensity above the vegetation, corrected for variety in light-sensitive surface of the two light sensors. Also temperature, using Ysi temperature sensors, and relative humidity of the air, using Vaisala MPH-21 probes, were measured in several height intervals in the vegetation canopy. These series of data were used to calculate the water vapour pressure saturation deficit (SD) according to Unwin (1980).

Results

Vegetation structure

In Table 3 the aboveground biomass, *LAI* and *LAR* values are presented. Lowest total biomass values are obtained in the Drenthe *Calthion* where the growing season is shortest, while the coastal *Calthion* reaches the highest values. As regards the values for the phanerogams these differences are particularly clear. The bryophytes and litter fractions in the D and C sites are comparable in absolute values, but relative to the total biomass the Drenthe *Calthion* shows a slightly higher value as compared to the coastal *Calthion*, whereas both score much higher than the SCE site (see also Table 4).

The SCE-August sample has a higher total biomass value than the SCE-July sample, mainly as a result of an increase in litter and bryophytes. This development probably results from the dying of *Crepis paludosa* rosettes in the period between the sampling dates, which increases the amount of litter, and at the same time improves the light condi-

Table 4. Growth form spectra of the *Calthion* communities in percentage of the total aboveground biomass (dry weight) using the growth form system of J. J. Barkman (unpubl.).

Location Date of harvesting	D 28-7	C 15-7	SCE-July 1-7	SCE-August 11-8
growth form:				
Graminoïds total	34.8	40.4	69.5	72.7
Schoenids	13.0	–	–	–
Airids	4.0	1.2	–	2.5
Anthoxanthids	16.9	12.6	2.9	1.1
Dactylids	0.9	26.6	66.6	69.2
Equisitids	7.9	4.7	0.1	0.1
Herbs total	15.3	20.6	19.6	4.3
Primulids	0.3	5.3	–	0.7
Digitalids	0.4	11.0	14.7	2.2
Epipactids	5.5	3.6	2.8	0.5
Ranunculids	8.9	0.7	2.1	0.9
Gypsophilids	0.2	–	–	–
Bryophytes	24.0	17.8	1.3	9.8
Standing litter	18.0	16.5	9.5	12.8

Table 5. Leaf size classification of Raunkiaer (1934) with subdivision according to Taylor (1975).

Size class	cm²		Small (s)	Medium (m)	Big (b)	
leptophyllous	0 –	0.25	0 – 0.056	– 0.12	– 0.25	
nanophyllous	0.25–	2.25	0.25 – 0.52	– 1.08	– 2.25	
microphyllous	2.25–	20.25	2.25 – 4.68	– 9.74	– 20.25	
mesophyllous	20.25–	182.25	20.25 – 42.09	– 87.68	– 182.25	
macrophyllous	182.25–	1640.25	182.25 – 378.82	– 789.13	–1640.25	
megaphyllous	>1640.25		1640.25 –3409.31	–7102.11	–∞	

tions near the soil surface, allowing the bryophytes to grow (see Fig. 4). This change in litter and bryophyte fractions has no parallel in the development of the total life biomass of phanerogams, though here the stem–leaf–inflorescence ratios change considerably (Table 3).

LAI is highest in the coastal *Calthion*, and lowest in the SCE-August sample, despite the fact that here the highest phanerogam biomass value is measured. Relative investment of aboveground biomass in leaf area (*LAR*) is highest in the Drenthian *Calthion*, however.

Growth form spectra show (Table 4) that at all sites various types of graminoids by far contribute most to the total biomass, reaching highest values in the SCE samples whereas equisitids are conspicuously scarce here. As regards the forbs the Drenthe *Calthion* contains mainly those with leafy stems (ranunculids and epipactids), while in the coastal *Calthion* and in the SCE-July sample rosette forbs (primulids and digitalids) are dominant (compare Table 2).

Biomass allocation per layer typically follows an exponential curve in all types (Fig. 2). However, whereas most samples show a rather gradual decrease in biomass for the strata above 30 cm, the Drenthe *Calthion* rapidly decreases from there upwards. Similarly this type is distinct in that it contains litter up to 30 cm above the surface.

Vertical distribution of *LAI* is rather variable (Fig. 2). SCE-July and SCE-August show the greatest similarity but a decrease in the *LAI* of the lowest

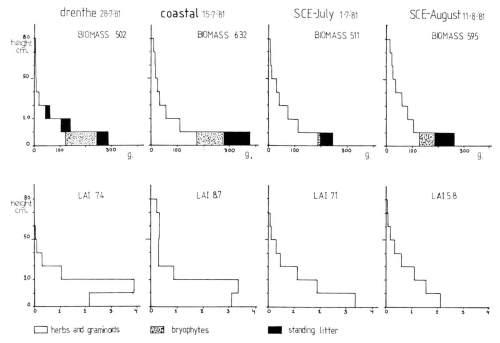

Fig. 2. Vertical distribution of aboveground biomass in g . m⁻² dry weight and *LAI* in m² . m⁻² of the three *Calthion* stands. The SCE site has been sampled on two occasions. Further explanation see text.

166

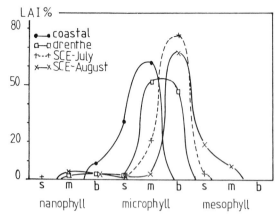

Fig. 3. Leaf size spectra of the *Calthion* communities as percentage of the *LAI*. Leaf size classes according to Raunkiaer (1934) with subdivision by Taylor (1975). Compare Table 5 for abbreviations and class limits.

layer between the dates of sampling is clearly discernible. The distribution pattern is largely the same, however.

Leaf size spectra (Fig. 3) of the three *Calthion* sites show that most leaves are microphyllous. There is, however, a clear shift in the maximal representation of the subdivisional classes of the microphyllous leaf size using the system of Taylor (1975): The coastal *Calthion* has the smallest leaves with its maximum in the medium microphyllous class; the Drenthe *Calthion* has a broader maximum and in the SCE *Calthion* the big microphylls are best represented. This shift in leaf sizes parallels a decreasing gradient in wind velocity (Table 1) and seems reciprocal to total *LAI* (Table 3).

Microclimate

In Figures 4 and 5 the vertical change in microclimates in the vegetation is illustrated. In all types

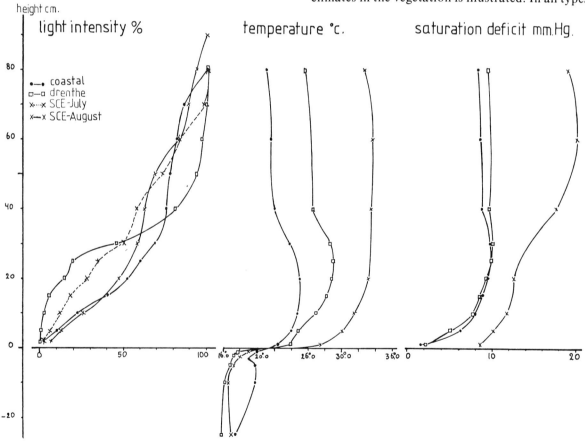

Fig. 4. Microclimatic profile in the vegetation canopy of the three *Calthion* communities: light intensity (%), temperature (°C) and saturation deficit (mm Hg).

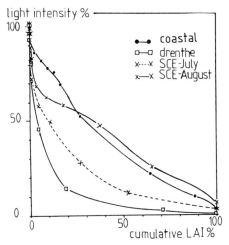

Fig. 5. Light intensity as a function of foliage area (cumulative *LAI*) for the three *Calthion* communities.

light intensities decrease at lower heights in the vegetation, and at a higher rate than cumulative *LAI* values increase. The decrease in light intensity is steepest just above the layer of high vegetation density in the Drenthe *Calthion* (20–40 cm) (cf. Fig. 2), while the other sites show a more gradual decrease in the uppermost dense vegetation layers. Both the SCE and the coastal *Calthion* have a deeper light penetration than the Drenthe *Calthion* (Fig. 5).

In all three vegetation stands the temperature within the vegetation initially shows an increase, that becomes stronger in the 30–40 cm layer, just above the densest vegetation layers, in the C and D sites. Temperature profiles, measured by Gisi & Oertli (1981) within production grassland (*Arrhenatheretum*) show similar shapes and also reach maximum values just above the densest concentration of biomass. Lower down in the vegetation (at about 20 cm) the temperature falls again, but in the coastal *Calthion* near the soil surface it is still slightly higher than the temperature above the vegetation. This is probably due to the high soil temperature in this *Calthion* stand. Comparing the temperatures at 2 cm above and 1 cm under the soil surface there is a smaller decline (of 2.8 °C) here than at the other sites where the declines are 6.1 and 8.5 °C, respectively for the D and the SCE sites (Fig. 4).

In all types the water vapour pressure deficit (*SD*) decreases in the dense vegetation layers, and varies over a range of 9 mm Hg. The *SD* minima are reached near the soil surface, and in the C and D

sites they approach saturated level. In the SCE site the large *SD* at soil surface may result from higher ambient temperature (Fig. 4) and from a reduced transpiring surface, as shown by the lower values for absolute *LAI* and for *LAR* (Table 3).

Conclusions

Macroclimatological factors are highly responsible for gross vegetation structure, though they affect it in a complex way. In various parts of the trajectory over which they range they have a different effect on plant organs. Apart from macroclimate, also nutrient availability, water regime and management practices ultimately determine the resulting vegetation structure. In our *Calthion* stands, these latter controlling factors are similar however: the *Calthion* is an azonal vegetation, on all three sites occupying similar peaty soils with a high groundwater table and a similar mowing regime. Height of the groundwater table strongly influences the rate of mineralization in wet soils, and this determines the nutrient availability for the plants. Balátová-Tuláčková (1968) and Grootjans (1980) emphasized the strongly determinant effects of this interaction between water table and mineralization upon the floristic composition and the peak standing crop in *Calthion* communities. The relatively small range in peak biomass values for our stands thus also reflect the similarity in soil conditions between the sites.

In our view the differences in vertical arrangement of plant biomass and *LAI*, as well as differences in leaf size, are expressions of differences in macroclimate.

The parallel between increasing wind velocity and diminishing average leaf size is as expected. Grace (1977) reviewed many references and discussed some of the underlying physiological characteristics to explain the finding. It is obvious that more and stronger wind imply a higher transpiration per unit of time (Gates, 1968; Taylor, 1975), which has 'aridifying' aspects in the leaf vicinity. Grace (1977) argued that the morphological modifications of leaves developed under windy conditions, including size reduction, may be of value in maintaining a positive leaf water balance, but are also adaptive in withstanding mechanical stress.

The shorter growth season going from the coastal, via the SCE, to the Drenthe *Calthion* is reflected in the decreasing peak biomass values for these sites. Furthermore, the relatively late start of the growth season in Drenthe apparently results in a sudden burst of growth of most plants which causes a stronger concentration of both total biomass and leaf area in the lower layers and very little above these, than in the other *Calthion* stands. This rhythm somewhat resembles the successive bursts of guilds of species in the undergrowth of deciduous woodlands in springtime (e.g. Bratton, 1976). This means that at least during this phase there will be a strong competition for light (and perhaps for other resources) between the individuals. Thus there is a strong concentration of leaf area in the 10–20 cm layer at Drenthe. This leaf area is to a large degree made up by horizontal leaves (of e.g. *Rhinanthus serotinus, Ranunculus acris, Rumex acetosa*), while in the other *Calthion* stands leaves are predominantly erect and erectopatent (cf. Barkman, 1979). This leaf orientation is significant since the light intensity at this point is already relatively low (Fig. 4) due to the high density of early flowering inflorescences of these species in the higher layers. The result is a strongly diminished light intensity in the lowest vegetation layer, and in the sense of Loomis & Williams (1961), an inefficient distribution of leaf area occurs as regards light supply (Fig. 5). In correspondence herewith digitalids (and primulids) with relatively much of their leaf area concentrated in the rosettes are conspicuously under-represented in this vegetation. *Rumex acetosa*, for example, a digitalid which occurs at all three sites, has virtually no rosette leaves at the peak flowering season in the Drenthe *Calthion*.

At the other *Calthion* sites, there is not such a strong concentration of particularly horizontally exposed leaf area in the layers of 0–20 cm. Light penetration in the vegetation is therefore deeper and growth form spectra are correspondingly (Table 4). This distribution of leaf area in relation to light supply also shows a regular, efficient pattern (Fig. 5, compare also Pümpel, 1977).

Though macroclimate thus seems salient in determining vegetation structure, vegetation structure in turn seems to develop in such a way that it smoothes out the climatic differences and brings about uniform or corresponding climatic conditions in the lower layers of stands of vegetation which, as regards their ecology, can be reasonably compared. In our study this is shown in very similar curves for temperature and *SD* within the vegetation canopy for all stands. Cernusca's (1976) results from a number of dwarf shrub communities growing on sites with strongly different macroclimate in the Alps seem to point in the same direction. Thus we may theorize that macroclimate determines various structural features of the vegetation. However, the net effect of this structural adaptation of the vegetation to macroclimate seems to be such, that in ecologically comparable vegetation with an approximately full ground cover this results in microclimate conditions within the canopy which are strongly similar.

References

Balátová-Tuláčková, E., 1972. Flachmoorwiesen in mittleren und unteren Opava-Tal (Schlesien). Vegetace ČSSR A4: 1–201. Praha.

Barkman, J. J., 1958. Phytosociology and ecology of cryptogamic epiphytes. van Gorcum, Assen.

Barkman, J. J., 1979. The investigation of vegetation texture and structure. In: M. J. A. Werger (ed.). The study of vegetation, p. 123–160. Junk, The Hague.

Barkman, J. J. & Westhoff, V., 1969. Botanical evaluation of the Drenthian District. Vegetatio 19: 330–388.

Bratton, S. P., 1976. Resource division in an understory herb community: responses to temporal and microtopographic gradients. Am. Nat. 110: 679–693.

Cernusca, A., 1976. Bestandesstruktur, Bioklima und Energiehaushalt von alpinen Zwergstrauchbeständen. Oecol. Plant. 11: 71–102.

Cowling, R. M. & Campbell, B. M., 1980. Convergence in vegetation structure in the mediterranean communities of California, Chile and South Africa. Vegetatio 43: 191–197.

Dolph, G. E. & Dilcher, D. L., 1980. Variation in leaf size with respect to climate in Costa Rica. Biotropica 12: 91–99.

Gates, D. M., 1968. Transpiration and leaf temperature. Ann. Rev. Plant Physiol. 19: 211–238.

Gates, D. M., 1976. Energy exchange and transpiration. In: O. L. Lange et al. (eds.). Water and plant life. Ecol. Studies 19: 137–147. Springer, New York.

Gisi, U. & Oertli, J. J., 1981. Oekologische Entwicklung in Brachland verglichen mit Kulturwiesen. IV. Veränderungen im Microklima. Oecol. Plant. 16: 233–249.

Givnish, T. J. & Vermey, G. J., 1976. Sizes and shapes of liane leaves. Am. Nat. 110: 743–778.

Grace, J., 1977. Plant response to wind. Acad. Press, London.

Grootjans, A. P., 1980. Distribution of plant communities along rivulets in relation to hydrology and management. In: O. Wilmanns & R. Tüxen (eds.). Epharmonie. Ber. intern. Symposien der I.V.V., 1979, p. 143–170. Cramer, Vaduz.

Heukels, H. & van Oostroom, S. J., 1977. Flora van Nederland. 19th ed. Wolters-Noordhoff, Groningen.

Loomis, R. S. & Williams, W. A., 1969. Productivity and the morphology of crop stands: patterns with leaves. In: J. D. Eastin et al. (eds.). Physiological aspects of crop yield, p. 27–47. Madison, Wisconsin.

Mennema, J., Quené-Boterenbrood, A. J. & Plate, C. L., 1980. Atlas of the Netherlands Flora. Part I. Junk, The Hague.

Mooney, H. A. (ed.), 1977. Convergent evolution in Chile and California: Mediterranean climate ecosystems. US/IBP Synthesis Series 5. Dowden, Hutchinson & Ross, Stroudsburg.

Orians, G. H. & Solbrig, O. T., 1977. A cost-income model of leaves and roots with special reference to arid and semi-arid areas. Am. Nat. 111:677–690.

Parkhurst, D. F. & Loucks, O. L., 1972. Optimal leaf size in relation to environment. J. Ecol. 60: 505–537.

Parsons, D. J., 1976. Vegetation structure in the mediterranean scrub communities of California and Chile. J. Ecol. 64: 435–447.

Pümpel, B., 1977. Bestandesstruktur, Phytomassevorrat und Produktion verschiedener Pflanzengesellschaften im Glocknergebiet. In: A. Cernusca (ed.). Alpine Grasheide Hohe Tauern, P. 83–101. Wagner, Innsbruck.

Raunkiaer, C., 1934. The life-forms of plants and statistical plant geography. Oxford Un. Press, Oxford.

Schimper, A. F. W., 1898. Pflanzengeografie auf physiologischer Grundlage. Fischer, Jena.

Soest, J. L. van, 1932. Plantengeografische districten in Nederland. In: H. Heukels & W. H. Wachter. Beknopte schoolflora voor Nederland. Noordhoff, Groningen.

Taylor, S. E., 1975. Optimal leaf form. In: D. M. Gates & R. B. Schmerl (eds.). Perspectives of biophysical ecology. Ecol. Studies 12: 73–86. Springer, New York.

Unwin, D. M., 1980. Microclimate measurements for ecologists. Biol. Tech. Series 3. Acad. Press, London.

Warming, E., 1895. Plantesamfund. Grundtraek af den økologiske plantegeografi. Philipsens, Copenhagen.

Werger, M. J. A., 1980. Structure and function. Inaugural Lecture University of Utrecht. Junk, The Hague.

Werger, M. J. A. & Ellenbroek, G. A., 1978. Leaf size and leaf consistence of a riverine forest formation along a climatic gradient. Oecologia 34: 297–308.

Accepted 24.4.1984.

Anthropogenous areal extension of central European woody species on the British Isles and its significance for the judgement of the present potential natural vegetation*

H. Dierschke
Systematisch-Geobotanisches Institut, Untere Karspüle 2, D-3400 Göttingen, F.R.G.

Keywords: Anthropogenous areal extension, British Isles, Postglacial vegetational history, Potential natural vegetation, Tree species

Abstract

Several central European species have failed to reach the British Isles because of the early separation from the continent. The two tree species *Fagus sylvatica* and *Carpinus betulus* reached southern England but were unable to spread much further. Other species, such as *Acer pseudoplatanus,* were only relatively recently introduced. Recent distribution maps for *Fagus* and *Acer* show an almost uninterrupted distribution in the whole British Isles. The beech must be considered to be an important element of the present potential natural vegetation; it has been planted widely and regenerates freely. The woodlands of Brittany may provide a model to enable us to visualize the possible species composition and appearance of these potential Atlantic beech woods.

Introduction

In addition to the yearly symposia and meetings of working groups the International Association for Vegetation Science organizes excursions yearly–two-yearly. Only few can participate and the majority of the Society is not informed on the results. Therefore more lectures on such results should be given at symposia. Our latest excursion was held 21–30 July, 1980, in Ireland and thanks to the local organizers (whose help is again acknowledged) the 34 participants from 9 countries got a good impression of flora and vegetation (see also White, 1982). This paper reports on a problem observed and discussed during the excursion and also related to the symposium subject.

Some chorological and ecological characteristics of woodlands on the British Isles

For a botanist of central Europe the islands along the northwestern fringe of Europe, especially Ireland, have unaccustomed climatic, soil ecological and vegetation historical characteristics (see also Lüdi, 1952). First of all the lack of near-natural woodlands or even woodlands at all is very striking. Ireland for example has only 5% woods, and even less than 1% near-natural ones (Neff, 1975). Also in Great Britain most of the woods are strongly influenced by man. Because of many centuries of grazing and coppicing it is difficult to judge the present potential natural vegetation (Bunce, 1982).

A second peculiarity, especially in Ireland, is the relatively poor woody flora. During the Atlanticum mixed broad-leaved forests occurred everywhere, with *Quercus* species, *Ulmus glabra* and *Alnus glutinosa,* and in England also *Tilia cordata* (Pennington, 1969). Already in the Subboreal these forests were influenced by man with a local development of

* Nomenclature of vascular taxa follows Ehrendorfer (1973).

Vegetatio 59, 171–175 (1985).

172

heathlands. *Fraxinus excelsior* moved in and extended.

In the Bronze Age woodlands had disappeared over large areas or become opened up (Pennington, l.c.). Most of the present woodlands are of secondary origin. At the beginning of the Subatlanticum, ca. 500 BC, *Fagus sylvatica* and *Carpinus betulus* became manifest in pollen spectra of southwest England. However, according to Pennington both species must have reached the area already before 5500 BC, when Great Britain became isolated in the Boreal. Ireland had become an isolated island still earlier.

The absence of natural *Fagus* in Ireland and the greater part of Great Britain is astonishing for the central European botanist. Beech still arrived at the southeast of England in a natural way but never became the dominant tree where one might expect this. The causes for this are not yet known, though there are some indications. Apparently *Fagus* could not enter easily under the Atlantic climatical conditions into the dense broad-leaved forests at that time covering Great Britain. According to Godwin (1975) the species could only become dominant secondarily after clearing and on abandoned fields, especially on limestone in south England. The increase in beech pollen in the early Iron Age can be related to the possibility of cultivation of deeper soils with the plough, which led to abandonment of former cultivated limestone areas (Pennington, 1969). Stojanoff (1931) has already pointed to the tendency of *Fagus* spreading in abandoned fields in a later period.

Another limiting factor for the spreading of beech may have been relatively low summer temperatures (bad fruit ripening) and frequent late frosts (damage to saplings and flowers) (Tansley, 1953; see also Pennington, 1969; Godwin, 1975). Also the development of raw humus under the influence of the cool-moist climate may have caused limited regeneration (Watt, 1931). Finally the early practising of coppicing may have prevented the extension of beech towards the west and the north (Bunce, 1982).

In Ireland it was *Fraxinus excelsior* which spread on base-rich soils whereas *Quercus* woods occurred on the poorer ones (Mitchell, 1976). While broad-leaved forests in southeast England only grow in moist habitats, outside the natural area of *Fagus* they also grow on dryer *Fagus*-like soils (Klötzli,

1970). So, the climax forests in Ireland, vicarious to the continental beech forest, are the *Corylo–Fraxinetum* and the *Blechno–Quercetum,* for the first time described by Tüxen (1950) and Braun-Blanquet & Tüxen (1952) (see also Moore, 1967; Kelly & Moore, 1974; Kelly, 1981; Kelly & Kirby, 1982).

For England Klötzli (1970) described a *Dryopterido-, Querco-,* and *Hyperico–Fraxinetum* in addition to the *Blechno–Quercetum.* These are partly vicarious to the *Stellario–Carpinetum* of central Europe.

Present situation of distribution areas of central European tree species

Besides *Fagus sylvatica* the central European *Acer* and *Tilia* species as well as *Carpinus betulus* never reached Ireland nor the major part of Great Britain. If we delete these species from Ellenberg's (1978) ecogram of the forest trees of central Europe large gaps arise in the mesic part which indeed have been filled up on the British Isles by *Quercus* and *Fraxinus* (see Fig. 1). Geographically this becomes clear from the maps in Meusel *et al.* (1965, 1978; see Fig. 2). The border of subspontaneous expansion of *Fagus* is to be seen in northern England whereas Ireland is excluded. *Acer pseudoplatanus* is entirely synanthropous. However, the maps in Perring & Walters (1976) are completely different (see Fig. 3). Today *Fagus sylvatica* is spread in northwestern Europe over large areas, as well as *Acer pseudoplatanus. Carpinus betulus,* however, which is supposed to have arrived in England at the same time as *Fagus,* still shows many gaps.

The unsaturated flora of the British Isles allows the establishment of many neophytes, especially of those which did not reach the islands for historical reasons. *Acer platanoides* for example is known for Ireland only recently (Webb, 1979). Only one really exotic species, *Rhododendron ponticum,* managed to establish in the vegetation of the British Isles (Tansley, 1953; Neff, 1975; Mitchell, 1976; Kelly, 1981; Cross, 1982).

All common European woody species which were not found originally have successfully been planted and their vigour indicates that historical reasons must have prevented those species to establish naturally. Whereas *Acer pseudoplatanus* plays a minor part (Watt, 1924/25; Klötzli, 1970; Kelly,

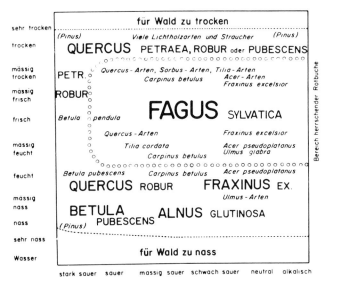

Fig. 1. Ecogram of woodland tree species of submontane Central Europe (on the left, from Ellenberg, 1978) and the natural competition situation in Ireland (on the right).

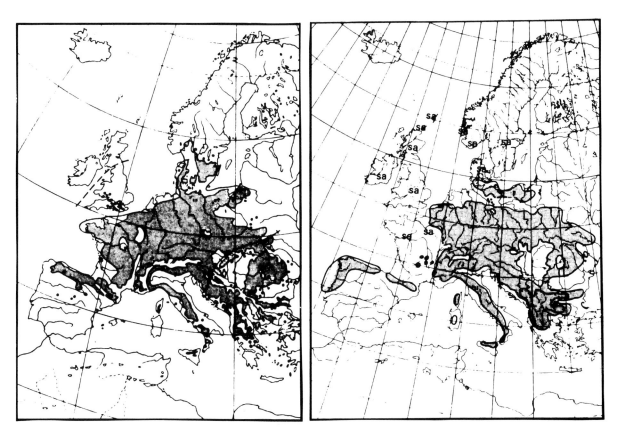

Fig. 2. Area maps of *Fagus sylvatica* and *Acer pseudoplatanus* (after Meusel *et al.*, 1965, 1978).

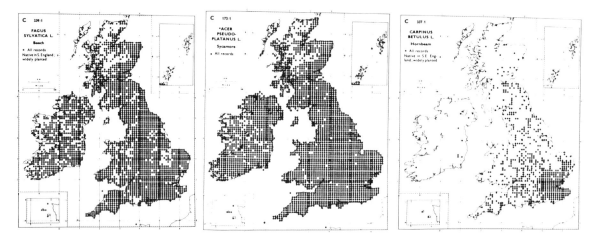

Fig. 3 Present distribution of *Fagus sylvatica, Acer pseudoplatanus* and *Varpinus betulus* on the British Isles (after Perring & Walters, 1976).

1981) especially the importance of *Fagus* for the present potential natural vegetation is of interest. The new vegetation map for the Council of Europe member states (Ozenda *et al.*, 1979) reflects the present state of knowledge. The potential natural vegetation outside the *Fagetum* area in southeast England is indicated as *Quercus* and *Fraxinus* forests.

According to the many references *Fagus sylvatica* has established without doubt on the British Isles outside its natural area with human help. Not only do planted stands thrive, but also a good regeneration was observed in several regions (Watt, 1924/25, 1931; Tansley, 1953, 1968; Klötzli, 1970; Neff, 1975; Mitchell, 1976; Webb, 1977; Kelly, 1981; Bunce, 1982). This means that *Fagus* has become part of the present potential natural vegetation over large areas. However, the species was still partly considered as an undesirable alien (see Neff, 1975) which is not logical in view of the equally anthropogenous character of the 'near-natural' woods. For example, the old trees in the present characteristic Atlantic oak woods of Ireland have often been planted in the 18th and 19th centuries according to Kelly & Moore (1974).

Today there is not the question *if* but *where* *Fagus* plays a part in the potential natural vegetation. The species will find optimal conditions on mesic soils with moderate moisture and a good base and nutrient status. Here *Fagus* is the strongest competitor to *Fraxinus*. So Klötzli (1970) judges many *Fraxinus* habitats as *Fagus*-like.

During the Ireland excursion we saw well-developed *Fagus* stands with a species-rich natural looking understorey, often not completely balanced (Dierschke, 1982). According to Kelly (1981) beech regenerates best on well-drained soils along a broad pH gradient, but not on moister sandstone sites. At least on limestone the species is able to invade the present woodlands. During the excursion we saw beech stands with a good regeneration within the area of the *Blechno-Quercetum*. Even at the wind-exposed Irish west coast with an eu-Atlantic climate *Fagus* is found at appropriate sites.

Ideas on the present potential natural vegetation

Since most *Fagus* stands on the British Isles are of human origin, the present potential natural vegetation in such areas must be judged in relation to climatically and geologically similar areas with natural occurrence of beech woods. As such the 'hygro-Atlantic zone' of the continent may be considered (Ozenda, 1979), i.e. the coastal zone from northwestern France to the Basque country, especially Brittany. Here we find beech woods which are called 'hyperhumid' by Ozenda. They are described by Durin *et al.* (1967). Though the *Rusco-Fagetum* has more submediterranean elements (for example *Buxus sempervirens, Ruscus aculeatus*) it can give a good impression of the potential beech woods of the British Isles with its evergreen species (*Ilex aquifolium, Taxus baccata* etc.), many Atlantic herb species, for example *Conopodium majus, Dryopteris aemula, D. pseudomas, Primula vulgaris, Scilla non-scripta*), and its richness in mosses and lichens.

The *Ilici–Fagetum* has links to the *Blechno–Quercetum* and to the south English beech woods (Klötzli, 1970; Géhu, 1975). Also Braun-Blanquet's (1967) differentiation of an acidic *Ilici–Fagion* and a more basic-rich *Scillo–Fagion* in the Basque country can be taken as a model.

In conclusion the anthropogenous extension of central European tree species to northwestern Europe leads to a change in the interpretation of the present potential natural vegetation of the British Isles. Consequently the natural conservation status and perspectives of these species for forestry planning must be reconsidered.

References

Braun-Blanquet, J., 1967. Vegetationsskizzen aus dem Baskenland mit Ausblicken auf das weitere Ibero-Atlantikum. II. Vegetatio 14: 1–126.

Braun-Blanquet, J. & Tüxen, R., 1952. Irische Pflanzengesellschaften. Veröff. Geobot. Inst. Rübel Zürich 25: 224–415.

Bunce, R. G. H., 1982. Some effects of man on the structure and composition of atlantic deciduous forests. In: H. Dierschke (ed.), Struktur und Dynamik von Wäldern. Ber. Internat. Sympos. IVV Rinteln 1981: 681–698. Cramer, Vaduz.

Cross, J. R., 1982. The invasion and impact of Rhododendron ponticum in native Irish vegetation. In: J. White (ed.), Studies on Irish Vegetation, pp. 209–220. Royal Dublin Society.

Dierschke, H., 1981. Internationale Vereinigung für Vegetationskunde (IVV). Bericht über die Jahre 1980/81. Phytocoenologia 9(3): 413–416.

Dierschke, H., 1982. The significance of some introduced European broad-leaved trees for the present potential natural vegetation of Ireland. In: J. White (ed.), Studies on Irish Vegetation, pp. 199–207. Royal Dublin Society.

Durin, L., Géhu, J.-M., Noirfalise, A. & Sougnez, N., 1967. Les hêtraies atlantiques et leur essaim climacique dans le nord-ouest et l'ouest de la France. Bull. Soc. Bot. Nord de la France, 20. Anniversaire: 59–89.

Ehrendorfer, F. (ed.), 1973. Liste der Gefässpflanzen Mitteleuropas. 2nd rev. ed. Fischer, Stuttgart. 318 pp.

Ellenberg, H., 1978. Vegetation Mitteleuropas mit den Alpen in ökologischer Sicht. 2nd ed. Ulmer, Stuttgart. 981 pp.

Géhu, J.-M., 1975. Aperçu sur les chênaies-hêtraies acidiphiles du sud de l'Angleterre, l'exemple de la New-Forest. Coll. Phytosoc. 3: 133–140.

Godwin, H., 1975. The History of the British Flora. A factual basis for phytogeography. 2nd ed. Cambridge University Press. 541 pp.

Kelly, D. L., 1981. The native forest vegetation of Killarney, South-West Ireland: an ecological account. J. Ecol. 69: 437–472.

Kelly, D. L. & Kirby, E. N., 1982. Irish native woodlands over limestone. In: J. White (ed.), Studies on Irish vegetation, pp. 181–198. Royal Dublin Society.

Kelly, D. L. & Moore, J. J., 1974. A preliminary sketch of the Irish acidophilous oakwoods. Coll. Phytosoc. 3: 375–387.

Klötzli, F., 1970. Eichen-, Edellaub- und Bruchwälder der Britischen Inseln. Schweizer. Ztschr. f. Forstwes. 121(5): 329–366.

Lüdi, W. (ed.), 1952. Die Pflanzenwelt Irlands (The Flora and Vegetation of Ireland). Veröff. Geobot. Inst. Rübel Zürich 25: 1–421.

Meusel, H., Jäger, E. & Weinert, E., 1965. Vergleichende Chorologie der zentraleuropäischen Flora I. Fischer, Jena. 583 + 258 pp.

Meusel, H., Jäger, E., Rauschert, S. & Weinert, E., 1978. Vergleichende Chorologie der Zentraleuropäischen Flora II. Fischer, Jena. 418 + 421 pp.

Mitchell, F., 1976. The Irish Landscape. Collins, London. 240 pp.

Moore, J. J., 1967. Zur pflanzensoziologischen Bewertung irischer nacheiszeitlicher Pollendiagramme. In: R. Tüxen (ed.), Pflanzensoziologie und Palynologie. Ber. Internat. Symposium IVV Stolzenau/Weser 1962, pp. 96–105. Junk, Den Haag.

Neff, M. J., 1975. Woodland conservation in the Republic of Ireland. Coll. Phytosoc. 3: 273–285.

Ozenda, P., 1979. Sur la correspondance entre les hêtraies médioeuropéennes et les hêtraies atlantiques et subméditerranéennes. Docum. Phytosoc. N.S. 4: 767–782.

Ozenda, P. et al., 1979. Vegetation map of the Council of Europe member states. Nature and Environment Ser. 16. Strasbourg. 99 pp.

Pennington, W., 1969. The History of British Vegetation. English Universities Press Ltd., London. 152 pp.

Perring, F. H. & Walters, S. M. (eds.), 1976. Atlas of the British Flora, 2nd ed. EP Publishing Ltd. 432 pp.

Stoyanoff, N., 1931. Notes on English beech woods. Bull. Soc. bot. Bulg. 4: 57–66.

Tansley, A. G., 1953. The British Isles and Their Vegetation, Vol. I, 3rd ed. Cambridge University Press. 484 pp.

Tansley, A. G., 1968. Britain's Green Mantle. Past, present and future, 2nd ed. George Allen & Unwin Ltd., London. 327 pp.

Tüxen, R., 1950. Observations in Irish woodland associations, with special reference to practical forestry. The Irish Naturalist's Journ. 10(4): 99–104.

Watt, A. S., 1924/25. The development and structure of beech communities on the Sussex Downs. J. Ecol. 12: 145–204; 13: 27–73.

Watt, A. S., 1931. Preliminary observations on Scottish beechwood. J. Ecol. 19: 137–157; 321–359.

Webb, D. A., 1977. An Irish flora, 6th rev. ed. Dundalgan Press, Dundalk. 277 pp.

Webb, D. A., 1979. Three trees naturalized in Ireland. Ir. Nat. J. 19(10): 369.

White, J. (ed.), 1982. Studies on Irish Vegetation. Contributions from participants in the vegetation excursion to Ireland, July 1980, organized by The International Society for Vegetation Science. Royal Dublin Society. 368 pp.

Accepted 19.1.1984.

Die *Fagus*-Sippen Europas und ihre geographisch-soziologische Korrelation zur Verbreitung der Assoziationen des *Fagion* s.l.*

J. Duty**
Tweel 11, 2500 Rostock, D.D.R.

Keywords: Beech woods, Distribution, Europa, *Fagion, Fagus*

Abstract

Studies of the beeches and beech woods of eastern central Europe revealed, that in the postglacial period not only *Fagus sylvatica* (L.) emend. reimmigrated from the refugial territories in the SE – as has generally been accepted, but also the 'transitional taxa', which originated from hybrids with *F. orientalis* Lipsky. The NW area limit of these intermediate taxa must be revised. The presence in central Europe of these taxa – which form own *Fagion* alliances and associations in the SE (*Fagus intermedia* ssp. *moesiaca* and ssp. *taurica*) as well as the presence of other southeastern species in central European beech woods shows, that their postglacial development is parallel to, but different from other areas. The taxon *Fagus intermedia* (ssp. *neglecta* and ssp. *transitus*) became differential taxa of a central European region of the *Fagion medioeuropaeum*. Plant sociologists are therefore requested to make new and critical analyses of the beech woods in Europe, with special attention to the *Fagus* taxa, in order to establish in detail the geographical distribution and phytosociological significance of *Fagus intermedia*.

The author offers to determine or revise *Fagus* material (herbarium collections).

Die Verbreitung der *Fagus*-Sippen Europas deckt sich im großen und ganzen mit derjenigen der *Querco–Fagetea europaea* incl. der *Fagetalia, Quercetalia, Fraxinetalia* und *Aceretalia* und ihren *Fagion*-Verbänden. Die Buchen bilden erstaunlicherweise (nach Ellenberg, 1963, p. 103) selbst an der N and NW Grenze ihrer Verbreitung kräftige Hochwälder und sind dominant, wo sie aus klimatischen

u.a. Gründen an der oberen Waldgrenze zu Krüppelwuchs gezwungen sind.

Von den Buchen wurden deshalb schon früh forstliche Standort-Formen und 'Rassen' beschrieben (Fig. 1). Auch die Zahl der beschriebenen Buchenwald-Gesellschaften ist außerordentlich groß, vgl. Bibliographie der *Querco–Fagetea*, Tüxen *et al.* (1980–1981). Als wichtig erscheint es hierbei zu betonen, daß arealgeschichtlich zwischen den west-, mittel-und nordeuropäischen Gesellschaftsentwicklungen und denen S und SO Europas (d.h. den nicht durch die Eiszeit total veränderten Gebieten) ein gesellschafts-entwicklungsgeschichtlicher, spürbarer Klima-, Raum- und Zeitunterschied existiert, der Artenspektrum und Sippenbildung, Bodenentwicklung und anthropogenen Einfluß bestimmte. So sind die mitteleuropäischen Gesellschaften postglaziale 'junge' Artenkombinationen gegenüber den seit dem Alt-Tertiär bestehenden

 * Die Sippennomenklatur richtet sich nach Ehrendorfer (73).
** *Danksagung.* Ich möchte meinen tiefsten und allzeitigen Dank besonders den Herren Prof. Dr. A. O. Horvat (Pécs), Prof. Dr. Ch. Moulopoulos, Prof. Dr. B. Jovanović, Prof. Dr. I. Dumitriu-Tataranu, Prof. Dr. R. Bornkamm, Prof. Dr. M. A. Kotschkin, Prof. Dr. P. Fukarek, Prof. Dr. K. Browicz für gewährte Unterstützung aussprechen und ganz besonders unserem unvergeßlichen verstorbenen Prof. Dr. Drs. h.c. R. Tüxen, der mir zur Fortführung dieser Studien Mut machte.

Fig. 1. Ökotypen (Klima-, Wuchs- und Standorttypen) von *Fagus sylvatica* L. nach Svoboda (1955); ohne Nr. entspr. Literaturangaben bei Dengler (1904), Schwappach (1911), Hartmann (1953), Galoux (1966), Krahl-Urban (1954), Horvat (1977). 1- *britannica*, 2-*scandinavica*, 3- *jutlandica*, 4- *celtica*, 5- *pyrenaica*, 6- *gallica*, 7- *alpina* (3 Regionen), 8- *helvetica* (2 Regionen), 9- *appenina*, 10- *sicilica*, 11- *suntalica*, 12- *hercynica*, 13- *pommeranica* (3 Regionen), 14- *carpatica* (3 Regionen), 15- *transsylvanica* (2 Regionen), 16- *balcanica* (3 Regionen).

S und. SE Europas (Horvat, 1977). Da aber über die Refugien der mitteleuropäischen Gehölze während der Eiszeit bezüglich Ausdehnung und Arten nur wenig Daten und Vorstellungen vorliegen (vgl. z.B. Firbas, 1949–1952) ist es heute sehr schwierig, die stark wechselnde Gehölzarten-Kombination im Bereich der Rückzugsgebiete und Randzonen anzugeben. Zumindest starben auch dort viele empfindliche Florenelemente aus. Von den 6 alten Buchensippen des Tertiärs (*F. attenuata* Goeppert und *F. feroniae* Ung., *F. deucalionis* Ung., *F. antipovii* Heer., *F. orient. fossilis* Palib.) (vgl. Goeppert, 1855) blieb bis auf *F. orientalis* Lipsky, deren Areal wohl weit nach Westen reichte, keine erhalten; doch wird angenommen, daß 'Hybridformen' der voreiszeitlichen Buchen mit *F. orientalis* existierten und aus diesen (Takhtajan, 1973) oder aus den Refugial-Hybriden erst die heute ganz Mittel-, West- und Süd-Europa besiedelnde *F. sylvatica* L. und somit ihre Gesellschaften hervorgingen (vgl. Lämmermeyer, 1923; Kolakovsky, 1960; Fukarek, 1980; Mišič, 1957).

Leider hat bisher niemand eine echte Prüfung der europäischen *Fagus*-Sippen genetisch vorgenommen. Die künstlich erzeugten Bastarde zwischen *F. orientalis* und *F. sylvatica* (Schaffalitzky de Muckadell & Nilson, 1954) brachten nur eine geringe Nachkommenschaft und keinen Formenschwarm, der *F. sylvatica* als Tochtersippe von *F. orientalis* erkennen ließ. Die atavistischen Formen brachten nach Mišič (1955) ebenfals keine Hinweise auf eine Hybridnatur von *F. sylvatica*. Eher zeichnen sich atavistische 'Altformen' ab, bzw. Öko- und Physiokline. So sind z.B. von *F. sylvatica* mehr als 120 Cultivare beschrieben (vgl. z.B. Wyman, 1964). Leider wurde aber auch sonst eine Trennung der Sippen von Vegetationskundlern, wie Horvat, Glavač & Ellenberg (1974, p. 315) betonen, kaum vorgenommen. Da nun neuere Ergebnisse zur Sippenerfassung in Mitteleuropa vorgelegt werden können, wonach der Kreis der zwischen *F. orientalis* und *F. sylvatica* existierenden Taxa und Transitussippen eine viel weitere und bis zur nördlichen Arealgrenze reichende Ausbreitung hat (Fig. 2), ist neben der systematischen, auch die Beurteilung der Taxabindung an die Gesellschaften des *Fagion im weitesten Sinne* zu überprüfen und bezüglich analoger Arteneinbindungen zu analysieren.

In arealgeographischer Hinsicht setzen sich nach Meusel (Meusel *et al.*, 1969) die mitteleuropäischen Buchenwaldgesellschaften in erster Linie aus süd- bis mitteleuropäisch verbreiteten Arten zusammen, die mehr oder weniger auch einen montanen Charakter und subatlantische bis zentraleuropä-

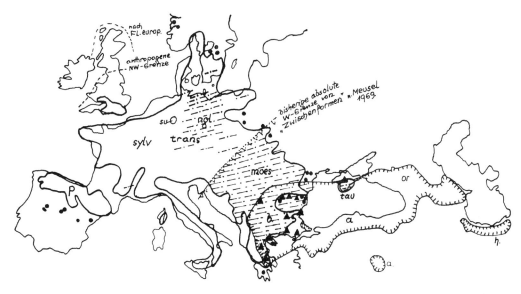

Fig. 2. Verbreitung der europäischen *Fagus*-Sippen. *Fagus sylvatica* (incl. subsp. *pyrenaica* - p, *sylvatica* - sylv, *suntalensis* - su, >200 Formen und Varietäten). ⌀ *Fagus intermedia* (incl. subsp. *transitus* - trans, *neglecta* - ngl, *moesiaca* - moes, *taurica* - tau, >50 Formen und Varietäten). *Fagus orientalis* (incl. subsp. *orientalis* - or, *hohenackeriana* - h, *anatolica* - a, *balcanica* - b, <20 Formen und Varietäten).

ische Verbreitungstendenz besitzen. Außerdem sind allgemein-boreomeridionale Arten beteiligt. Nur wenige kommen südlich bis in die Tropen vor (z.B. *Sanicula*) oder entstammen typischen alten tropischen Verwandschaftskreisen (wie *Ilex*). In S Europa treten dann submediterrane und eine große Zahl von S und SE Relikttypen hinzu, so daß die *Fagion*-Verbände stark differenzierbare Artenkombinationen besitzen (vgl. Soó, 1964), darunter auch *F. intermedia* ssp. *taurica*, *F. intermedia moesiaca* sowie *F. sylvatica* ssp. *pyrenaica*.

Nach dem vorläufigen Vergleich (Fig. 3) sind es 7–9 Regional-Verbände im südlichen Raum, wenn man Soó (1964) zustimmt, daß die floristisch differenzierenden Artengruppen der Krim und der Kaukasus-Buchenwälder ausreichen, eine eigene Klasse (*Fagetea orientalis*) zu begründen. Im W Teil (spanische und Pyrenäen-Vorkommen), ebenfalls durch viele eigene zum Teil endemische Arten abgrenzbar, ist es das *Scillo-Fagion*. Über die korsischen und süd-französischen Buchenwälder ist bislang keine eigene Abgrenzung gegeben worden. Die mittel- und süd-italienischen Buchenwälder, einschließlich der sizilianischen, stellt Soó (o.c.) zum *Fagion austro-italicum*. der jugoslavische W Teil ergibt in Übereinstimmung mit Fukarek (1977) u.a. das *Fagion illyricum* (mit

den UV *Primulo-Fagion*, *Lonicero-Fagion* und *Collurno-Fagion*). Nordöstlich angrenzend wird ein *Fagion circumpannonicum* und ein *Ostryo-Fagion* differenziert. Vom Karpatenraum nach Süden bis Bulgarien und Jugoslavien das *Fagion dacicum* (mit *Symphyto-Fagion* und *Acerion dacicum*) dem aber Fukarek in S Jugoslavien noch das *Fagion moesiacum* zwischenschaltet und südlich das *Fagion scardo-pindicum* anschließt.

Das nördlich liegende Großgebiet des *Fagion medioeuropaeum* ist vorerst nicht weiter differenziert, obwohl auch hier verschiedene Vorstellungen vorliegen. Durch die Untersuchungen zu den *Fagus*-Sippen zeigt sich, daß der östliche Gebietsraum durch das häufige Vorhandensein von *Fagus intermedia* eine Gliederung möglich erscheinen läßt.

Für die soziologische Gliederung der Buchenwälder ist die breite ökologische Amplitude der bestandbildenden Baumarten, sogar von *Fagus* selbst, erschwerend. Die Differenzierung erfolgte deshalb meist nur auf der ökologisch spezifischen und besser differenzierenden Bodenvegetation (Makrophyten, Bryophyten etc.), sowie deren geographische Verbreitung. Trotzdem verblüfft, daß zur Benennung und Kennzeichnung der 'Fageten' 22 Gehölze herangezogen werden, auch wenn es oft die Kontaktbereiche besonderer Standorte sind.

180

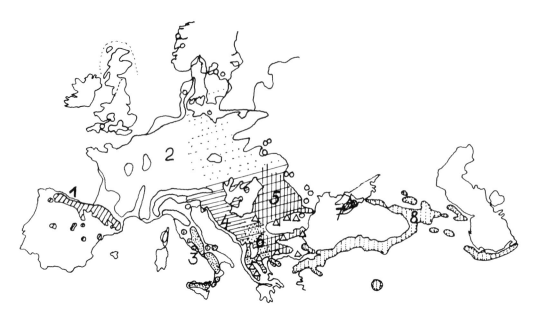

Fig. 3. Verbreitung der Verbände der *Fagetalia sylvaticae* (et *intermediae*) und der *Fagetalia orientalis*. 1- *Fagion pyrenaicum* (=*Scillo-Fagion*), 2- *Fagion medio-europaeum* (mit *Eu-Fagion, Cephalanthero-Fagion, Aceri-Fagion, Fraxino-Fagion, Luzulo-Fagion, Deschampsio-Fagion*), 3- *Fagion austro-italicum*, 4- *Fagion illyricum* (mit *Ostryo-Fagion, Lonicero-Fagion, Collurno-Fagion, Primulo-Fagion, Fagion-circumpannonicum*), 5- *Fagion dacicum* (*Symphyto-Fagion*), 6- *Fagion moesiacum* (*Fagion scardo-pindicum*), 7- *Fagion tauricum*, 8- *Fagetalia orientalis*, △ *Fagion orientalis*.

Die nicht immer mit 'floristischer Treue' nach Braun-Blanquet differenzierten 'geographischen Gesellschaften', zeigen ebenfalls die recht beachtlichen Schwierigkeiten der soziologischen wie ökologischen Amplitude. Hinzu kommt, daß durch die historisch-anthropogenen Einflüsse auf die Gehölzartenkombination (vgl. bei Pniower, 1954) aufgenommene Bestände nicht ohne weiteres einfach zugeordnet werden können, da heute von Buchen beherrschte Bestände durchaus nicht zwangsläufig zu Gesellschaften des *Fagion* gehören.

Als interessant erwies sich nun, daß bei der Buche und mit ihr korrelierten Arten z.B. *Abies, Acer, Tilia, Quercus, Fraxinus* u.a. eine deutliche W–SE Differenzierung in der 'Sippengliederung' durch taxonomische Untersuchungen ermittelt wurde (vgl. Flora Europaea, Hegi: Flora von Mitteleuropa; Ehrendorfer, 1973, usw.); diesbezüglich ist eine weitere intensive, kritische Bearbeitung anzustreben. Ich fasse bei *Fagus* die wohl einst hybridogen (und mutativ (?)) entstandenen Übergangssippen zwischen *F. orientalis* Lipsky und *F. sylvatica* L. emend. unter dem Namen *Fagus intermedia* nov. spec. coll. zusammen und gliedere nach den bisher erfaßten Sippen.

Da diese meist weit geographisch differenziert sind, gelten sie als Subspecies (nach ihren Autoren sind es Arten!). Es bleibt verwunderlich, wieso forstbotanisch eine so umfassende 'Rassen'-und Ökoklinen-Kenntnis entstand, ohne dabei taxonomisch zu erkennen, daß *F. intermedia* in den meisten mitteleuropäischen Wäldern Populationen besitzt, d.h. daß eigentlich zwei Buchenarten in unseren Wäldern wachsen. Auch Mattfield (1936) suchte die Zwischenformen nur auf dem Balkan und in Griechenland, d.h. im Bereich der Areal-überschreidung, ohne zu ahnen, daß hunderttausende solcher Bäume in seiner Heimat wuchsen.

Die Pflanzengeographen und Arealkundler sowie die Bearbeiter der *Fagus*-Sippen des Balkan (z.B. Stojanoff, 1955; Mišić, 1957; Dumitriu-Tataranu, 1950; Moloupoulos, 1965, u. a. m.) vermuteten und erwogen nicht die Möglichkeit einer Einwanderung der Zwischen-Sippen nach Mitteleuropa. Erst seit 1962, während der Tagung der Tschechoslowakischen botanischen Gesellschaft in Prag, verdanke ich Herrn Prof. Dr. A. O. Horvat (Pécs) die Anregung, auf Buchensippen zu achten. Tatsache ist, daß die mitteleuropäischen *Fagion*-Gesellschaften in ihrem östlichen Bereich eine weitere Klassenkennart als

Differentialart enthalten, verstärkt durch das Hinzutreten einiger Taxa, die ebenfalls im Südosten den Schwerpunkt ihrer Verbreitung besitzen, z.B. *Acer pseudoplatanus* ssp. *villosum*, *Fraxinus excelsior* ssp. *angustifolia*, *Tilia cordata* ssp. *rubra* (+*T. cordifolia* Bess.). *Glechoma hirsuta* L. u. a. Auf eine ganz analoge Situation im Südosten machte bereits Fukarek (1977) aufmerksam. Vergleicht man die modernen kritisch bearbeiteten Floren, so wurden für den Balkan und den Kaukasus viele mitteleuropäisch bekannte Sippen als selbständige, vikariante Arten beschrieben oder als Subspecies fixiert. Wohl sind diese Arten meist nicht allzu scharf getrennt, doch sinkt die Zahl der mit Mitteleuropa gemeinsamen Arten beachtlich. Es zeigt sich eine häufige Dreiteilung in mitteleuropäische, südosteuropäische, kaukasisch-taurische Sippen, auch bei rund 30% der im europäischen Buchenwald auftretenden Arten der Gehölz und Bodenvegetation. Einige Beispiele: *Acer campestre*, *A. platanoides*, *A. pseudoplatanus*, *Fraxinus excelsior*, *Abies alba*, *Tilia cordata*, *F. platyphyllos*, *Carpinus betulus*, *Hepatica nobilis*, *Primula vulgaris*, *Arum maculatum*, *Symphytum tuberosum*, *Trifolium medium*, *Melampyrum nemorosum*, *Aconitum variegatum*, *Primula veris*, *Poa nemoralis*, *Glechoma hederacea*. Inwieweit nun historisch bedingte oder natürliche Verbreitungslücken bestehen, müßte analysiert werden. Arten sind oder erscheinen heute unvollständig verbreitet, da sie oft nur in den Gesellschaften auftreten, wo sie früher entsprechende Einwanderungsbedingungen vorfanden, oder fehlen durch Veränderungen (Erlöschen). Das Auftreten der SE Sippen in mitteleuropäisch-östlichen Buchenwald-Gesellschaften zeigt, daß die Gesellschaftsentwicklung zu 'Vorposten' eines mit *Fagus intermedia* Populationen charakterisierten Regional-Bereiches führte. *Fagus intermedia* und die erwähnten Begleitsippen werden so zu Trennarten. Sicher ist diese Differenzierung kein Zwang und ohne die exakte Kenntnis der Gesamtverbreitung der Zwischenbuche und klare soziologische Analysen nicht möglich. Es bleibt ein zu lösendes Problem, da in anderen Klassen, Ordnungen und Verbänden sowie Associationen *Fagus intermedia*-Taxa, nämlich *moesiaca* und *taurica*, zur *Fagion* Differenzierung herangezogen wurden, d.h. zur namensgebenden Art der Gesellschaft und des Verbandes benutzt wurden.

Die Buchenwald-Gesellschaften mit *F. intermedia* sind, je weiter sie vom Balkan über Podolien, Karpaten, Slowakei und Böhmen nach Norden, z.B. in die DDR reichen, in der Regel auch immer Arten- und SE Sippen-ärmer, so daß mitunter nur die Sub- oder Voll-dominanz von *Fagus intermedia* (ssp. *neglecta* und ssp. *transitus*) vorhanden ist. Dies erscheint mir jedoch nicht mehr vergleichbar mit dem analogen Auftreten von *F. intermedia* ssp. *moesiaca* auf dem Balkan.

Da häufig 60–70% der Arten einer Aufnahme zu Verbands- bis Klassenarten gestellt werden, mag das Hinzutreten von Kleinsippen und *F. intermedia* vielleicht auch ohne gesellschaftssystematische Bedeutung sein. Von den höheren Einheiten – etwa vom Verband an – verlangen wir aber, daß sie mit Kennarten überregionaler Bedeutung ausgestattet sind. Als differenzierend sind dann jedoch nur die atlantischen, borealen, kontinentalen oder süd(ost)europäischen Arten deutlich. Aus geographischen, vor allem aber aus ökologischen Gründen scheint es sinnvoller, die Regional-Verbände durch entsprechende Differentialarten-Gruppen zu untergliedern. *Fagus intermedia* hätte dann bei uns eine solche Bedeutung.

Als Letztes sei nach dem Indigenat von *Fagus intermedia* gefragt. Ist sie vielleicht eine verschleppte, anthropogen ausgebreitete Sippe? D.h. könnte es wie bei Kiefer, Lärche, Fichte und auch anderen Arten zu einer allgemeinen Verbreitung in natürlich ehemals nicht besiedelte Bereiche gekommen sein? Haben vielleicht Kreuzfahrer, Mönche oder Kaufleute Saatgut transportiert? Nun, das letztere geschah nach forstwissenschaftlichen Untersuchungen viel später, etwa nach 1800, als es im mittleren Europa zu mehrfachem Ausfall guter Mastjahre kam. Zu früheren Terminen wurden zwar Obstgehölze und Arzneipflanzen sowie besonders Exoten, Juwelen u.a. mitgebracht, doch kaum Samen von Gehölzen, gegen die man im Heimatland noch Rodungskampagnen durchführte (Pniower, 1954). Es ist im Gegenteil selbst im 12.–14. Jahrhundert anzunehmen, daß Arten, die der damals sehr wichtigen Waldmast dienten (Eichen, Buchen), in bestimmtem Umfang geschont wurden. Die extremen Verluste in den Kriegen (z.B. 30-jährigen Krieg) konnten, da der Handel sich nur sehr langsam erholte, nicht aus den SE europäischen Ländern großräumig abgedeckt werden. Es erscheint unwahrscheinlich, daß aus den wenigen, örtlich möglichen Saatguteinbringungen daraus schon nach ein bis zwei Baumgenerationen eine ganz allgemeine Verbreitung resultiert, wie sie heute vorliegt.

Es scheint auch interessant, daß zum Teil die heute ältesten Buchen im Untersuchungsgebiet (250–300 Jahre alt) zu *F. intermedia* ssp. *neglecta* gehören. Auch die wenigen aufgefundenen 'subspontanen' Vorkommen echter *Fagus orientalis* im Gebiet sind viel zu jung (Saatgutverteilung 1904, nach Wilhelm, MDDG, 1909), um eine so allgemeine Verbreitung von Zwischenformen zu begründen. Erschwerend zeigen künstlich erzeugte Bastarde eine etwas verringerte Fertilität, wobei diese nicht vergleichbar mit den Jahrtausenden der natürlichen Vegetationsentwicklung sind. Trotz experimentell oft nicht erreichter Bastarde belegt Schwarz (1937, 1962, 1964) z.B. für *Quercus* viele natürliche Bastardpopulationen. Ebenso belegt Schwarz eine 'parallele' Zwischenformen- und Subspecies-Serie für *Q. pubescens* s.l. und *Q. petraea*. Auch Menizkii (1971) belegt Hybriden für die SE europäischen und Kaukasus Eichen.

Die reinen taxonomischen Probleme und die der recht komplizierten Nomenklatur sollen hier unberührt bleiben und nur einige Hinweise zur Beobachtung, Materialnotwendigkeit folgen.

Der einzige systematisch wichtige Kriterienbereich sind die Blüten und Fruchtmerkmale der Arten (vgl. Appendix). Es gibt für den erfahrenen Beobachter zwar auch Blatt-, Rinden- und Wuchs-Merkmale, die in einer recht unterschiedlichen Korrelation zu den Fertil-Kriterien stehen. Leider ist das Erlangen von blühenden oder fruchtenden Zweigen bei den häufig sehr hohen Bäumen geschlossener Bestände nicht leicht und so bleibt, zum Beurteilen, nach der frisch abgefallenen Cupulae zu suchen.

Mit etwas Erfahrung lassen sich so auch die zugehörigen Stämme ermitteln, da die Cupulae fast stets 'individuell' ausgebildet sind. Man kann so relativ rasch ermitteln, ob es sich um eine 'Reine' oder 'Misch-Population' der Arten handelt. Interessant erscheint aus den bisherigen Beobachtungen, daß *F. intermedia* häufig an exponierten Standorten, so Felspartien, Süd-Hänge, Steilhanglagen und in nördlichen Arealteilen auf Sand und Torfböden auftrat. Sogar nasse Standorte werden besiedelt. In den Optimalbereichen, bei 'Hallenwald' verrät sich ein Teil der Sippen durch den 'rissigen'-Rindentyp, der von fraxinoid bis quercoid be-

zeichnet wurde, und den Forstleuten durch die Härte der Rinde, als 'Steinbuchen' bekannt sind. Dazu kommt fast immer eine nicht glatt-zylindrische, sondern unregelmäßige bis spannrückig erscheinende Stamm-Form. Die Zwischenbuche zeigt auch häufig eine Winkel- bis Knorrastigkeit, die bis zu Renkformen geht, doch gibt es an guten Standorten auch Sippen mit völlig geraden, wipfelschäftigen Stämmen. Die südosteuropäischen Taxa besitzen ein sehr großes 'Stockausschlagvermögen' und die Fähigkeit der Astbewurzelung. Bei *Fagus intermedia* ssp. *neglecta* und noch mehr bei ssp. *transitus* verlieren sich diese Eigenschaften mehr und mehr.

Die unteren Schattenblätter sind meist mehr als 9-nervig und zeigen an der Basis häufig beidseitig eine schwache Einschnürung (Eindellung) des Blattrandes. Sie sind meist auch stärker behaart. Die Variabilität ist jedoch so groß, daß es kurzzeitig nicht möglich ist, sondern erst mit aufwendigen statistischen Verfahren (vgl. bei Podgorska, 1955) eine vegetative Sippenabgrenzung nach Blättern sicher vorzunehmen. Die Kurztriebe zeigen an den Blattansatzstellen oft eine noch nach dem Blattfall erkennbare Behaarung. Die größeren Astansätze am Stamm sind häufig konisch verlängert und nicht kurz wie bei *F. sylvatica*. Insgesamt gibt es also eine beachtliche Zahl von Merkmalskorrelationen um *Fagus intermedia* s.l. in den Buchenwäldern aufzufinden und zu determinieren.

Literatur

Čelakovsky, C., 1887. Über die morphologische Bedeutung der Cupula bei den echten Cupuliferen. Sitzungsber. Böhm. Wiss. Sitzung v. 12. XI. 1886, Prag, in Jahrb. Wiss. Bot. 21, 1890.

Czeczottova, H., 1933–1936. A study on the variability of the leaves of Beeches, F. orientalis Lipsky, F. silvatica L. and intermediate forms. Ann. Soc. Dendrol. Pologne, 1–5: 45–121, 1933, II–6, 1935.

Dengler, A., 1904. Untersuchungen über die natürlichen und künstlichen Verbreitungsgebiete einiger forstlich und pflanzen-geographisch wichtigen Holzarten in Nord und Norddeutschland. Neudamm.

Dimitriu-Tataranu, I. & Ocskay, S., 1950. Contributiuni la studiul Fagului din R.P.R. Anal. Acad. R.P.R., Ser. A, Mem. 4: 76–90.

Dobzhansky, Th. et al., 1980. Beiträge zur Evolutionsforschung, 'Genetik' Beitrag 10. Fischer Verlag, Jena.

Doing Kraft, H. & Westhoff, V., 1958. De Plaats van de Beuk (Fagus sylvatica) in het Midden- en West-europese Bos. Jaarb. Dendrol. Ver., 1958: 226–234.

Domin, K., 1933. On the variability of the beech. Rozpr. 2. Tř. Čes. Akad., 13 (14): 1–24; 66–74.

Ehrendorfer, F., 1973. Liste der Gefäßpflanzen Mitteleuropas. 2. ed. Stuttgart.

Ellenberg, H., 1963. Vegetation Mitteleuropas mit den Alpen. Stuttgart.

Firbas, F., 1949–1952. Spät- und nacheiszeitliche Waldgeschichte Mitteleuropas nördlich der Alpen. Vol. 1, 1949, Vol. 2, 1952.

Fukarek, P., 1977. Die Verbreitung der Buchenwälder in dem südlichen Raum Pannoniens. Stud. Phytosoziol. Festschr. 1977: A. O. Horvat, 33–37.

Galoux, A., 1966. La variabilité génélogique de Hêtre commun (F. sylvatica L.) Trav. Stat. Rech. Eau Forest. Separ. A 11: 117 pp.

Goeppert, R., 1855. Die Tertiärflora von Schlesien. Görlitz.

Hartmann, F. K., 1953. Die Buchenwälder Europas. Veröff. Geobot. Inst. Rübel Zürich, 8.

Hegi, G., et al., 1958. Flora von Mitteleuropa III. Stuttgart.

Hjelmquist, H., 1940. Studien über die Abhängigkeit der Baumgrenzen von den Temperaturverhältnissen unter besonderer Berücksichtigung der Buche und ihrer Klimarassen. Diss. Lund.

Horvat, I., Glovac, V. & Ellenberg, H., 1974. Die Vegetation Südosteuropas. Stuttgart.

Kolakowsky, A. A., 1960. Zur Geschichte der Buche in Eurasien. Arbeit. Internat. Ges. Naturforsch. Moskau.

Krahl-Urban, J., 1954. Buchenrassenstudien. Forstw. Centralbl., 1954/9–10.

Lämmermayr, L., 1923. Die Entwicklung der Buchenassoziation seit dem Tertiär. Feddes Repert. Spec. Nov. Regni Veget. 24. 100 pp.

Mattfeld, W., 1936. Die Buchen der Chalkidike. Bull. Soc. Bot. Bulg. 7: 63–73.

Menizkij, Y. L., 1971. Die Eichen des Kaukasus (Дуδbr. Kabka3a). Leningrad (in Russian).

Meusel, H., 1942. Der Buchenwald als Vegetationstyp. Bot. Arch. 43: 305–321.

Meusel, H., Jäger, E. & Weinert, E., 1969. Vergleichende Chorologie der zentraleuropäischen Flora, Vol. 1. Jena.

Mišić, V., 1955. Ancestral manifestation on leaves of the Balkan beech in Jugoslavia. Arch. Sci. Biol. 7 (1–2): 115–120.

Mišić, V., 1957. Variabilitet, ekologija bukov v Juguslaviji Monographie, Vol. 1. Beograd.

Moulopoulas, Ch., 1965. The beech woods of Greece. Thessaloniki.

Pniower, G., 1954. Über die Entwicklungsgeschichte und landeskulturelle Bedeutung der Dendrologie, Jena et Leipzig. (pp. 13–140).

Podgorska, L., 1955. Materiali do studium nad geograficzna zmiennośti liśći bukve w Polsce. Acta Soc. Bot. Polon. 24 (1): 1–80.

Poplovskaja, H., 1938. Die Buche der Krim und ihre Variabilität. Österr. Bot. Z. 27:

Pott, R., 1981. Der Einfluß der Niederwaldwirtschaft auf die Physiognomie und floristisch-soziologische Struktur von Kalk-Buchenwäldern. Tuexenia NS 1: 239 Ser.

Rameau, J.-C., 1982. Chorologie et différenciation de quelques groupements du Carpinion et du Quercion robori-petraeae dans le nord-est de la France. Ref. 26 th. Internat. Sympos. ISV, Prag 5.-8.4. 1982.

Rübel, E., 1932. Die Buchenwälder Europas. Veröff. Geobot. Inst. Rübel Zürich. 8.

Schaffalitzky de Muckadell, M. & Nilson, F., 1954. Flower observations in Fagus. Z. Forstgenetik 'Forstpflanzenzüchtung' 3. Frankfurt/M.

Schwappach, A., 1911. Die Rotbuche. Neudamm.

Schwarz, O., 1936. Monographie der Eichen Europas und des Mittelmeergebietes mit Atlas der Blattformen. Sonderbeih. Feddes Repert. Spec. Nov. D. 1.4 et D. 5.

Schwartz, O., 1962. Die Populationen mediterraner Eichen in Mitteleuropa nördlich der Alpen-Karpaten-Schranke. Drudea 2: 1–4.

Soó, R., 1964. Die regionalen Fagion-Verbände und Gesellschaften Südosteuropas. Stud. Biol. Hung., 1954: 1–104.

Stojanoff, V., 1955. Über die Buchenarten in den Wäldern Bulgariens. Sitz.-Ber. DAL DDR, 4/7: 1–16.

Svoboda, P., 1955. Lesní dřeviny a jejich porosty. Část 2. Praha.

Takhtajan, A., 1973. Evolution und Ausbreitung der Blütenpflanzen. Jena.

Tutin, T. G. et al., 1964–1976. Flora Europaea. Cambridge.

Tüxen, R. et al., 1962. International colloquium on the systematics of European beech woods. 12.–14. 4. Stolzenau/Weser.

Tüxen, R. et al., 1960–1963. Bibliographie der Querco-Fageeta. Excerpta Bot. Bd. 1–5. Jena.

Wilhelm, F., 1909. Mitteilungen der Deutschen Dendrologischen Gesellschaft. Berlin.

Wymann, D., 1964. Registration list of cultivar names of Fagus L. Arnoldia 24/1: 1–24.

Accepted 27 February 1984.

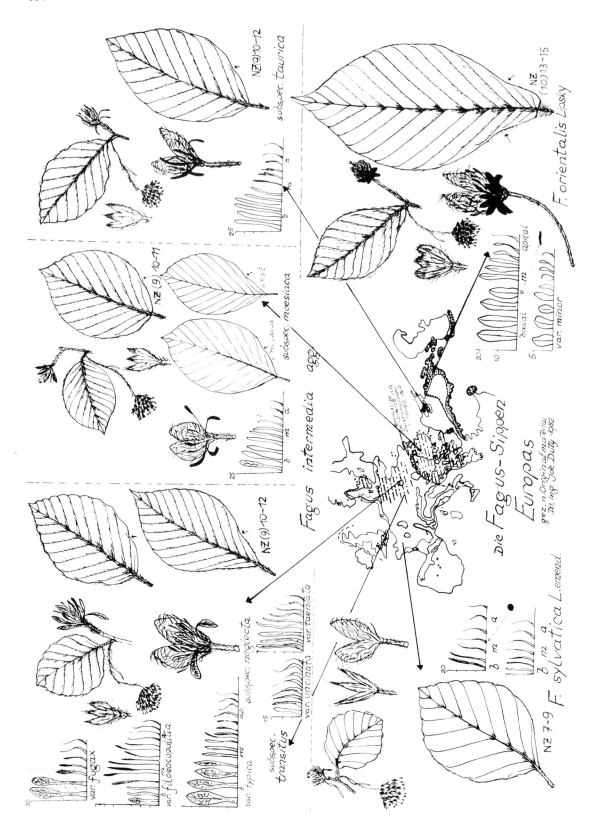

Appendix. Einige Merkmale der *Fagus*-Sippen Europas.

Die zönologischen Verhältnisse der dazischen und dazisch-balkanischen Arten aus dem rumänischen Karpatenraum*

N. Boşcaiu & F. Täuber
Acad. RSR, Subcomisia Monumentelor Naturii, Str. Republicii 9, 3400 Cluj – Napoca, Rumania

Keywords: Dacian-Balkan phytoelement, Dacian phytoelement, Phytochorology in the Carpathian-Pontic-Danubial area, Rumanian Carpathian Mountains, Syntaxonomy

Abstract

The initiatives adopted in Rumania for conserving the native flora pointed out the importance of identifying syntaxa which contain endemic and relict species. The Dacian endemic plants, form the most representative phytoelement in the flora of the Rumanian Carpathians. They include the species whose extension is limited to the Carpathian-Pontic-Danubial area (Pax, 1898).

On the basis of caryological information some considerations concerning the origin of certain Dacian plant species as well as the vicarious relationship with species in other mountain-mass of Eurasia are presented. In order to contribute to the protection of the Dacian and Dacian-Balkan phytoelement within the natural reserves, the sociological affinity of these species and the specification of the alliances where they find their sociological optimum, are persued.

Allgemeine Betrachtungen

Die in Rumänien adoptierten Initiativen zur Konservierung des autochthonen Florenbestandes haben die Notwendigkeit des Studiums der endemischen und reliktären Arten hinsichtlich ihres Schutzes in einem ausgedehnten Netz von Naturreservaten, erneut in den Vordergrund gerückt (Boşcaiu, 1979; Täuber, 1980; Täuber & Wollmann, 1979). Die Schutzinteressen dieser Pflanzenarten sind umsomehr begründet, als sie genetisch und auch in ihren zönologischen Verhältnissen die höchstmögliche Quantität von biogeographischer und biohistorischer Information enthalten. Die biogeographische Bedeutung der Endemiten hat als lebendige Zeugen florogenetischer Archive be-

sonders infolge der rapiden Zerstörung der natürlichen Standorte, an denen sie bis heute überdauerten, steigendes Interesse gefunden (Favarger, 1974).

Tatsächlich verbleiben die Phytoendemiten als jene 'semantophore' Arten im biogeographischen Rahmen Rumäniens, welche die reichste Information über die phytohistorischen Ereigniße liefern, welche die gegenwärtige Florenvielfalt des karpatisch-pontisch-danubischen Raumes charakterisieren. Die Diversität der Geoelemente, welche die Flora Rumäniens bilden, erklärt sich durch die geographische Lage dieses Landes, das in der Übergangszone zwischen den Gebirgsmassiven Zentraleuropas und der Balkanhalbinsel liegt, wie auch durch die vielseitige, temperiert-kontinental-submediterrane Klima-Interferenz und vor allem durch die schwächeren Auswirkungen der pleistozänen Vereisung, welche die Erhaltungsmöglichkeiten einiger Pflanzenarten 'in situ' und die frühzeitige Rückwanderung der Waldvegetation aus ihren glazialen Refugien zuließen.

* Die Sippennomenklatur richtet sich nach Flora Europaea, Vol. 1–5, Cambridge University Press (1964–1980). Flora Republicii Populare Romîne, Vol. 1–10, Edit. Acad. R.P. Romîne (1952–1965). Flora Republicii Socialiste România, Vol. 11–13, Edit. Acad. R.S. România (1966–1976). Index Kewensis, Vol. 1–2, Oxford Clarendon Press (1960).

Vegetatio 59, 185–192 (1985).

Der Endemismus ist ein relativer Begriff, wie es schon Pax (1898) hervorgehoben hat, der von der Ausbreitung des in Betracht genommenen Territoriums abhängt, wobei die Anzahl der Endemiten-Taxa gleichfalls mit dessen Vergrößerung zunimmt. In unseren Betrachtungen beziehen wir uns auf das dazische Phytoelement, welches das repräsentativste Element der rumänischen Karpatenflora darstellt, wenn es auch nicht immer mit der strengen Auffaßung des Endemiten-Begriffs übereinstimmt. Einige dazische Arten, die wir vorstellen, können nicht immer als echt endemisch in engstem Sinn betrachtet werden, da die geographische Verbreitung ihrer Sippen nicht exklusiv auf den rumänischen Karpatenbogen beschränkt ist.

Das dazische Element

In der vorliegenden Arbeit betrachten wir das dazische Element in seiner klassischen Auffassung von Pax (1898), der als berühmter Phytogeograph des Karpatenraums darunter diejenigen Pflanzenarten verstand, deren Verbreitung im Norden der Balkanhalbinsel, im karpatisch-pontisch-danubischen Raum liegt und welche einige Ausstrahlungen in südlichere Balkan-Massiven oder zu den West-Alpen aufweisen können. Nach dieser Auffassung entspricht das Areal des dazischen Phytoelements dem Südostkarpatenraum. Zum Unterschied zu den dazischen Arten sind jene, welche den gesamten Karpatenraum bewohnen, als karpatische Arten (pankarpatische) aufgefaßt. Unter dieser Auffassung wurde das dazische Element sowohl von Borza (1931), Soó (1933), Máthé (1940), wie auch vom Großteil der rumänischen, deutschen und ungarischen Phytogeographen benützt. Es muß noch bemerkt werden, daß sich das dazische Element nach Pax's Auffassung mit dem illyrischen nach Gajewski (1937) überlagert, wie auch mit dem nord-balkanischen nach Popov (1949). Zu dem in dieser Weise aufgefaßten dazischen Element rechnen wir noch einige Arten mit folgendem Areal: dazisch-balkanische, dazisch-balkanisch-pannonische, dazisch-balkanisch-pontische und sogar dazisch-balkanisch-kaukasische (zusammenhängende oder konnektive) Elemente.

Obwohl im gewöhnlichen Sprachgebrauch der Phytogeographie ein klarer Unterschied zwischen den arealgeographischen, historischen und genetischen Elementen gemacht wird, ist in der Tat jeder Hinweis zu einem Geoelement auch mit gewissen Angaben (wenn sie auch noch so vag erscheinen) zu ihrer Entstehung (Genese) und Geschichte verbunden. Dank dieser Umstände wird jedes phytogeographische Spektrum immer Auskünfte vermitteln, in denen sich arealographische Informationen mit ökologischen und historischen in verschiedenen Beziehungen verbinden.

Die statistische Bewertung der Häufigkeit der dazischen Endemiten in den rumänischen Karpaten-Massiven bereitet vorläufig noch größere Schwierigkeiten, sowohl durch die nicht einheitliche Abschätzung der Rangstufe mancher Taxa, als auch durch die öfters lückenhaften Kenntniße über ihre Verbreitung in verschiedenen Gebieten. Borza (1931) schätzte z.B. die Anzahl der endemischen Taxa (Arten und Infrataxa) in Rumänien auf bis zu 283 Einheiten. Dazu muß aber bemerkt werden, daß in diese Vielzahl auch zahlreiche apogame Taxa *(Hieracium)*, wie auch einige Art-Untereinheiten eingereiht wurden, deren taxonomischer Status durch spätere Forschungsergebnisse widerlegt wurde. In einer kritischen Neubetrachtung, die sich auf den taxonomischen Standard des 'Conspectus florae romaniae regionumque affinium' (Borza, 1948) stützt, rechnet der Autor zu den 3 339 aus Rumänien bekannt gewordenen Gefäßpflanzen 148 endemische Arten und 38 wichtigere endemische Unterarten. Diese Anzahl schloß aber auch die extrakarpatischen Endemiten, also auch jene aus der Dobrudscha ein. Wir erwähnen noch, daß Máthé (1940) in der Liste des dazischen Elements 112 Arten aufgezählt hat. Pawlowski (1970) hingegen führt 123 Endemiten (einschließlich Subendemiten) im gesamten Karpatenbogen an, darunter 25 pankarpatische Arten und 85 Arten, die nur den Südostkarpaten (dazisch!) eigen sind; für den rumänischen Karpatenraum werden also 110 Arten angegeben. Nach vergleichendem Studium der Flora aus den wichtigsten Gebirgsmassiven Rumäniens erwähnt Beldie (1967) 98 Endemiten in de Rumänischen Karpaten, zu denen er auch noch 76 dazische Arten (karpatisch-balkanische) hinzufügt. Eine von Morariu & Beldie (1967) veröffentlichte Liste enthält über 127 endemische Arten (auch einige Unterarten einbezogen) für den Gesamtraum Rumäniens, eine neuere Statistik von Stefureac & Tăcină (1978) 130 endemische Taxa (Arten, Unterarten, seltener Varietäten).

Typen von Endemiten

Erwähnenswert ist, daß häufig Betrachtungen hinsichtlich der Beziehung zwischen Häufigkeit und Floren-Alter folgendermaßen angestellt wurden: umso älter die Flora eines Territoriums ist, umso mehr Endemiten – die in geographischer und systematischer Sicht isoliert erscheinen – enthält sie (Pawlowski, 1970). Diese axiomatische Aussage enthält, bis zu einer gewissen Grenze, eine beträchtliche Wahrheitsdosis. Es wurde bereits unterstrichen (Küpfer, 1974), daß das Endemiten-Vorkommen nicht immer ausschließlich die konservative Komponente der Flora widerspiegelt. Die Anwendung von cytotaxonomischen und cytogenetischen Methoden in der Pflanzengeographie ermöglichte es, innerhalb einer Flora verschiedenaltrige Elemente zu unterscheiden und somit den passiven Endemismus (Paläoendemismus) vom aktiven Endemismus (Neoendemismus) zu trennen (Favarger & Contandriopoulos, 1961).

Als Beispiele aus der Endemiten-Flora Rumäniens können wir *Lychnis nivalis (=Polyschemone nivalis)* und eventuell auch *Symphytum cordatum* erwähnen, worunter die letzte Art sogar gegenüber *Symphytum grandiflorum* aus dem Kaukasus ziemlich isoliert bleibt, obwohl sie zu dieser Art einige Ähnlichkeiten aufweist.

In der Flora Rumäniens treten schizoendemische Taxa viel häufiger auf, die durch eine stufenweise Differenzierung aus einer gemeinsamen Ur-Sippe hervorgingen. Zu dieser Endemitenkategorie gehören, durch ihr disjunktes Areal und dieselbe Chromosomen-Anzahl, folgende vikariierende Taxa: *Syringa josikaea* (Westgebirge) – *Syringa emodi* (Himalaja Gebirge), *Potentilla haynaldiana* (Parîng und Pirin Gebirge) – *Potentilla valderia* (Meeresalpen), *Melandrium zawadzkii* (Ostkarpaten) – *Melandrium elisabetae* (zwischen Lago de Como und Mont Baldo), *Primula leucophylla* (Ostkarpaten) – *Primula ruprechtii* (Kaukasus), *Asperula capitata* (Südkarpaten u. Westgebirge) – *Asperula hexaphylla* (Meeresalpen), *Veronica bachofenii* (Südkarpaten) – *Veronica dahurica* (China, Japan, Ostsibirien). Wahrscheinlich könnten zukünftige cytotaxonomische Forschungen zu dieser Kategorie noch folgende vikariierende Arten einbeziehen: *Saxifraga carpathica* (Südostkarpaten) – *Saxifraga sibirica* (Sibirien), *Gypsophila petraea* (Südostkarpaten) – *Gypsophila uralensis* (Ural Gebirge),

Galium bailloni (Vîlcei Gebirge, Südkarpaten) – *Galium valantoides* (Kaukasus), *Hepatica transsilvanica* (Südostkarpaten) – *Hepatica henryi* (China), *Aconitum moldavicum* (Karpaten) – *Aconitum septentrionale* (Nord-Eurasien). Die phytohistorischen Implikationen der zitierten Beispiele heben das hohe Alter einiger dazischer und dazisch-balkanischer Schizoendemiten hervor, wie auch die Komplexität der florogenetischen Vorgänge, die zur Genese der Karpatenflora beitrugen.

Aus der Reihe der apo-endemischen Taxa, die durch Auto- oder Allopolyploidie entstanden sind, können Arten wie *Draba dorneri, D. simonkaiana, Campanula abietina* und eine große Anzahl von apomyktischen Mikrospezies der Gattungen *Alchemilla, Rubus* und *Hieracium* aufgezählt werden, welche den aktiven Anteil des dazischen und dazisch-balkanischen Endemismus darstellen.

Geographische Verteilung der Endemiten

Die geographische Verteilung dieser Endemiten in den Südostkarpaten-Massiven ist sehr verschieden. Neben den pankarpatischen und dazischen Elementen, die in allen Massiven vorkommen, existieren auch einige Arten mit eng begrenzter Verbreitung. Im allgemeinen ist die Frequenz der Endemiten in den Südkarpaten größer als in den Ostkarpaten. Trotzdem sind einige der wichtigsten Endemiten, die von südlicheren Sippen ausgingen, wie *Andryala levitomentosa, Melandrium nemorale* und *Heracleum carpaticum* ausschließlich nur im Norden der Ostkarpaten verbreitet und fehlen in den Südkarpaten. Sowohl Pax (1898), wie auch Pawlowski (1970) haben ebenfalls festgestellt, daß die endemitenreichsten Gebiete nicht im Innenraum der Karpaten-Massive liegen, wie in den Alpen, sondern am Aussenrand, in ihrer Peripherie. Ebenso hat Pawlowski die Beobachtung vermerkt, daß die ökologische und phytosoziologische Amplitude der Karpaten-Endemiten weiter ausgedehnt ist, als die der Alpen-Endemiten. Zum Unterschied aber zu den Waldökosystemen der Alpen, die nur wenige Endemiten beherbergen, zählen wir die Karpaten-Buchenwälder zu den endemitenreichsten Wäldern, die zahlreiche dazische Elemente enthalten.

Die weite Disjunktion der Vikarianten einiger nemoraler Endemiten aus der montanen Stufe der

Südostkarpaten zeigt, daß sie bezüglich der floro-genetischen Etappen, in welchen sie entstanden, älter sind. In dieser Hinsicht könnten die Betrachtungen von Favarger (1975) über die Genese der orophilen Alpenflora auch auf die Entstehung der Karpatenflora ausgeweitet werden. Tatsächlich ist die montane und subalpine Flora in den Karpaten viel reicher an Endemiten und Tertiärrelikte – deren entsprechende Vikarianten oft weit entfernt liegen – als ihre alpine Flora, wie es auch schon Meusel (1968) in den Südkarpaten beobachtet hat. Diese Naturbegebenheit ist erklärbar, wenn wir auf die Bemerkung von Favarger (1975) Rücksicht neh-men, die besagt, daß die Karpatenflora wie auch die Alpenflora zur heutigen Höhenlage erst gelangte, nachdem die nemorale miozäne Flora bereits dif-ferenziert war. Diesbezüglich haben sich höchst-wahrscheinlich, ebenso wie im Fall der Alpenflora, zahlreiche nemorale Oreophyten (und Hemioreo-phyten) schon im Miozän aus der planaren ark-tisch-tertiären Flora differenziert, welche durch orogenetische Hebungen in höhere Lagen gelang-ten, wo neue ökologische Bedingungen einwirkten.

Das xerotherme Klima aus der Pontian-Zeit hat seinerseits das Eindringen der Steppenpflanzen aus Asien begünstigt, deren Sippen ebenso altitudial durch pliozäne orogenetische Gebirgsfaltungen emporgehoben wurden und dadurch zur Genese einiger Gesteins- und Geröll-Endemiten (Astraga-lus, Oxytropis, Helictotrichon) beitragen konnten. Die ökologischen Argumente plädieren für die Hypothese, daß ursprüngliche Steppen-Arten in höhere Stufen nur in einer xerothermen, waldlosen oder waldlückigen Periode gelangen konnten. Da sich die Flora des karpatisch-pontisch-danubischen Raumes an der Ausklangsgrenze der zentraleuro-päischen Provinz befindet, war sie den orientali-schen Einflüßen mehr ausgesetzt als die Alpenflora.

Gestützt auf die phytohistorischen Analogien mit der Alpenflora und anhand der taxonomischen Affinität mit aktuellen Vikarianten können wir auch unter den Karpaten-Oreophyten einen meri-dionalen Flora-Stamm und einen borealen unter-scheiden, wie ihn schon frühzeitig Diels (1910) erkannte. Sowohl die paläogeographischen Unter-lagen, wie auch jene cytotaxonomischen zeigen, daß die aus den meridionalen Pflanzen hervorge-gangenen Endemiten viel älter sind als jene, die dem borealen Stamm entsprangen. Trotzdem kann auch die Hypothese wahrscheinlich bestehen bleiben,

daß einige aus dem Nord-Stamm hervorgegangene Endemiten sich auch schon am Ende des Tertiärs trennen konnten (Cochlearia borzeana = Cochlea-ria pyrenaica ssp. borzeana); der Ursprung des Großteils bleibt aber an die Wirkung der pleisto-zänen Vergletscherung gebunden. Die polyploiden Sippen, die unter der Wirkung der Vereisung ent-standen sind, haben dank ihrer ökologischen Agres-sivität ein eurizönotisches Verhalten. Insbesondere die apomyktischen Arten (Alchemilla, Hieracium) werden in sehr unterschiedlichen phytozönotischen Komplexen angetroffen. Die phytosoziologischen Betrachtungen, die von der 'stenozönotischen' Spe-zialisierung einiger Endemiten ausgehen, die vom borealen Oreophyten-Zweig abstammen, ergeben die Notwendigkeit einer kritischen Überprüfung der Hypothesen hinsichtlich der Wanderungs-ströme und des floristischen Austausches, die in der Gletscherzeit stattgefunden haben (Merxmüller & Poelt, 1954; Favarger, 1975).

Schlußfolgerung

Die vorangestellten Betrachtungen ermöglichen auf florengeschichtlicher Basis eine klarere Unter-scheidung der dazischen und karpatischen Phyto-elemente. So kann z.B. festgestellt werden, daß im Rahmen des dazischen Elements der Anteil der aus dem meridionalen Stamm entsprungenen Arten miozänen Ursprungs aus der oreophilen Flora Europas überwiegt, indem auch ein hoher Bestand von balkanogenen Arten teilnimmt. Im Fall des karpatischen Elements herrschen die Arten borea-len Ursprungs vor, die dem nördlichen Stamm ent-stammen, einen alpigenen Ursprung zeigen und sich wahrscheinlich später während des Pliozäns und Pleistozäns differenzierten.

Hinsichtlich eines wirksamen Schutzes der da-zischen, karpatischen und dazisch-balkanischen Florenelemente der Rumänischen Karpaten in Naturreservaten, die sich auf phytosoziologische Kriterien stützen, haben wir die Festlegung ihres soziologischen Anschlusses als notwendig gefun-den, indem die Syntaxa bestimmt werden, in wel-chen sie ihr soziologisches Optimum finden. Die soziologische Amplitude der Mehrheit der unter-suchten Pflanzenarten zeigt, daß die aussage-kräf-tigste syntaxonomische Einheit, auf welche das soziologische Optimum der dazischen, karpatischen und dazisch-balkanischen Phytoelemente bezogen werden kann, der Verband darstellt. Eine statistische Übersicht der Zuteilung der Florenelemente Rumäniens zu verschiedenen Verbänden wurde schon früher veröffentlicht (Boşcaiu, 1979).

Nachfolgend stellen wir in analytischer Weise die syntaxonomischen Zugehörigkeiten der dazischen und dazisch-balkanischen Elemente zu den Verbänden dar, in welchen sie ihr soziologisches Optimum vorfinden. Wir unterstreichen, daß jede einzelne Pflanzenart einem einzigen Verband zugeteilt wurde, der wahrscheinlich ihr soziologisches Optimum im rumänischen Karpatenraum ausdrückt, wenngleich sie auch in anderen Assoziationen erscheinen kann.

Es bleibt eine Aufgabe für zukünftige Forschungen, ausführlicher auch die Zugehörigkeit dieser Pflanzenarten im Rahmen einer umfassenden Übersicht der Phytoassoziationen der Rumänischen Karpaten festzulegen und widerzugeben.

Literatur

Beldie, Al., 1967. Endemismele şi elementele dacice din flora Carpaţilor României. Comunicări de botanică. A V-a Consfătuire de geobotanică: 113–129. Bucureşti.

Borza, Al., 1931. Die Vegetation und Flora Rumäniens. Guide de la sixième Excursion Phytogéographique Internationale Roumanie: 1–55. Cluj.

Borza, Al., 1948. Une nouvelle statistique de la flore roumaine. Bull. Soc. Bot. Fr. 95: 289–293.

Boşcaiu, N., 1979. L'integration phytosociologique du génofond végétal (plantes vasculaires) de la Roumanie. Docum. phytosociol. ser. n. 4: 87–109.

Boşcaiu, N., 1979. Integrarea fitocenotică şi constituirea rezervaţiilor botanice. Ocrot. Nat. Med. Inconj. 19: 17–21.

Diels, L., 1910. Genetische Elemente in der Flora der Alpen. Beibl. Bot. Jahrb. 102: 7–46.

Favarger, C., 1969. L'endemisme en géographie botanique. Scientia 681–682: 1–16.

Favarger, C., 1972. Endemism in the Montane Floras of Europe. In: D. H. Valentine (ed.), Taxonomy, Phytogeography and Evolution, pp. 191–204. Academic Press, London, New York.

Favarger, C., 1974. Progrès récents dans l'étude de l'endémisme végétal en Europe. Lavori Soc. Ital. Biogeogr. ser. n. 5: 1–29.

Favarger, C., 1975. Cytotaxonomie et histoire de la flore orophile des Alpes et de quelques autres massifs montagneux d'Europe. Lejeunia ser. n. 77: 1–45.

Favarger, C. & Contandriopoulos, J., 1961. Essai sur l'endémisme. Ber. Schweiz. bot. Ges. 71: 384–408.

Gajewski, W., 1937. Elementy flory polskiego Podola. (Elements of the flora of Polish Podole). Planta Polonica 5.

Hendrych, R., 1981. Bemerkungen zum Endemismus in der Flora der Tschechoslowakei. Preslia 53: 97–120.

Küpfer, P., 1974. Recherches sur les liens de parenté entre la flore orophile des Alpes et celle des Pyrénées. Boissera 23: 1–322.

Kuzmanov, B. A., 1969. Some aspects of the origin of the Bulgarian flora. Publ. de la Universidad de Sevilla. V Simp. de Flora Europaea (20–30 le majo de 1967): 133–147. Sevilla.

Máthé, I., 1940. Magyarorszag növényzetének flóraelemei. Tisia 4: 116–147.

Merxmüller, H. & Poelt, J., 1954. Beiträge zur Florengeschichte der Alpen. Ber. Bayer. Bot. Ges. 30: 91–101.

Meusel, H., 1968. Geobotanische Beobachtungen in den Südost-Karpaten. Arch. Naturschutz u. Landschaftforsch. 8: 175–210.

Morariu, I. & Beldie, Al., 1976. Endemismele din flora R. S.R. In: Flora Republicii Socialiste Romania XIII: 97–133, Edit. Academiei Bucureşti.

Pawlowski, B., 1970. Remarque sur l'endémisme dans la flore des Alpes et des Carpates. Vegetatio 21: 182–243.

Pax, F., 1898. Grundzüge der Pflanzenverbreitung in den Karpathen. I. Band. Leipzig. 270 pp.

Popov, M. G., 1949. Ocherk rastitelnosti i flory Karpat. Studie über Vegetation und Flora von Karpaten. Verlag Izdatelstvo Moskovskogo obschestva ispytatelej prirody. Moskva.

Soó, R., 1933. Analyse der Flora des historischen Ungarns (Elemente, Endemismen, Relicte). Magy. Biol. Inst. Munkai 4.

Stefureac, T. I. & Tăcină Aurica, 1978. Unele consideraţii asupra endemismelor şi corologia taxonilor endemici în România. Stud. şi cetc. Biol. Ser- Biol. veg. 30: 85–92.

Szafer, W., 1975. General plant geography. PWN, Polish Scientific Publishers, Warszawa. 430 pp.

Szücs, L., 1943. A Keleti Karpátok endemikus növényfajai I (Die endemischen Pflanzenarten der Ostkarpathen I). Acta Geobot. Hung. 5: 185–240.

Täuber, F., 1980. Preocupări pentru conservarea florei autohtone. Ocrot. Nat. Med. Inconj. 24: 111–114.

Täuber, F. & Wollmann, S., 1979. Syntaxonomische Verhältnisse einiger Pflanzenendemiten aus den rumänischen Karpaten an ihrer Fundorten (I Ost-Karpaten). Documents phytosociologiques NS 4: 917–922.

Walter, H., 1975. Über ökologische Beziehungen zwischen Steppenpflanzen und alpinen Elementen. Flora 164: 339–346.

Accepted 25.2.1984.

Appendix

Syntaxonomische Eingliederung der dazischen und dazisch-balkanischen Phytoelemente aus dem rumänischen Karpatenraum

ASPLENIETEA RUPESTRIA Br.-Bl. 34
A n d r o s a c e t a l i a   v a n d e l i i Br.-Bl. 34
Silenion lerchenfeldianae Horv. et Pawl. 49
D: *Dianthus henteri* Heuff., *Draba dorneri* Heuff., *Draba si-*

monkaiana Jáv., Senecio glaberrimus (Roch.) Simk., *Silene dinarica* Spreng., *Silene nutans* L. ssp. *dubia* (Herb.) Zapal.
Carp: *Poa rehmanii* (A. et G.) Woloszczak
D-Balc: *Potentilla haynaldiana* Janka, *Silene lerchenfeldiana* Baumg.
D-Balc-Anat: *Symphyandra wanneri* (Roch.) Heuff.
D-Balc-Pont: *Veronica bachofenii* Heuff.

Potentilletalia caulescentis Br.-Bl. et Jenny 26
Moehringion muscosae Horv. et H-ić ap. Horv. 62
Carp: *Campanula carpatica* Jacq.

Gypsophilion petraeae Borhidi et Pocs 58
D: *Andryala levitomentosa* (E. I. Nyár.) P. D. Sell., *Eritrichium nanum* (L.) Schrad. ssp. *jankae* Simk., *Saxifraga mutata* ssp. *demissa* (Schott. et Kotschy) D. A. Webb., *Thesium kernerianum* Simk.
D-Balc: *Gypsophila petraea* (Baumg.) Rchb.

Micromerion pulegii Boşcaiu 71
D: *Micromeria pulegium* (Roch.) Benth.
D-Balc: *Sesleria filifolia* Hoppe, *Silene saxifraga* L. ssp. *petraea* (W. et K.) Guşul.

THLASPIETEA ROTUNDIFOLII Br.-Bl. et al. 48
Androsacetalia alpinae Br.-Bl. et Jenny 26
Androsacion alpinae Br.-Bl. et Jenny 26
D: *Dianthus glacialis* Haenke ssp. *gelidus* (Schott., Nym. et Kotschy) Tutin, *Lychnis nivalis* Kit.
Carp: *Aquilegia transsilvanica* Schur, *Poa granitica* Br.-Bl. ssp. *disparilis* (E. I. Nyár.) E. I. Nyár., *Poa nyaradyana* Nannf.
Carp-Balc: *Saxifraga carpathica* Rchb.
D-Balc: *Doronicum carpaticum* (Gris. et Sch.) Nym. *Saxifraga pedemontana* All. ssp. *cymosa* (W. et K.) Engler, *Veronica baumgartenii* Roem. et Schult.

Thlaspietalia rotundifolii Br.-Bl. et Jenny 26
Papavero-Thymion pulcherrimi I. Pop 68
Carp: *Papaver corona-sancti-stephani* Zapal., *Thymus bihorensis* Jalas, *Thymus pulcherrimus* Schur.
Carp-Balc: *Poa cenisia* All. ssp. *contracta* (E. I. Nyár.) E. I. Nyár.
D-Balc: *Cerastium arvense* L. ssp. *lerchenfeldianum* (Schur) A. et G.

Stipion calamagrostis Jenny-Lips 30
D: *Aquilegia nigricans* Baumg. ssp. *subscaposa* (Borb.) Soó

MONTIO-CARDAMINETEA Br.-Bl. et Tx. 43
Montio-Cardaminetalia Pawl. 28
Cardamino-Montion Br.-Bl. 25
D: *Barbarea lepuznica* Nyár.
Carp: *Cardamine opizii* J. et C. Presl.
D-Balc: *Saxifraga heucherifolia* Gris. et Sch.

Cratoneurion commutati W. Koch 28
Carp: *Cochlearia borzeana* (Com. et Nyár.) Pobed. (= *Cochlearia pyrenaica* DC. var. *borzeana* Com. et Nyár.)

BETULO-ADENOSTYLETEA Br.-Bl. 48
Adenostyletalia Br.-Bl. 31
Adenostylion alliariae Br.-Bl. 25
D: *Heracleum transsilvanicum* Schur, *Hesperis oblongifolia* Schur.
Carp: *Alopecurus laguriformis* Schur, *Hesperis nivea* Baumg.

Carp-Balc: *Salix silesiaca* Willd.
D-Balc: *Aconitum toxicum* Rchb., *Melandrium nemorale* Heuff.

Calamagrostion Luq. 26
D: *Hieracium paltinae* Jáv. et Zahn.
Carp: *Festuca carpatica* Dietz, *Festuca porcii* Hack., *Heracleum carpaticum* Porc., *Phyteuma vagneri* A. Kern., *Saussurea porcii* Deg.

SCHEUCHZERIO-CARICETEA FUSCAE (Nordh. 37) Tx. 37
Caricetalia davallianae Br.-Bl. 49
Caricion davallianae Klika 34
D: *Armeria maritima* (Miller) Willd. ssp. *barcensis* (Simk.) P. Silva

Caricetalia fuscae Koch 26
Caricion canescenti-fuscae Koch 26
Carp: *Euphorbia carpatica* Wol.
D-Balc: *Carex dacica* Heuff., *Dactylorhiza cordigera* (Fries) Soó, *Pedicularis limnogena* A. Kern., *Pseudorchis frivaldii* (Hampe) P. Hunt.

SEDO-SCLERANTHETEA Br.-Bl. 55
Sedo-Scleranthetalia Br.-Bl. 55
Alysso-Sedion Oberd. et Müller 61
D-Balc: *Alyssum wierzbickii* Heuff.

FESTUCO-BROMETEA Br.-Bl. et Tx. 43
Brachypodio-Chrysopogonetalia Boşcaiu 71
Chrysopogoni-Danthonion Kojič 57
D: *Dianthus giganteus* D'Urv. ssp. *banaticus* (Heuff.) Tutin
D-Balc: *Dianthus giganteus* D'Urv. ssp. *giganteus*
D-Balc-Pann: *Dianthus pontederae* Kern. ssp. *giganteiformis* (Borb.) Soó

Stipo-Festucetalia pallentis I. Pop 68
Asplenio-Festucion pallentis Zolyomi 36
D: *Centaurea coziensis* Nyár.
D-Balc-Pann: *Minuartia hirsuta* (Bieb.) Hand-Mazz. ssp. *frutescens* (Kit.) Hand-Mazz.

Seslerio-Festucion pallentis Klika 31
D: *Centaurea phrygia* L. ssp. *rătezatensis* (Prod.) Dost. *Festuca pachyphylla* Deg., *Thymus comosus* Heuff.
Carp: *Centaurea reichenbachii* DC., *Knautia kitaibelii* (Schult.) Borb.
Carp-Balc: *Ligularia carpatica* (Sch. N. et Ky) Pojark.
Carp-Balc-Pann: *Chamaecytisus ciliatus* (Wahlenb.) Rothm.
D.-Balc: *Centaurea triniifolia* Heuff., (?) *Cirsium grecesui* Rouy, *Iris reichenbachii* Heuff., *Jovibarba heuffeli* (Schott) A. et D. Löve, *Potentilla tommassiniana* F. Schult, *Phleum montanum* K. Koch
D-Pann: *Ferula sadleriana* Ledeb., *Scabiosa columbaria* L. ssp. *pseudobanatica* (Schur) Jáv.

Bromo-Festucion pallentis Zolyomi 66
D: *Delphinium simonkaianum* Pawl.
D-Balc: *Cerastium banaticum* (Roch.) Heuff., *Chamaecytisus glaber* (L. fil.) Rothm., *Chamaecytisus rochelii* (Wierzb.) Rothm., *Delphinium fissum* W. et K., *Minuartia setacea* (Thuill.) Hay. ssp. *banatica* (Heuff.) Prod.
D-Balc-Pann: *Chamaecytisus heuffelii* (Wierzb.) Rothm.
D-Balc-(Pann): *Silene flavescens* W. et K.

Festucetalia valesiacae Br.-Bl. et Tx. 43

Festucion rupicolae Soó 40

Trans: *Cephalaria radiata* Gris., *Jurinea mollis* (Torn.) ssp. *transsilvanica* (Spreng.) Hayek, *Salvia transsilvanica* Schur.

Trans-Pont: *Centaurea trinervia* Stephen, *Iris humilis* M.B.

D-Balc: *Centaurea rutifolia* Sibth. et Sm. ssp. *jurineifolia* (Boiss.) Nyman

MOLINIO-ARRHENATHERETEA Tx. 37

Molinietalia W. Koch 26

Molinion coeruleae W. Koch 26

D: *Peucedanum rochelianum* Heuff.

Arrhenatheretalia Pawl. 28

Arrhenatherion elatioris Roch.

D-Balc: *Rhinanthus rumelicus* Velen.

D-Pann: *Centaurea banatica* Roch.

Polygono–Trisetion Br.-Bl. 48

Carp: *Centaurea phrygia* L. ssp. *carpatica* (Porc.) Dostál, *Phyteuma tetramerum* Schur, *Ranunculus carpaticus* Herb., *Viola declinata* W. et K

SALICETEA HERBACEAE Br.-Bl. 47

Salicetalia herbaceae Br.-Bl. 26

Salicion herbaceae Br.-Bl. 26

D-Balc-Anat: *Plantago gentianoides* Sibth. et Sm.

ELYNO-SESLERIETEA Br.-Bl. 48

Seslerietalia Br.-Bl. 26

Seslerion bielzii Pawl. 35

D: *Centaurea pinnatifida* Schur. *Dianthus callizonus* Sch. et Ky., *Dianthus tenuifolius* Schur, *Gentiana cruciata* ssp. *phlogifolia* (Schott. et Kotschy) Tutin

Carp: *Achillea oxyloba* (DC.) Schultz-Bip ssp. *schurii* (Schultz-Bip) Heimerl, *Campanula kladniana* (Schur) Witas, *Draba haynaldi* Stur, *Erysimum wittmannii* Zawad. ssp. *transsilvanicum* (Schur) P. W. Ball, *Leontodon montanus* Lam. ssp. *pseudotaraxaci* (Schur) Finch et P. D. Sell, *Oxytropis carpatica* Uechtr., *Primula wulfeniana* Schott ssp. *baumgarteniana* (Deg. et Moesz.) Lüdi

Carp-Balc: *Erysimum wittmannii* Zawad. spp. *wittmannii*

D-Balc: *Bupleurum diversifolium* Roch., *Centaurea kotschyana* Heuff., *Linum perenne* L. ssp. *extraaxillare* (Kit.) Nyman, *Sesleria bielzii* Schur, *Sesleria coerulans* Fries, *Sesleria rigida* Heuff. ssp. *haynaldiana* (Schur) Beldie.

Seslerion rigidae Zolyomi 39

D: *Astragalus pseudopurpureus* Guşul, *Astragalus römeri* Simk., *Athamantha turbith* (L.) Brot. ssp. *hungarica* (Borbás) Tutin, *Helictotrichon decorum* (Janka) Henrad, *Linum uninerve* (Roch.) Borb., *Melandrium zawadzkii* (Herb.) A. Br., *Pedicularis baumgartenii* Simk., *Primula elatior* (L.) Hill ssp. *leucophylla* (Pax) H.-Harrison, *Saxifraga marginata* Sternb. var. *rocheliana* (Sternb.) Engl. et Irm., *Viola jooi* Janka

Carp: *Centaurea marmarosiensis* (Jáv.) Czerep., *Hieracium pojoritense* Wol., *Trisetum macrotrichum* Hack.

D-Balc: *Asperula capitata* Kit., *Carduus kerneri* Simk. *Centaurea atropurpurea* W. et K., *Cephalaria laevigata* (W. et K.) Schrad., *Dianthus petraeus* W. et K. ssp. *simonkaianus* (Péterfi) Tutin, *Dianthus petraeus* W. et K. ssp. *petraeus*,

Dianthus spiculifolius Schur, *Erysimum comatum* Pancic, *Festuca xanthina* Roem. et Schult. *Hypericum umbellatum* A. Kern., *Lilium jankae* A. Kern, *Saxifraga luteoviridis* Sch. et Ky, *Scrophularia heterophylla* Willd ssp. *laciniata* (W. et K.) Maire et Petitmengin, *Seseli gracile* W. et K., *Seseli rigidum* W. et K. ssp. *rigidum, Sesleria rigida* Heuff.

D-Balc-Pann: *Centaurea mollis* W. et K., *Draba lasiocarpa* Roch., *Sempervivum marmoreum* Gris.

CARICETEA CURVULAE Br.-Bl. 48

Caricetalia curvulae Br.-Bl. 26

Caricion curvulae Br.-Bl. 25

Carp-Balc: *Senecio abrotanifolius* L. ssp. *carpaticus* (Herb.) Nyman, *Poa media* Schur

NARDO-CALLUNETEA Prsg. 40

Nardetalia (Oberd. 49) Prsg. 29

Potentillo ternatae–Nardion Simon 57

D-Balc: *Potentilla aurea* ssp. *chrysocraspeda* (Lohm.) Nym., (?) *Viola dacica* Borb., *Gentianella bulgarica* (Velen.) J. Holub.

Nardo–Agrostion tenuis Sillinger 33

Carp: *Armeria pocutica* Pawl.

TRIFOLIO-GERANIETEA SANGUINEI Th. Müll. 61

Origanetalia Th. Müll. 61

Trifolion medii Th. Müll. 61

D: *Centaurea trichocephala* Bieb. ssp. *simonkaiana* (Hayek) Dostál.

Carp-Balc: *Laserpitium archangelica* Wulfen

D-Balc: *Centaurea degeniana* Wagn.

Geranion sanguinei Tx. 60

D-Balc: *Dianthus puberulus* (Simk.) Kern.

QUERCETEA PUBESCENTI-PETRAEAE Jakucs 60

Orno-Cotinetalia Jakucs 60

Orno-Cotinion Soó 60

Carp-Balc: *Coronilla elegans* Panč.

D-Balc: *Allium paniculatum* L. ssp. *fuscum* (W. et K.) Archangeli

Syringo–Carpinion orientalis Jakucs 59

D: *Dianthus giganteus* D'Urv. ssp. *banaticus* (Heuff.) Tutin

D-Balc: *Syringa vulgaris* L.

Quercion farnetto Horvat 54

D:Balc: *Campanula sparsa* Friv., *Dianthus trifasciculatus* Kit.

Quercetalia pubescentis Br.-Bl. 31

Quercion petraeae Zolyomi et Jakucs 60

D: Centaurea indurata Janka

D-Pann: *Paeonia officinalis* L. ssp. *banatica* (Roch.) Soó

CARPINO-FAGETEA Jakucs 60

Fagetalia Pawl. 28

Alno-Padion Knapp 42

D: *Syringa josikaea* Jacq. f.

D-Balc-(Pann): *Oenanthe banatica* Heuff.

Symphyto–Fagion Vida 59

(incl. *Lathyro–Carpinenion* (Boşcaiu 79) et *Moehringio–Acerenion* (Boşcaiu 79)

D: *Hepatica transsilvanica* Fuss, (?) *Galium baillonii* Brandza, *Galium kitaibelianum* Schultes et Schultes fil., *Lathyrus hal-*

lersteinii Baumg., *Symphytum cordatum* W. et K.

Carp: *Aconitum moldavicum* Hacquet, *Lathyrus transsilvanicus* (Spr.) Rchb., *Pulmonaria filarszkyana* Jáv.

D-Balc: *Cardamine glanduligera* O. Schwartz, *Crocus banaticus* Gay., *Pulmonaria rubra* Schott, *Verbascum lanatum* Schrad.

D-Pann: *Helleborus purpurascens* W. et K.

D-Balc-Pann: *Melampyrum bihariense* A. Kern.

Carp-Balc-Cauc: *Wadsteinia geoides* Willd., *Waldsteinia ternata* (Stephen) Fritsch

VACCINIO-PICEETEA Br.-Bl. 39
Vaccinio - Piceetalia Br.-Bl. 39
Chrysanthemo-Piceion Krajina 33
Carp: *Leucanthemum waldsteinii* (Schultz-Bip.) Pouzar
D-Balc: *Anthemis macrantha* Heuff.

Junipero - Pinetalia mugi Boşcaiu 71
Rhododendro-Vaccinion Br.-Bl. 26
D-Balc: *Rhododendron kotschyi* Simk.
Junipero-Bruckenthalion (Horv. 49) Boşcaiu 71
Carp: *Melampyrum saxosum* Baumg.
D-Balc: *Campanula abietina* Gris. et Sch.
D-Balc-Anat: *Bruckenthalia spiculifolia* (Salisb.) Rchb.

Ozeanische Florenelemente in aquatischen Pflanzengesellschaften der D.D.R.*

H.-D. Krausch
Wilhelm-Pieck Strasse 32, 1500 Potsdam, D.D.R.

Keywords: Aquatic plant community, Oceanic, Phytosociology, Plant geography, Suboceanic

Abstract

Aquatic macrophytes with oceanic and suboceanic distribution occur in the G.D.R. mainly in *Littorelletea, Isoeto–Nanojuncetea* and *Montio–Cardaminetea* communities. Within the *Potametea* they are almost entirely confined to the *Ranunculetum fluitantis* and to the group of *Batrachium* communities within the *Nymphaeion*. Within the territory of the G.D.R. several of these species are reaching the eastern border of their distribution areas. Nearly all of them are more or less endangered or even threatened by extinction.

Allgemeine Vorbemerkungen

Innerhalb der in Mitteleuropa vorkommenden höheren Wasserpflanzen stellen die Arten mit mehr oder weniger starker Ozeanitätsbindung ihrer Areale eine zahlenmässig große Gruppe dar. Dabei verstehen wir unter dem Begriff 'Ozeanisches Florenelement' hier solche Sippen, welche sich innerhalb der temperaten Zone auf die Ozeanitätsstufen I und II nach Jäger (1968) beschränken, also die Ozeanitätsindices oz_1, oz_{1-2} und $oz_{1(-3)}$ nach Meusel, Jäger & Weinert (1965) aufweisen, bzw. nach anderer Definition als Atlantiker bzw. Subatlantiker bezeichnet werden.

Diese ozeanischen bzw. subozeanischen Arten zeigen hinsichtlich ihrer zonalen Bindung keine Einheitlichkeit. Neben solchen Arten, die sich im wesentlichen auf die temperate Zone beschränken (atlantische und subatlantische Arten, z.B. *Deschampsia setacea, Hottonia palustris*) gibt es Arten, die mehr oder weniger weit in die südlich angrenzenden submeridionalen und meridionalen Zonen vordringen, in Europa also auch im Mittelmeergebiet vorkommen (atlantisch-mediterrane bzw. subatlantisch-submediterrane Arten, z.B. *Ludwigia palustris*). Ihnen stehen Arten mit geringeren Wärmeansprüchen gegenüber, die im westlichen Europa meist nicht sehr weit nach Süden gehen, dafür aber im Norden bis in die boreale, mitunter sogar bis in die arktische Zone reichen (nordisch-atlantische bzw. subatlantische Arten, z.B. *Myriophyllum alterniflorum* und *Littorella uniflora*).

Die Ursachen der Ozeanitätsbindung derartiger Sippen liegen wohl eindeutig bei den Klimaverhältnissen. Zum einen bewirken die höheren Niederschläge an den atlantischen Küsten Europas, verbunden mit fehlenden oder kaum ausgeprägten Perioden sommerlicher Austrocknung und winterlicher Froststarre, eine stärkere Auswaschung des Bodens und dadurch die Entstehung zahlreicher elektrolytarmer, sauer-oligotropher Gewässer, an die viele der ozeanisch verbreiteten Wasserpflanzen angepaßt sind. Nicht wenige dieser Arten zeigen zugleich ein amphibisches Verhalten, sie vermögen sowohl submerse Formen als auch auf zeitweise trockenfallenden Ufern, Teich-, Tümpel- und Grabenböden emerse Formen zu entwickeln. Häufige Niederschläge und hohe Luftfeuchtigkeit sind für diese Arten unentbehrlich, für die ein zu starkes

---

* Die Sippennomenklatur richtet sich nach Casper & Krausch (1981, 1982).

Austrocknen der litoralen Standorte offensichtlich ungünstige Faktoren darstellen.

Von besonderer Wichtigkeit für die Verbreitung der ozeanischen Wasserpflanzen sind die Temperaturverhältnisse. Es handelt sich bei ihnen um Arten, die an einen ausgeglichenen Gang der Jahres- und Tagestemperaturen angepaßt sind, keine höheren Wassertemperaturen während des Sommers benötigen, dafür aber gegen langanhaltende und starke Winterkälte, welche eine längere Eisbedeckung und ein tiefes Durchfrieren der flacheren Gewässer zur Folge haben, empfindlich sind. Sie besitzen meist keine oder nur unzureichend ausgebildete Überwinterungsorgane, eine ganze Reihe von ihnen ist wintergrün. Ebensowenig sind sie offensichtlich an eine stärkere und länger andauernde sommerliche Austrocknung des Gewässers angepaßt. Im ozeanischen Bereich werden alle diejenigen Wasserpflanzen ausgeschaltet oder gehemmt, die für Keimung, Wachstum und Samenerzeugung auf höhere sommerliche Wassertemperaturen angewiesen sind. Die direkt oder indirekt durch das ozeanische Klima bedingten sauer-oligotrophen Gewässer, die in den westeuropäischen Landschaften verbreitet sind, lassen die konkurrenzkräftigen Eutraphenten nicht zur Entwicklung kommen, sodaß sich die ozeanischen Oligotraphenten ungestört entfalten können. Wie empfindlich und konkurrenzschwach derartige ozeanisch verbreitete Arten meist sind, zeigt sich heute überall dort mit großer Deutlichkeit, wo vordem oligotrophe Gewässer durch zivilisatorische Einflüsse eutrophiert werden.

Verbreitung und Einordnung der ozeanischen Arten innerhalb der Wasservegetation

Littorelletea

Die ozeanischen Arten häufen sich innerhalb der D.D.R. in bestimmten Gesellschaften in erster Linie der ozeanische Klasse der *Littorelletea*. Zu ihren Komponenten gehören neben eigentlichen Wasserpflanzen viele Arten, die sowohl Wasser- als auch Landformen ausbilden können, und eine Reihe von Helophyten. Die meisten der hierhin gehörenden Gesellschaften zeigen eine so enge Bindung an die Gebiete mit hoher Ozeanität, daß sie das Territorium der D.D.R. nicht mehr oder nur noch in Fragmenten erreichen, wobei diese Vor-

postenvorkommen einzelner Kennarten, z.B. von *Lobelia dortmanna, Isoetes lacustris, Subularia aquatica, Hypericum elodes*, heute meist erloschen sind. Die in der D.D.R. auftretenden Gesellschaften dieser Klasse konzentrieren sich in der Hauptsache auf das küstennahe Gebiet an der Ostsee und das Altmoränengebiet, das sich von der Altmark, Südwest-Mecklenburg und der Prignitz über den Fläming weiter nach der Nieder- und nördlichen Oberlausitz und Niederschlesien hinzieht. Ein besonderes Häufungszentrum ozeanischer Wasserpflanzen liegt dabei in der an Fischteichen und neuerdings auch an Tagebaurestgewässern reichen Nieder- und Oberlausitz. Verschiedene dieser Arten reichen über Oder und Neiße hinweg weiter nach Osten, was darauf hindeutet, daß ihre Vorkommen weniger durch klimatische als vielmehr durch edaphische Faktoren (arme ausgelaugte Böden, saueroligotrophe Gewässer) bestimmt werden.

Die meisten der in den genannten Gebieten anzutreffenden *Littorelletea*-Gesellschaften gehören dem Verband *Hydrocotylo–Baldellion* Tx. et Dierßen apud Dierßen 72 an. Das *Eleocharitetum multicaulis* All. 22 wächst an Heideseen und Fischteichen der Lausitzen, wo auch *Deschampsia setacea* als weitere Kennart der Gesellschaft vorkommt. Dagegen endet das Areal von *Ranunculus ololeucos* als der dritten Kennart schon weit vor unserem Gebiet im Bremer Raum.

Von der drei Kennarten des *Hyperico–Potametum oblongi* (All. 21) Br.-Bl. et Tx. 52 ist *Hypericum elodes* verschwunden; *Eleogiton fluitans* wächst zusammen mit dem gleichfalls ozeanisch verbreiteten *Apium inundatum* in einem Moorgraben und an ähnlichen Standorten, während das frühere Vorkommen in Heidegräben inzwischen erloschen ist; *Potamogeton polygonifolius* wächst hier und da in sauer-oligotrophen Heidegewässern, mit reichlichem Vorkommen von *Littorella uniflora* und anderen ozeanischen Arten. Auch bildet er z.B. in verwachsenen Torfstichen eine artenarme Gesellschaft, die wohl als Fragment des *Hyperico–Potametum oblongi* aufzufassen ist.

Auch das *Pilularietum globuliferae* Tx. 55 ex Müller et Görs 60 mit der ozeanischen bis subozeanischen *Pilularia globulifera* ist in den Lausitzer Teich- und Heidegebieten noch anzutreffen. Zwar sind viele frühere Vorkommen erloschen, doch hat sich die Art auch mehrfach neu in Ausstichen, Grubenseen und Entwässerungsgräben eingestellt. Pietsch (1974) nennt nicht weniger als 72 Fundstellen. Neben anderen Ausbildungsformen wurde stellenweise, so in Gräben des nördlichen Spreewaldgebietes, auch die Subassoziation von *Apium inundatum* angetroffen, die in Westeuropa weit verbreitet ist. Häufiger findet sich das *Pilularietum* in den Lausitzen in der Subass. von *Eleocharis acicularis*, die im atlantischen Bereich nur sehr selten vorkommt, sowie in der Typischen Subassoziation. In ähnlicher Vergesellschaftung trat *Pilularia globulifera* früher auch in der Prignitz und in Mecklenburg auf (Fischer, 1959; Pankow & Rattey, 1963), doch wurden die dortigen Vorkommen seit langem nicht mehr bestätigt und dürften erloschen sein.

Eine weitere Gesellschaft des *Hydrocotylo-Baldellion* ist das *Samolo-Littorelletum* Westhoff 47. Die Küstenpflanze *Samolus valerandi* hat ein weiteres Areal als die anderen hier genannten Arten, denn sie geht ostwärts bis Mittelasien und im Süden bis Mittelafrika; sie ist vor allem im Mediterrangebiet verbreitet. Gleichwohl verhält sie sich in der temperaten Zone wie eine subozeanische Art, indem sie sich auf die Küstengebiete West- und Mitteleuropas beschränkt und im Norden bis Dänemark, Mittel-Schweden und Süd-Finnland reicht. Es handelt sich in der Hauptsache um eine Art lockerer Salzröhrichte und tritt als solche auch an Binnensalzstellen der D.D.R. auf. Stellenweise trifft man sie aber auch zusammen mit *Littorella uniflora* und *Baldellia ranunculoides* in einer Vergesellschaftung an, die dem *Samolo-Littorelletum* Westhoff 47 entspricht oder ihm nahekommt. Die dortige Ausbildung faßt Dierßen (1975) als Subass. von *Eleocharis acicularis,* sie ist stark mit Arten der *Phragmitetea*, der *Isoeto-Nanojuncetea* und der *Agrostietea stolonifera* durchsetzt. Auffällig ist auch hier das Vorkommen auf oligohalinen Standorten z.T. in Nachbarschaft von Binnensalzstellen. Während aber *Samolus valerandi* entsprechend seiner Gesamtverbreitung relativ weit verbreitet ist und innerhalb der DDR auch in ausgesprochenen Wärme- und Trockengebieten vorkommt, verhält sich *Baldellia ranunculoides* als typisch ozeanische Art ganz anders. Von Nordwesten ragt ihr Areal über das westliche Mecklenburg, die Altmark und das Havelland bis in den Berliner Raum, im Norden reicht es entlang der Ostseeküste bis in das Odermündungsgebiet. In den Lausitzen fehlt diese Art. Nahezu überall tritt *Baldellia ranunculoides* dabei

im *Samolo-Littorelletum* bzw. in verarmten Ausbildungen dieser Gesellschaft (ohne *Littorella*) auf. Eine abweichende Vergesellschaftung zeigt die Art an ihrem östlichen Vorkommen im mitteleuropäischen Binnenland. Dort wächst sie als submerse Form in Reinbeständen, die nur spärlich mit *Nuphar lutea* und *Equisetum fluviatile* durchmischt sind, in 0.6–1.2 m Tiefe in einem meso- bis oligotrophen Waldsee.

Myriophyllum alterniflorum ist im westlichen Europa weitgehend an Gesellschaften der *Littorelletea* gebunden und gilt vielfach als Kennart der Klasse bzw. der Ordnung *Littorelletalia*. Auch in der D.D.R. kommt diese Art stellenweise in *Littorelletea*-Gesellschaften vor; auch wächst sie zusammen mit Submers-Formen von *Littorella uniflora* und *Chara*-Arten in einer '*Myriophyllum alterniflorum-Littorella uniflora*-Gesellschaft', die Jeschke (1959) noch den *Littorelletalia* zurechnet. Dieselbe Ansicht vertritt auch Dierßen (1975), bezeichnet sie indessen als *Chara aspera-Myriophyllum alterniflorum*-Gesellschaft.

Eine *Ranunculion fluitantis*-Gesellschaft mit *Myriophyllum alterniflorum*, wie sie als *Callitricho-Myriophylletum alterniflori* (Steusloff 39) Weber-Oldecopp 67 aus unverschmutzten, kühlen und nährstoffarmen Perlmuschelbächen aus der Lüneburger Heide und der Oberpfalz bekannt geworden sind, wurde in der D.D.R. bisher noch nicht angetroffen. *Myriophyllum alterniflorum* greift jedoch im nordbrandenburgisch-mecklenburgischen Seengebiet vielfach in ausgesprochen kalk-oligotrophe Gewässer über. Während es in eutrophen Seen fehlt, kommt es in den kalk-oligotrophen Seen dieses Gebietes zusammen mit *Chara*-Arten in ärmeren Ausbildungen von *Potametea*-Gesellschaften vor oder bildet Dominanzbestände, die ebenfalls den *Potametea* und nicht mehr den *Littorelletea* anzuschließen sind.

Von den übrigen Kennarten der *Littorelletea* (bzw. *Littorelletalia*) kommt *Luronium natans* in der D.D.R. nur sehr zerstreut vor, wobei die meisten der früheren Vorkommen erloschen sind. Sie lagen vor allem in Mecklenburg und in den Lausitzen. Ein 1981 entdecktes Vorkommen zeigt die Art als Bestandteil einer armen *Potametea*-Gesellschaft eines Entwässerungsgrabens. Auch trat *Luronium natans* in ärmeren Ausbildungen von *Potametea*-Gesellschaften auf, dort allerdings zusammen mit *Eleogiton fluitans* und *Utricularia ochroleuca* (Freitag *et al.*, 1958).

Littorella uniflora hat in der D.D.R. ein ähnliches Areal wie *Pilularia globulifera*. Die Art kommt in verschiedenen *Littorelletea*-Gesellschaften in und an Fischteichen und oligotrophen bis mesotrophen Seen und Weihern vor, vor allem im *Eleocharitetum multicaulis* und in der Subass. von *Littorella uniflora* des *Eleocharitetum acicularis* (Baumann 11) W. Koch 26.

Die weiteste Verbreitung der *Littorelletea*-Arten innerhalb der D.D.R. besitzt *Juncus bulbosus*. Er findet sich in fast allen *Littorelletea*-Gesellschaften und bildet stellenweise, vor allem in sauren Grubengewässern und Heideseen, auch artenarme Dominanzbestände, die Dierßen (1975) als '*Juncus bulbosus*-Gesellschaft' bezeichnet. *Juncus bulbosus* hat sich in den Lausitzen in sauren Grubengewässern außerordentlich stark ausgebreitet und bildet sowohl im Wasser als auch (z.T. in der subsp. *kochii*) im litoralen Inundationsbereich nicht selten Massenbestände.

Insgesamt bleibt festzuhalten, daß die *Littorelletea*-Gesellschaften in der D.D.R. die aquatischen Makrophytengesellschaften mit dem höchsten Anteil an ozeanischen Arten sind. Durch ihre Bindung an nährstoffärmere, überwiegend saure Gewässer sind ihre Vegetationseinheiten heute in besonderem Maße gefährdet. Viele ihrer früheren Vorkommen sind bereits erloschen. Die Erhaltung der noch vorhandenen Bestände ist ein wichtiges Anliegen des Naturschutzes (Pietsch, 1977).

Isoëto–Nanojuncetea

Auch in den Gesellschaften der Isoëto—Nanojuncetea gibt es einige Arten mit mehr oder weniger starker Bindung an küstennahe Gebiete wie z.B. *Elatine hexandra, Lindernia dubia* und *Illecebrum verticillatum*. Hier soll lediglich auf *Ludwigia palustris* eingegangen werden, eine submediterranatlantische Art, die in der D.D.R. lediglich in der Niederlausitz und im westlich angrenzenden Elstergebiet gefunden wurde (Krausch, 1974). Zwar sind die meisten Vorkommen erloschen, doch konnten einige noch in neuerer Zeit bestätigt werden. Bezeichnenderweise liegen die meisten Fundorte in Gebieten mit höherer Sommerwärme, in denen u.a. auch *Trapa natans* und *Salvinia natans* vorkommen, während die Heidegebiete mit Häufung von *Littorelletea*-Arten gemieden werden. Stellenweise gibt es freilich auch Überschneidungen; so wurde die Art in einem Elster-Altwasser in submerser Form zusammen mit *Eleogiton fluitans* und *Luronium natans* in einem ärmeren *Myriophyllo–Nupharetum* angetroffen. Sonst tritt *Ludwigia palustris* in *Isoëto–Nanojuncetea*-Gesellschaften offener Sand- und Schlammufer auf und greift von diesen Stellen bei hohen Wasserständen in flutender Form auch in das offene Wasser aus.

Potametea

Unter den *Potametea*-Gesellschaften enthält vor allem der Verband *Ranunculion fluitantis* ozeanische Elemente, wenn auch zumeist solche von weiterer Verbreitung (subozeanische bzw. subatlantische Arten). Es handelt sich meist um wintergrüne Arten des fließenden Wassers, die offenbar vor allem gegen Eisbedeckung empfindlich sind. Von ihnen erreichen die beiden subatlantisch-submediterranen Arten *Potamogeton coloratus* und *Groenlandia densa* das Territorium der D.D.R. nur noch mit ganz vereinzelten, z.T. auf Verschleppung mit Fischbrut beruhenden und sämtlich wieder verschwundenen Fundorten. Weit verbreitet in der D.D.R. ist dagegen *Ranunculus fluitans*, wenngleich die Verbreitungskarte (Fukarek *et al.*, 1978) offensichtlich zahlreiche Verwechslungen mit *Ranunculus penicillatus* bzw. *R. peltatus* enthält, die aber ebenfalls, soweit bisher erkennbar, zu den Arten mit subozeanischer (subatlantischer) Verbreitung zu zählen sind. Es läßt sich erkennen, daß *Ranunculus fluitans* stets größere Fließgewässer (Flüsse) besiedelt, *Ranunculus penicillatus* dagegen überwiegend kleinere (Bäche). Der Schwerpunkt von *Ranunculus fluitans* liegt im Raum der Mittelgebirge und ihrer vorgelagerten Hügelländer, wo er relativ saubere, klare, sommerkühle und sauerstoffreiche Flüsse bewohnt. Die Art ist Verbands-Kennart und zugleich Hauptbestandteil des *Ranunculetum fluitantis*, das je nach Nährstoffgehalt des Gewässers in verschiedenen Ausbildungen auftritt. In nährstoff- und kalkarmen, oligosaproben Gewässern bzw. Gewässerabschnitten findet sich eine moosreiche Ausbildung mit *Fontinalis antipyretica* und *Hygrohypnum*-Arten sowie mit der ebenfalls subatlantischen *Callitriche hamulata*, in schwach belasteten ß-mesosaproben dagegen eine Ausbildung mit *Potamogeton pectinatus, Potamogeton crispus, Myriophyllum spicatum* und anderen Arten.

Zu den Kennarten des *Ranunculetum fluitantis*

zählt in Süddeutschland der subatlantisch-submediterrane *Potamogeton nodosus*. Im mediterranen Bereich (Südfrankreich) bildet diese Art zusammen mit *Vallisneria spiralis* das *Potamo-Vallisnerietum*. Im Gebiet der D.D.R. ist *Potamogeton nodosus* selten. Manche angeblichen Fundorte beruhen auf Verwechslung mit Fließwasserformen von *Potamogeton natans*, doch liegen auch einige sichere Belege vor. Soweit sich bisher erkennen läßt, kommt bzw. kam die Art hier meist im *Ranunculetum fluitantis* vor. Indessen sind die meisten Vorkommen heute erloschen.

Neben *Callitriche hamulata* enthält der Verband *Ranunculion fluitantis* noch zwei weitere *Callitriche*-Arten. Die subatlantische *Callitriche platycarpa* ist in ihrer Submersform vielfach Bestandteil von *Ranunculion fluitantis*-Gesellschaften, während die Schwimmformen in Gesellschaften der *Lemnetea* und der *Potametea* sowie des *Sparganio-Glycerion* auftreten. Die ähnliche, subatlantisch-submediterrane *Callitriche stagnalis* kommt ebenfalls vielfach in *Ranunculion*-Gesellschaften vor, besonders im *Veronico-Callitrichetum stagnalis*, das in höheren Lagen von Silikatgebirgen auftritt.

Die zweite Gruppe von *Potametea*-Gesellschaften mit subatlantischen Sippen ist die der Wasserhahnenfuß-Gesellschaften, die von einigen Autoren als eigener Verband (*Ranunculion aquatilis* Passarge 64 bzw. *Callitricho-Batrachion* den Hartog et Segal 64) aufgefaßt, sonst aber dem *Nymphaeion* zugewiesen werden. Es handelt sich um Gesellschaften flacher Tümpel, Wiesengräben und Fischteiche, die vor allem in den küstennahen Weidelandschaften weit verbreitet sind, aber teilweise auch noch in den Teichgebieten des Berg- und Hügellandes vorkommen. Von ihnen ist das *Ranunculetum baudotii* Br.-Bl. 52 in der D.D.R. heute ganz auf die Küstengebiete der Ostsee beschränkt, wo es Brackwassertümpel und -gräben besiedelt. Seine Kennart *Ranunculus baudotii* ist eine subatlantische Art, die auch im mediterranen Bereich eine weite Verbreitung besitzt, andererseits aber bis Archangelsk vordringt. Sie überwintert unter Eis als grüne Pflanze. In den Niederlanden rechnet man auch die atlantisch-mediterrane *Callitriche obtusangula* zu den Kennarten des *Ranunculetum baudotii*, die anderswo freilich auch in anderen Wasserpflanzengesellschaften auftritt. Sie gehört zu denjenigen ozeanischen Arten, die das Territorium der D.D.R. nicht mehr erreichen.

Sehr viel häufiger tritt in der D.D.R. das *Ranunceletum peltati* (Sauer 47) Segal 67 in Erscheinung. Sein Schwerpunkt liegt in den Grünland- und Weidelandschaften des Flachlandes, z.B. Mecklenburgs, der Altmark und des Havellandes, doch gibt es auch Vorkommen in Teichlandschaften des Berglandes im Süden, z.B. um Plothen. Im Thüringer Wald werden Höhen von 750 m erreicht.

Oft in enger Nachbarschaft mit dieser Assoziation, meist jedoch an etwas ärmeren Standorten, wächst das *Hottonietum palustris* Tx. 37 (*Callitricho-Hottonietum* (Tx. 37) Segal 65, *Ranunculo-Hottonietum* (Tx. 37) Oberd. *et al.* 67). Seine Kennart *Hottonia palustris* beschränkt sich auf die temperate Zone Europas und reicht hier von der Atlantikküste ostwärts bis ins obere Wolgagebiet. Auch andere wesentliche Komponenten dieser Gesellschaft (*Callitriche platycarpa, C. hamulata, Ranunculus peltatus, R. aquatilis* s.str., *Potamogeton acutifolius*) weisen subozeanische Verbreitung auf, sodaß man das *Hottonietum* sicher zurecht als subozeanische Gesellschaft bezeichnen kann.

Von den übrigen *Potamogeton*-Arten Mitteleuropas besitzt lediglich *Potamogeton nitens* – nach dem bisherigen Kenntnisstand – ein engeres Areal (oz_{1-2}) und muß, da sie vor allem in nördlichen Europa häufig ist, als nordisch-subatlantisches Florenelement eingestuft werden. In der D.D.R. wächst die Art in der Litoralzone klarer, mesotropher Seen, vor allem im Gebiet der nordbrandenburgisch-mecklenburgischen Seenplatte.

Eigenartig ist das Verhalten der Schwimmblattpflanze *Nymphoides peltata*, Kennart des *Nymphoidetum peltatae*. Die Art gehört zu den wärmeliebenden Wasserpflanzen vom *Trapa*-Typ (Jäger, 1968) und hat eine weite Verbreitung im gemäßigten bis kühlen Eurasien mit Schwerpunkt in sommerwarmen Gebieten des kontinentalen Bereichs ($k_{(1)-3}$ nach Jäger, 1968). Abweichend von diesem Gesamtareal beschränkt sich *Nymphoides peltata* in der D.D.R. auf das Küstengebiet und die Unterläufe von Elbe, Havel und Oder und verhält sich damit wie eine subatlantische Art. In den Gebieten mit einem Tagesmittel der Lufttemperatur von 10 °C und darüber an mindestens 160 Tagen im Jahr, an welche *Trapa natans* ganz deutlich gebunden ist, fehlt *Nymphoides peltata*, sodaß sich beide Arten hier weitergehend ausschließen. Die Gründe für dieses unterschiedliche Verhalten bedürfen noch der näheren Klärung. Offenbar sind die Wärmeansprüche besonders für die Zeit des Frühlings und des Frühsommers bei der ausdauernden und sich relativ spät entwickelnden *Nymphoides peltata* geringer als bei der annuellen *Trapa natans*, welche zur Keimung Wassertemperaturen von mindestens 12 °C benötigt.

Montio–Cardaminetea

Ozeanische Elemente finden sich dann noch in den Quellfluren (*Montio–Cardaminetea*). So besitzen alle einheimischen *Montia*-Sippen subozeanische Verbreitung (oz$_{1-2}$). Die in klaren Quellbächen der Mittelgebirge, seltener der Moränenketten des Flachlandes wachsenden Unterarten von *Montia fontana* (ssp. *fontana, variabilis* und *amporitana*) haben durch Gewässerverschmutzung, Meliorationen und Grundwasserabsenkungen überall einen starken Rückgang erfahren. Am häufigsten findet man gegenwärtig noch die subsp. *chondrosperma* (syn. *Montia minor* C. Gmel.), die nur selten in Quellfluren, meist jedoch in ackerbewohnenden Gesellschaften der *Isoëto–Nanojuncetea* auftritt. Infolge hochgradiger Herbicid-Empfindlichkeit ist auch diese Sippe stark gefährdet.

Eine typisch ozeanische Art der Quellfluren ist ferner *Chrysosplenium oppositifolium*, die in der D.D.R. teilweise die Ostgrenze ihres Areals erreicht, im Südosten aber weiter ostwärts vordringt und auch im mittleren Polen noch vorgeschobene Fundorte besitzt. Sie wächst in mehr oder weniger beschatteten Quellfluren, an Bachufern, an überrieselten Felsen, in Quell-Erlenbrüchen und Erlen-Eschenwäldern, insgesamt and Standorten mit hoher Luftfeuchtigkeit, wobei ihr Verbreitungsschwerpunkt im Bergland liegt. An den gleichen Standorten und in ähnlicher Vergesellschaftung findet sich auch die ebenfalls subozeanische *Lysimachia nemorum*.

Während die meisten Autoren auch das *Ranunculetum hederaceae* (Tx. et Diemont 36) Libbert 40 zu den *Montio–Cardaminetea* rechnen, wird es in den Niederlanden dem dortigen *Callitricho–Batrachion* (*Potametea*) zugezählt. Die Assoziation besiedelt Quellstellen, Quelltümpel, Quellbäche und Gräben mit kühlem und klarem, kalk- und nährstoffarmem Wasser. Der ozeanische *Ranunculus hederaceus* erreicht in der D.D.R. die absolute Ostgrenze seines Areals und galt hier bereits als ausgestorben. In den letzten Jahren konnten jedoch in der Altmark noch einige Vorkommen entdeckt werden. Da die Art an verschiedenen Fundstellen in der Altmark schon in der Mitte des vorigen Jahrhunderts bereits wieder erloschen war, wird man bei dem Rückgang der Art nicht nur Gewässerverschmutzungen und Meliorationen, sondern auch klimatische Gründe in Rechnung stellen müssen. Insgesamt gilt *Ranunculus hederaceus* für das Gebiet der D.D.R. als eine vom Aussterben bedrohte Art.

Literatur

Casper, S. J. & Krausch, H.-D., 1981, 1982. Pteridophyta und Anthophyta. Süßwasserflora von Mitteleuropa 23 u. 24. VEB. G. Fischer-Verlag. 942 pp.

Dierßen, K., 1975. Littorelletea uniflorae. Prodromus der europäischen Pflanzengesellschaften 2. Vaduz. 149 pp.

Fischer, W., 1959. Pflanzenverbreitung und Florenbild in der Prignitz. Wiss. Z. Päd. Hochsch. Potsdam, Math.-nat. R. 5: 49–84.

Freitag, H., Markus, Chm. & Schwippl, I., 1958. Die Wasser- und Sumpfpflanzengesellschaften im Magdeburger Urstromtal südlich des Fläming. Wiss. Z. Päd. Hochsch. Potsdam, Math.-nat. R. 4: 65–92.

Fukarek, F., Knapp, H.-D., Rauschert, St. & Weinert, E., 1978. Karten der Pflanzenverbreitung in der D.D.R. Hercynia N.F. 15: 229–320.

Jäger, E., 1968. Die pflanzengeographische Ozeanitätsgliederung der Holarktis und die Ozeanitätsbindung der Pflanzenareale. Feddes Repert. 79: 157–335.

Jeschke, L., 1959. Pflanzengesellschaften einiger Seen bei Feldberg in Mecklenburg. Feddes Repert., Beih. 138: 161–214.

Krausch, H.-D., 1974. Ludwigia palustris (L.) Ell. in der Niederlausitz. Niederlaus. Flor. Mitt. 7: 23–32.

Meusel, H., Jäger, E. & Weinert, E., 1965. Vergleichende Chorologie der zentraleuropäischen Flora. Bd. 1. Jena. 583 u. 258 pp.

Meusel, H., Jäger, E., Rauschert, St. & Weinert, E., 1978. Vergleichende Chorologie der zentraleuropäischen Flora. Bd. 2. Jena. 418 pp. u. S. 259–421.

Müller-Stoll, W. R. & Krausch, H.-D., 1959. Verbreitungskarten brandenburgischer Leitpflanzen. 2. Reihe. Wiss. Z. Päd. Hochsch. Potsdam, Math.-nat. R. 4: 105–150.

Müller-Stoll, W. R., Fischer, W. & Krausch, H.-D., 1962. Verbreitungskarten brandenburgischer Leitpflanzen. 4. Reihe. Wiss. Z. Päd. Hochsch. Potsdam, Math.-nat. R. 7: 95–150.

Pankow, H. & Rattey, F., 1963. Verbreitungskarten zur Pflanzengeographie Mecklenburgs. 2. Reihe. Wiss. Z. Univ. Greifswald, Math.-nat. Reihe 12: 359–376.

Pietsch, W., 1974. Zur Verbreitung und Soziologie des Pillenfarns (Pilularia globulifera L.) in der Lausitz. Niederlaus. Flor. Mitt. 7: 11–22.

Westhoff, V. & Held, A. J. den, 1969. Plantengemeenschappen in Nederland. Zutphen. 324 pp.

Accepted 22. 2. 1984.

Contributions to the sociology and chorology of contrasting plant communities in the southern part of the 'Wienerwald' (Austria)*,**

G. Karrer***
Institute of Botany, University of Vienna, Rennweg 14, A-1030 Vienna, Austria

Keywords: *Asperulo–Fagetum*, Austria, Climax, Disjunction quotient, *Euphorbio saxatilis–Pinetum nigrae, Fumano–Stipetum eriocaulis, Mesobromion,* Permanent community, *Querco–Carpinetum* s.l., Spectra of area types, Vegetation relevés

Abstract

Five plant communities contrasting in successional status and human impact from the southern part of the 'Wienerwald' (Austria) are analyzed using vegetation relevés, spectra of area types and a newly proposed disjunction quotient. A climax community (*Asperulo–Fagetum*), a subclimax community (*Querco–Carpinetum* s.l.), an anthropogenous substitute community (*Mesobromion*) and two natural, non-climax permanent communities (*Euphorbio saxatilis–Pinetum nigrae* and *Fumano–Stipetum eriocaulis*) are recognized.

The disjunction quotient is defined as the number of partial (discontinous) areas divided by the size of the total area of distribution of a species.

In particular, the average disjunction quotients of the species in the first two communities reflect relatively stable environments only slightly influenced by man, with many ancient, stable taxa. These communities are characterized by species with well-delimited, stable distribution areas. The species in the *Mesobromion* community have very low average disjunction quotients as its component species are widely and continuously distributed and are often promoted by man. In contrast to these communities, the species linked to the natural permanent, non-climax communities of extreme habitats, have high distribution quotients i.e. small, discontinuous areas; this illustrates the relic character of these plant communities and of the eastern edge of the Alps as a whole.

Using the highly variable disjunction quotient of all species and communities examined, the concepts of climax and permanent communities (of different origin) are discussed with regard to European conditions.

Introduction

Within the scope of a research program undertaken at the Institute of Botany, University of Vienna, some contrasting plant communities in the southern part of the 'Wienerwald' (about 30 km SSW of Vienna resp. W of Bad Vöslau) were investigated from different points of view. The contrast lies in both successional position, ecological conditions and degree of human impact.

The following communities were selected:
– communities equivalent or at least closely connected with the zonal climax of the southern part of the 'Wienerwald' (*Asperulo–Fagetum, Querco–Carpinetum* s.l.);

* Nomenclature for vascular plants follows Ehrendorfer (1973), for mosses Rabenhorst (1884–1918) and for lichens Poelt (1966) and Poelt & Vězda (1977 + 1981).

** Preliminary results from a dissertation guided by Dr F. Ehrendorfer at the University of Vienna.

*** *Acknowledgement.* I should like to thank Dr H. Niklfeld (Vienna) for his permanent interest in my work, Dr F. Ehrendorfer for proposing the topic and for his interesting and manifold suggestions, Dr J. Saukel (Vienna) for the naming of the mosses and Mr W. Brunnbauer (Vienna) for reviewing some of the lichens; I am also very grateful to Dr H. Meusel (Halle) and to Dr H. W. Luftensteiner (Vienna) for their valuable help, discussions and joint excursions; the English manuscript was reviewed by Miss D. E. Mantell (Vienna) and Dr H. Niklfeld.

– an anthropogenous substitute community (= Ersatzgesellschaft sensu Braun-Blanquet, 1964; Ellenberg, 1976) on a climax site (*Mesobromion* community);
– natural permanent communities (= azonal communities sensu Ellenberg, 1976) on special sites with shallow soils (*Euphorbio saxatilis-Pinetum nigrae, Fumano–Stipetum eriocaulis, Pinus nigra* stone heath).

The aspects of dispersal biology (Luftensteiner, 1978, 1979, 1980, 1981) and ploidy spectra (Ehrendorfer, 1980) have already been considered and discussed. Detailed sociological data can be found in Karrer (in prep.). Based on these exhaustive analyses of well-chosen permanent sample plots, chorological aspects are also discussed comprehensively. The differentiation of area forms and area dimensions is stressed. Until now, only a few authors have taken these rather quantitative aspects of plant areas into account (i.e. Dansereau, 1957; Kornaś, 1980).

Methods

Relevés of the permanent sample plots and adjacent sites were made according to Braun-Blanquet (1964) and Ellenberg (1956). The typology of distribution areas largely follows Ehrendorfer (1972). Mosses and lichens were not taken into account. Distribution areas of the species involved were taken from maps in Hultén (1958, 1962, 1970), Meusel, Jäger & Weinert (1966), and Meusel, Jäger, Rauschert & Weinert (1978).

The degree of areal fragmentation, referred to as 'area form', was taken as a main characteristic. Whether this is possible or not depends very much on the matter of representation, on the scale, as well as on the density of single data, and on the taxonomical significance of the respective taxon, etc. (cf. also Straka, 1970).

In order to record quantitative characters of areas, the absolute sizes of the total areas were determined in arithmetically easy-to-use units. Different scales and non-congruent map projections (i.e. Mercator's projection in Meusel, Jäger & Weinert, 1966) were corrected. In this way it was possible to obtain the absolute size of the total areas, measured in cm^2 at a scale of 1 : 50000000 as is shown by the map of Europe in Meusel, Jäger &

Weinert (1966) (see e.g. Fig. 179b, *Aethionema saxatile* in Meusel, Jäger & Weinert, 1965).

The number of partial areas was determined by counting as such – continuous areas, islands, isolated point signs or groups of point signs, and gaps within continuous areas (i.e. the Pannonian lowland gap in *Hordelymus europaeus* or the central parts of the Alps in *Rosa arvensis*). Then a disjunction quotient (DQ) was calculated as:

$$\frac{\text{number of partial areas}}{\text{total area size on the map}}$$

For example, the area of *Aethionema saxatile* has 60 partial areas and a total size of 5.03 cm^2, the DQ is therefore 60/5.03 = 11.93. This species, distributed over the meridional and submeridional zone of Europe and northwest Africa, combines a relatively high disjunction with a limited total area of (mainly montane) distribution. Other examples are: *Bromus erectus* with a DQ of 36/10.43 = 3.45, *Veronica arvensis* with a DQ of 56/22.30 = 2.51, *Minuartia setacea* with a DQ of 52/3.28 = 15.85 and *Thlaspi montanum* (with pronounced disjunct distribution area!) with a DQ of 64/1.13 = 56.69.

The DQ values for the character-species, differential species, dominants and companions (with presence degree II-V) of the examined communities were calculated – as far as good distribution maps were available (see Table 2).

Results and discussion

Sociology

Climax and subclimax communities. In the southern part of the 'Wienerwald', beech forests are developed as the climax vegetation on calcareous soils at montane altitudes (cf. Zukrigl, 1973; Mayer, 1974, 1977). At the Peilstein (about 30 km SW of Vienna), an *Asperulo–Fagetum* H. May. 64, growing on Terra fusca (calcareous brown loam soil) in the middle of the eastern slope and only in a few areas changed by man into an oak–hornbeam forest, is found. However, on the more rocky parts of the slope, thoroughly site-adequate, natural oak–hornbeam forests are to be found, which are not sufficiently considered in phytosociological literature but can be assigned to the *Querco–Carpinetum* s.l. (sensu Tüxen 37).

Table 1. Sociological characters of 3 plant communities from the southern part of the 'Wienerwald'; 1 = *Euphorbio saxatilis–Pinetum nigrae* Wendelb. 62 (subass. of *Cyclamen purpurascens* Wendelb. 62), 2 = *Pinus nigra* stone heath, 3 = *Fumano–Stipetum eriocaulis* Wagner 41 corr. Zólyomi 66 *minuartietosum setaceae* Karrer 84; a = constancy, b = average abundance (in % of total relevé area).

Plant community	1		2		3	
Number of relevés	12		7		11	
Average species number	38,3		32,2		36	
Average exposition	NW-N		W-SSW		SW-S	
Constancy + average abundance	a	b	a	b	a	b
Tree layer:						
Pinus nigra	V	(61)	V	(61,1)	I	(15)
Sorbus aria	II	(10,2)				
Shrub layer:						
Amelanchier ovalis	V	(22,5)	V	(21,1)	III	(4,6)
Pinus nigra	III	(3,9)	V	(1,2)	III	(1,3)
Sorbus aria	IV	(1,8)	II	(0,5)		
Cotoneaster tomentosus	III	(7,5)				
Berberis vulgaris	III	(3,8)				
Sorbus torminalis	I	(0,5)				
Herb and moss layer:						
Leucanthemum maximum	V	(0,6)				
Thesium alpinum	V	(0,5)				
Scabiosa lucida f. badensis	IV	(0,5)				
Thlaspi montanum	IV	(0,5)				
Euphrasia salisburgensis	III	(0,5)				
Gentianella austriaca	III	(0,4)				
Primula auricula	III	(1)				
Goodyera repens	III	(0,2)				
Pyrola rotundifolia ssp. rotundifolia	III	(0,4)				
Daphne cneorum	V	(1,6)				
Pimpinella saxifraga	V	(0,6)				
Erica herbacea	V	(31,7)	III	(32,5)		
Campanula glomerata	V	(0,4)	I	(0,5)		
Phyteuma orbiculare	V	(0,8)	IV	(0,4)	III	(0,3)
Polygala chamaebuxus	V	(15,7)	V	(3,7)	II	(0,5)
Thalictrum minus	IV	(0,5)	IV	(0,5)	II	(0,1)
Hieracium bifidum	IV	(0,5)	II	(0,5)	II	(0,5)
Galium austriacum	IV	(0,7)	I	(0,5)		
Scleropodium purum	III	(0,8)	I	(0,5)		
Hypnum cupressiforme s.l.	III	(4,1)	I	(0,5)		
Polygala amara ssp. amara	III	(0,4)				
Pleurozium schreberi	III	(3,3)				
Hylocomium splendens s.l.	II	(0,5)				
Fragaria vesca	II	(0,5)				
Plantanthera bifolia	I	(0,5)				
Gymnadenia conopsea	I	(0,1)				
Coronilla vaginalis	I	(1,3)				
Allium montanum	IV	(0,7)	V	(0,8)	V	(0,4)
Carex humilis	IV	(9,4)	V	(24)	III	(3,3)
Asperula cynanchica	V	(0,9)	V	(0,5)	V	(0,5)
Leontodon incanus	III	(0,4)	V	(0,8)	V	(0,4)
Thymus praecox ssp. praecox	IV	(4,4)	V	(1,9)	V	(2,7)
Dorycnium germanicum	III	(0,7)	V	(1,1)	V	(0,5)
Galium lucidum	III	(0,5)	V	(0,5)	V	(0,4)

A (groups: Shrub layer Cotoneaster tomentosus – Sorbus torminalis)
B (groups: Leucanthemum maximum – Pyrola rotundifolia)
C (groups: Daphne cneorum – Coronilla vaginalis)
D (groups: Allium montanum – Galium lucidum)

Table 1. (Continued).

Plant community		1		2		3	
Number of relevés		12		7		11	
Average species number		38,3		32,2		36	
Average exposition		NW-N		W-SSW		SW-S	
Constancy + average abundance		a	b	a	b	a	b
D	Globularia cordifolia	II	(1,8)	V	(5,7)	V	(0,6)
	Teucrium montanum	I	(1,5)	V	(3,3)	V	(0,9)
	Festuca stricta	III	(1,2)	V	(1,6)	III	(2,9)
	Thesium linophyllon	I	(0,3)	V	(0,5)	III	(0,4)
	Scabiosa canescens	I	(0,5)	III	(0,5)	II	(0,5)
	Centaurea scabiosa ssp. badensis	II	(0,5)	III	(0,5)	IV	(0,7)
	Acinos alpinus	I	(0,5)	III	(0,5)	II	(0,5)
	Pulsatilla grandis	III	(0,3)	III	(0,5)	I	(0,3)
E	Euphorbia saxatilis	V	(0,8)	V	(0,5)	V	(0,6)
	Hieracium glaucum (incl. H. bupleuroides)	III	(0,4)	V	(0,5)	V	(0,4)
	Minuartia setacea			I	(0,5)	IV	(0,5)
	Aethionema saxatile					IV	(0,3)
	Cardaminopsis petraea			I	(0,5)	IV	(0,3)
	Erysimum sylvestre			I	(0,5)	II	(0,2)
	Alyssum montanum			I	(0,5)	II	(0,5)
F	Onosma visianii			III	(0,5)	V	(0,3)
	Jurinea mollis			III	(0,5)	III	(0,4)
	Fumana procumbens					III	(1,1)
	Campanula sibirica					III	(0,7)
G	Seseli austriacum	I	(0,1)	III	(0,4)	V	(0,4)
	Jovibarba hirta			II	(0,5)	V	(0,7)
	Potentilla arenaria			V	(0,4)	IV	(0,4)
	Helianthemum canum			III	(1)	V	(0,7)
	Toninia coeruleonigricans			II	(0,5)	V	(0,5)
	Lecidea decipiens			I	(0,5)	IV	(0,5)
	Seseli hippomarathrum			III	(0,5)	V	(0,4)
	Linum tenuifolium			III	(0,5)	IV	(0,4)
	Melica ciliata			II	(0,5)	IV	(0,5)
	Allium sphaerocephalon					III	(0,8)
	Squamarina crassa					III	(0,5)
	Hieracium echioides					II	(0,2)
	Asplenium ruta-muraria					II	(0,3)
	Tortella tortuosa			I	(0,5)	I	(1,5)
H	Sesleria varia	V	(38,3)	V	(28,2)	V	(5)
	Anthericum ramosum	V	(1,5)	IV	(1,3)	V	(0,5)
	Biscutella laevigata ssp. austriaca	V	(0,4)	II	(0,3)	III	(0,3)
	Genista pilosa	IV	(0,8)	III	(1,2)	II	(0,3)
I	Campanula gentilis	II	(0,3)				
	Rhamnus saxatilis	II	(0,9)				
	Acer pseudoplatanus juv.	II	(0,3)				
	Sorbus aria juv.	III	(0,5)	II	(0,5)		
	Amelanchier ovalis juv.	III	(0,4)	III	(1,2)	I	(0,5)
	Euphorbia cyparissias	I	(0,3)	II	(0,5)	III	(0,3)
	Pinus nigra juv.	II	(0,3)	II	(0,5)	III	(0,4)
	Scorzonera austriaca	I	(0,3)	II	(0,5)	II	(0,4)
	Polygonatum odoratum	I	(0,1)	III	(0,5)	I	(0,5)

Table 1. (Continued).

Plant community	1		2		3	
Number of relevés	12		7		11	
Average species number	38,3		32,2		36	
Average exposition	NW-N		W-SSW		SW-S	
Constancy + average abundance	a	b	a	b	a	b
Senecio jacobea	I	(0,1)	II	(0,5)	II	(0,3)
Cytisus ratisbonensis	I	(0,5)	I	(0,5)	I	(0,5)
Campanula rotundifolia agg.	III	(0,4)			III	(0,7)
Teucrium chamaedrys	I	(2,5)			I	(0,5)
Reseda lutea					II	(0,1)

Sociological species groups: A = differential species of the *Euphorbio saxatilis–Pinetum nigrae*; B = enlarged group of character species (at least of local value) of the *Euphorbio saxatilis–Pinetum nigrae*; C = differential species of the *Euphorbio saxatilis–Pinetum nigrae* compared with the *Fumano-Stipetum eriocaulis* and the *Cotino–Quercetum pubescentis* Jakucs 61; D = species from the *Festuco-Brometea* and the *Quercetea pubescentis-petraeae*, having their ecological optimum in the *Pinus nigra* stone heath; E = differential species of the *Fumano–Stipetum eriocaulis minuartietosum setaceae*; F = character species of the typical *Fumano-Stipetum eriocaulis*; G = species of open vegetation types mostly from the *Festucetalia valesiacae* Br.-Bl. et Tüxen 43 and the *Seslerio-Festucion pallentis*; H = suboceanic species group with their sociological emphasis in the *Pine* forests in general; I = companions and accidental species.

Anthropogenous substitute community. In the neighbourhood of the above climax permanent sample plots, relevés were made in enclosured permanent sample plots of a rough meadow ('Magerwiese') on similar soils. Karrer (in prep.) classifies them as *Mesobromion* communities closely connected with the *Bromus erectus–Lathyrus pannonicus* ass. prov. Wagner 41. For further details see Karrer (in prep.) and Luftensteiner (1981).

Natural, non-climax permanent communties. The Lindkogel-massif near Bad Vöslau (about 30 km SSW of Vienna), built up on limestone and dolomite, is especially on its southern and western slopes copiously covered with landscape-forming *Pinus nigra* forests. These natural stands of black pine belong sociologically to the *Euphorbio saxatilis–Pinetum nigrae* Wendelb. 62, which is typically developed as dense forest on north-facing slopes. For detailed sociological data see Karrer (in prep.).

Distinct mosaic differentiation of border habitats of forests occur on the southern slopes of the Hauerberg (Lindkogel-massif). These slopes are intensely furrowed by erosion and because of this, soil development scarcely reaches the protorendsina's stage. Rough, open dolomite mounds and rocks vary with a quite open and low *Pinus nigra* forest type (= '*Pinus nigra* stone heath'). Rock vegetation

belongs to the alliance *Seslerio–Festucion pallentis* Klika 31 corr. 37 s.l., and within this alliance to the so-called 'limestone group' (= *Seslerio–Festucion pallentis* Klika em. Zólyomi 66) corresponding to Niklfeld (1979). The stands in the vicinity of the Lindkogel can be regarded as a montane subassociation of the *Fumano–Stipetum eriocaulis* Wagner 41 corr. Zólyomi 66 situated on dolomite rocks and detritus (cf. Karrer, in prep.). This subassociation of montane dolomite rock vegetation is characterized positively by a distinct block of differential species (cf. Table 1) and negatively by the almost complete absence of many spring annuals that had been noted by Wagner (1941) in his typical association.

The *Pinus nigra* stone heath intercedes sociologically between the *Euphorbio saxatilis–Pinetum nigrae* and the *Fumano–Stipetum eriocaulis*. The latter and the *Pinus nigra* stone heath form together a complicated mosaic pattern of vegetation. Here, the species of the *Festuco-Brometea* Br.-Bl. et Tüxen 43 – especially from the *Seslerio–Festucion pallentis* Klika 31 corr. 37 s.l. – and of the *Quercetea pubescentis-petraeae* Jakucs 61 have a clear sociological emphasis. The sociological position of this community is still to be determined (Karrer, in prep.).

Chorology

Area type spectrum (ATS, compare Fig. 1)

The *ATS* of the *Asperulo-Fagetum* shows a 74.5% European and central European species component; this includes a portion of 21% advancing further to the south. The remaining 15% is comprised of Eurasian species, which show oceanic distribution patterns connected with a characteristic Europe–eastern Asia disjunction; these species can be referred to the deciduous forest element throughout. In contrast, among the 26% of the Eurasian species in the *Querco–Carpinetum* s.l. relevés, there are already some species with continental affinities and without large disjunctions. With a little over 50%, the EUR + MEUR group is represented less in this community than in the climax community; however, the group includes a relatively large part (30%) of more southern species. The percentage of submediterranean species numbers 14.5% (20% resp.) in the *Querco–Carpinetum* s.l. (this figure increases in the oak–hornbeam forests on the 'Thermenlinie' – the dry and warm edge of the Alps against the Viennese basin – to about 50%). In every case, the examined sample plots show affinities to the Illyric and Carpathian oak–hornbeam forests. The differences in *ATS* between the probably natural *Querco–Carpinetum* and the *Querco–Carpinetum* influenced by man are few.

47% of the species occurring in the rough meadow on limestone (*Mesobromion* community) are of European–central European distribution; among these, 20.5% belong to the southern group. Central European species (10%) decrease in contrast to an increase in the widespread (European) species (37%). The Eurasian species show more continental affinities, almost throughout their entire areas. The semi-dry grasslands at the eastern edge of the Alps show a more continental influence (9.5% dis-

tribution areas with an oceanic tendency compared with 26% with a continental tendency in the examined relevé). Compared to this, the *Meso-* and *Xerobromion* communities of the Swiss Jura contain more species with a submediterranean–atlantic distribution (see Zoller, 1954).

The *ATS* of the *Euphorbio saxatilis–Pinetum nigrae* shows 45% submediterranean species of which the main part, i.e. 37%, has a montane affinity. Furthermore, 28% is comprised of European–central European species (those with southern distributions – 11%), 15% of dealpines, 9% of species with Eurasian distributions and 3% of species that are distributed over the whole of the northern hemisphere. In comparison the *Pinus nigra* stone heath has a montane group of only 24% within the 40% of submediterranean species; the Pontic (3%), Pannonian (10%) and Eurasian (16%) species, concentrating in *Quercetea pubescentis-petraeae* forest steppes, are more numerous than the European and central European species (18%), which also have an eastern or southern tendency here.

Two types of rock vegetation (*Fumano–Stipetum eriocaulis*) are connected spatially with these two *Pinus nigra* community types. On southern slopes, the *Pinus nigra* stone heath alters with the *Fumano–Stipetum eriocaulis minuartietosum setaceae;* on northern slopes, the *Euphorbio saxatilis–Pinetum nigrae* neighbours an impoverished *Fumano–Stipetum* type. Also as to their *ATS*, the connected communities coincide more than the two types of *Fumano–Stipetum eriocaulis*. For example, the *Euphorbio saxatilis–Pinetum nigrae* and the rock vegetation type of the northern slopes show a high percentage of submediterranean montane species, only a very small percentage of the Eurasian and Pontic-Pannonian species, and a small percentage of European-central European species (mostly eastern montane species).

Fig. 1. Area type spectra of typical relevés from plant communities discussed in the text and shown here as percentages of respective total species numbers.

NHEM = distributed in the northern hemisphere or all over the world; EURAS = Eurasian distribution; EUR = European: region of deciduous forest and northern (resp. montane) coniferous forest; incl. western Siberia and some parts of southwest Asia; outposts in the Mediterranean region (possibly to northwest Africa); MEUR = central European distribution; SMED = submediterranean: northern edge of the Mediterranean region, almost lacking sclerophyllous woody plants; also southern high mountains with altitudinal belts in which European species still dominate; SMED GEB = montane areas of the submediterranean region; MED = Mediterranean: region of evergreen sclerophyllous plants; TUR = Oriental-Turanean: semi-deserts of southwest and central Asia; PONT = Pontic: grass steppes in which woody plants are rare or completely absent, extending from middle Asia to the southeastern parts of central Europe; PANN = Pannonian distribution; DEALP = dealpine (incl. demontane): descending from alpine (montane) altitudes; ↓ = species advancing further to the south although essentially having European or central European distributions.

Asperulo-Fagetum (Karrer, in prep.)

NHEM
EURAS
EUR + MEUR/↓
SMED GEB/SMED
MED + TUR
PONT + PANN
DEALP

0   50   100%

Querco-Carpinetum s.l., influenced by man
(Karrer, in prep.)

NHEM
EURAS
EUR + MEUR/↓
SMED GEB/SMED
MED + TUR
PONT + PANN
DEALP

0   50   100%

Querco-Carpinetum s.l., natural (Karrer, in prep.)

NHEM
EURAS
EUR + MEUR/↓
SMED GEB/SMED
MED + TUR
PONT + PANN
DEALP

0   50   100%

Mesobromion community (Karrer, in prep.)

NHEM
EURAS
EUR + MEUR/↓
SMED GEB/SMED
MED + TUR
PONT + PANN
DEALP

0   50   100%

Euphorbio saxatilis-Pinetum nigrae (Karrer,
in prep.)

NHEM
EURAS
EUR + MEUR/↓
SMED GEB/SMED
MED + TUR
PONT + PANN
DEALP

0   50   100%

Fumano-Stipetum eriocaulis (pure type; Karrer,
in prep.)

NHEM
EURAS
EUR + MEUR/↓
SMED GEB/SMED
MED + TUR
PONT + PANN
DEALP

0   50   100%

Pinus nigra stone heath (Karrer, in prep.)

NHEM
EURAS
EUR + MEUR/↓
SMED GEB/SMED
MED + TUR
PONT + PANN
DEALP

0   50   100%

*Fumano-Stipetum eriocaulis minuartietosum
setaceae* (Karrer, in prep.)

NHEM
EURAS
EUR + MEUR/↓
SMED GEB/SMED
MED + TUR
PONT + PANN
DEALP

0   50   100%

Fumano-Stipetum eriocaulis (characteristic type;
Wagner, 1941: Table 1, relevé No. 12)

NHEM
EURAS
EUR + MEUR/↓
SMED GEB/SMED
MED + TUR
PONT + PANN
DEALP

0   50   100%

*Fumano-Stipetum eriocaulis laserpitietosum
sileris* (Niklfeld, 1979)

NHEM
EURAS
EUR + MEUR/↓
SMED GEB/SMED
MED + TUR
PONT + PANN
DEALP

0   50   100%

In contrast to the typical *Fumano–Stipetum eriocaulis* (Wagner, 1941, Table 1), the *ATS* of the montane subassociation *Fumano–Stipetum minuartietosum setaceae* shows a major component of dealpine, but less pure submediterranean, species (cf. Fig. 1). This tendency continues in relevés of the dolomite rock vegetation analyzed in the more montane and inner parts of the Alps (i.e. in a relevé from the Grillenberg Valley, Niklfeld, 1979, Table 9). There, already 17% of the species are dealpine, 16% of the species are European-central European, and 67% of the species are submediterranean (of which 42% are distributed in the submediterranean montane regions).

The *ATS* of the *Fumano–Stipetum eriocaulis laserpitietosum sileris* (Niklfeld, 1966) consists of only a few species with submediterranean montane distributions but has in contrast 23% Eurasian and 22% European-central European species. The dealpines are rarer than the Pontic and Pannonian species (6%). The distinction of the special syntaxon, at least at the subassociation level, is strengthened by this. Differences between the two montane subassociations are also caused by different soils; the *F.–S. minuartietosum setaceae* is situated on dolomite, whereas the *F.–S. laserpitietosum sileris* is situated on hard limestone and has a different soil development.

Disjunction quotient (DQ)

The values of *DQ* in this paper range from 0.68 for *Rumex acetosa* s.l. to 150 for *Euphorbia saxatilis*, which is an endemic of the northeast Alps and has a very small distribution area. A survey of *DQ* values for the examined communities are shown in Table 2.

For the *Querco–Carpinetum* s.l. (representing a near-climax community) the average $DQ = 5.49$. This is twice as large as that calculated for the anthropogenous *Mesobromion* community, whose average *DQ* is in turn exceeded by the *DQs* of the natural non-climax permanent communities.

According to Bach (1957, cit. in Braun-Blanquet, 1964), a climax is defined as the result of natural vegetation development under prevailing climatic and soil conditions, excluding all impeding outside influences like abnormal drainage, erosion or accumulation, without any successional dynamics at all. These conditions can be assumed to hold for the *Asperulo–Fagetum* and, also to a great extent for the *Querco–Carpinetum* s.l. On both sites, species that have been able to adapt to the overall climatic conditions (and are therefore generally widespread) are predominant (*DQ*-size group 0-5: 59%). However, species with smaller and/or more disjunct distributions (15% with size group 10-20) are also to be found. This can be explained by the history of the sites. Some species, that are at present disjunctly distributed, were restricted to the mountains of the meridional and submeridional zone during the glacial periods and invaded, in a sociologically significant manner, the climax and subclimax communities of the eastern edge of the Alps (e.g. *Staphylea pinnata, Euonymus latifolia, Cyclamen purpurascens*).

Gradually species that had been able to adapt to the supra-regional overall climate were incorporated into the central sociological groups of the climax forest. Temporal and regional deviations in soil and climate are balanced by the plants' ecological plasticity potential. For the most part these comprise genetically stable diploids and polyploids (Ehrendorfer, 1980) with an average sum of total area equivalent to the extension of overall climatic-geo-

Table 2. Percentage of species with *DQ* values in different *DQ* intervals and average *DQ* value for contrasting plant communities in the S Wienerwald area.

Size groups of *DQ*	0–5	5–10	10–20	20–40	>40	average *DQ*
Querco–Carpinetum	59	26	15	.	.	5.49[a]
Mesobromion community	91	9	.	.	.	2.46
Euphorbio saxatilis–Pinetum nigrae	.	26	32	21	21	27.05
Fumano–Stipetum eriocaulis minuartietosum setaceae	5	29	24	30	12	26.04

[a] For the *Asperulo–Fagetum* the value is 7.16

graphic units. In addition, their distribution areas are characterized by a degree of disjunction which is determined by their low tolerance to changes in altitude. (All the above conclusions are applicable to Europe only.)

The anthropogenous substitute community (*Mesobromion*) is characterized by widespread species with uniform areas. The incisive and limiting human influences (mowing, and to a certain extent, grazing) created habitats in which, on the one hand, species that had already existed in the area before human cultural activities began, were 'pre-fitted' to the changed ecological conditions. (These species, according to Ehrendorfer, 1980, comprise 47.2% genetically stable diploids and polyploids.) On the other hand, the aggravating changes in the ecology (of the area) and competitive conditions allowed newly developed – or developing – diploid and polyploid taxa to establish and disperse themselves over a wide area. Once a taxon has established itself in a meadow, for example, it is often distributed and promoted systematically by man. Thus the occurrence of the greater part of the more or less continuously, widespread species (91% with *DQ* below 5) in the *Mesobromion* community can be accounted for.

As to explain the *DQs* in the natural permanent communities, e.g. the *Euphorbio saxatilis–Pinetum nigrae* – characterized sociologically by Wendelberger (1962) and Zimmermann (1972), and by Ehrendorfer (1980) with reference to ploidy conditions – and the *Fumano–Stipetum eriocaulis minuartietosum setaceae*, two aspects are important.

1. Such communities are subject to permanent ecological stress (i.e. a poor water supply, stagnating soil development, erosion, and a high concentration of magnesium in the soil). Identical combinations of site conditions occur in scattered patterns only; the species of these sites are stenoecic and weak 'competitors' and are in most cases disjunctly distributed. Disjunct distributions and small sizes of total area cause a high value of *DQ* for many species of natural permanent communities. The genetically stable palaeo- and mesopolyploids, especially, have often retreated into and accumulated in such azonal permanent communities (cf. Ehrendorfer, 1980). The more or less complete absence of competition in these areas may lead to the 'freezing of characters'. And thus the chances of such species to expand their distribution range are decreased.

2. At the eastern edge of the Alps, a well-known relic region (cf. Niklfeld, 1972, 1973, 1974, 1979; Zimmerman, 1972, 1976), some plant communities occur that 'shelter' endemics and subendemics and many taxa, with disjunct distributions. Here, the *Euphorbio saxatilis–Pinetum nigrae*, besides the montane rock vegetation and the relic *Pinus sylvestris* forests (Gams, 1930; Schmid, 1936; Wendelberger, 1962; Zimmermann, 1972, 1976), occupies a central position. Within this community and the *Fumano–Stipetum eriocaulis minuartietosum setaceae* a great number (24.7% and 37.5% resp. according to Ehrendorfer, 1980) of diploid species with endemic or disjunct distributions and a basic systematic position can be found; they are partly considered to be 'parental taxa' of polymorphous polyploid species (Ehrendorfer, 1949, 1962; Polatschek, 1966); for example: *Biscutella laevigata* ssp. *austriaca*, *Galium austriacum* ssp. *austriacum*, *Erysimum sylvestre*, *Cardaminopsis petraea* and *Amelanchier ovalis*. In the *Fumano–Stipetum eriocaulis minuartietosum setaceae* such 'basic' diploids predominate over the group of 'stable' diploids (25.7% by Ehrendorfer, 1980). A small and disjunct (or endemic) area as well as a high value of *DQ* is common to all these relic species.

Conclusions

Spectra of area types and the values of average *DQ* for the communities which are compared in this study differ depending on the degree of human influence (cf. Fig. 1 and Table 2). The *ATS* of the montane climax community (*Asperulo–Fagetum*) shows its distinct middle-European character. 11% Eurasian species occur probably in this community because of the relatively continental all-over climate of the eastern edge of the Alps (with its transition to the Pannonian region). Especially, this portion of species increases in the *Querco–Carpinetum* s.l., which is promoted by man. In particular, clear cuttings, coppice woods or middle forests enable the invasion of thermo- and heliophilous plants. Their middle-European occurrences are only peripheral localities, whereas their distribution is wider and more continuous in the light forests of eastern Europe and Siberia. The lower average *DQ* of 5.49 in the *Querco–Carpinetum* s.l. (as against 7.16 in the *Asperulo–Fagetum*) also indicates the invasion of such widespread species. Furthermore the lower

208

DQ is caused by the changed ratio of only middle-European species to widespread European species (in % of total number of species at a sample plot). Widespread European species like *Clinopodium vulgare, Campanula rapunculoides, Crataegus monogyna* and *Sambucus racemosa*, can be easily brought in by man or his grazing animals.

In the anthropogenous *Mesobromion* substitute community, continuously and spaciously distributed European and Eurasian species predominate, whereas the portion of middle-European species is low.

In contrast to the climax and to the substitute community, the azonal permanent communities show a high degree of disjunctly distributed species and accordingly the average DQ is higher. The species that are sociologically relevant to those communities, concentrate in the submediterranean (resp. Illyric) regions. Neighbouring hardwood forests (e.g. *Cotino–Quercetum pubescentis* Jakucs 61) – in consequence of the greatly varied relief, mostly of azonal character – and usually secondary semi-dry grasslands also give shelter to many submediterranean elements (see also Niklfeld, 1964, 1973; Wagner, 1956). Thus the 'Thermenlinie' can really be regarded as a submediterranean island.

References

Braun-Blanquet, J., 1964. Pflanzensoziologie. Springer Verlag, Wien, New York, 865 pp.

Dansereau, P., 1957. Biogeography. The Ronald Press Company, New York, 394 pp.

Ehrendorfer, F., 1949. Zur Phylogenie der Gattung Galium I. Polyploidie und geographisch-ökologische Einheiten in der Gruppe des Galium pumilum MURRAY (Sekt. Leptogalium LANGE sensu ROUY) im österreichischen Alpenraum. Österr. Bot. Z. 96: 110–138.

Ehrendorfer, F., 1962. Cytotaxonomische Beiträge zur Genese der mitteleuropäischen Flora und Vegetation. Ber. Dt. Bot. Ges. 75: 137–152.

Ehrendorfer, F., 1972. Arealtypen in der Wiener Flora. In: F. Ehrendorfer et al. (eds.), Naturgeschichte Wiens, Band 2. Jugend und Volk, Wien-München.

Ehrendorfer, F., 1973. Liste der Gefäßpflanzen Mitteleuropas. Fischer, Stuttgart, 318 pp.

Ehrendorfer, F., 1980. Polyploidy and distribution. In: W. H. Lewis (ed.), Polyploidy: Biological Relevance. Plenum Publ. Corp., New York.

Ellenberg, H., 1956. Grundlagen der Vegetationsgliederung. I. Teil. Aufgaben und Methoden der Vegetationskunde. In: H. Walter (ed.), Einführung in die Phytologie IV. Ulmer, Stuttgart.

Ellenberg, H., 1976. Vegetation Mitteleuropas mit den Alpen in ökologischer Sicht, 2nd ed. Ulmer, Stuttgart.

Gams, H., 1930. Über Reliktföhrenwälder und das Dolomitphänomen. Veröff. Geobot. Inst. ETH Rübel 6: 32–80.

Hultén, E., 1958. The Amphi-Atlantic plants and their phytogeographical connections. K. Svensk. Vet. Akad. Handl. 4, 7.

Hultén, E., 1962. The circumpolar plants I. Vascular cryptogams, conifers, monocotyledons. K. Svensk. Vet. Akad. Handl. 4, 8.

Hultén, E., 1970. The circumpolar plants II. Dicotyledons. K. Svensk. Vet. Akad. Handl. 4, 13.

Jakucs, P., 1961. Die phytozönologischen Verhältnisse der Flaumeichen-Buschwälder Südosteuropas. Monographie der Flaumeichen-Buschwälder I. Akadémiai Kiadó, Budapest.

Karrer, G., (in prep.). Arealkundliche Untersuchungen kontrastierender Pflanzengesellschaften im südlichen Wienerwald. Diss. Phil. Fak. Univ. Wien.

Kornaś, J., 1972. Corresponding taxa and their ecological background in the forests of temperate Eurasia and North America. In: D. H. Valentine (ed.), Taxonomy, Phytogeography and Evolution. Academic Press, New York.

Luftensteiner, H. W., 1978. Experimentelle Untersuchungen zur Reproduktions- und Verbreitungsbiologie an vier Pflanzengemeinschaften des niederösterreichischen Alpenostrandes. Diss. Phil. Fak. Univ. Wien. 240 pp.

Luftensteiner, H. W., 1979. The eco-sociological value of dispersal spectra of two plant communities. Vegetatio 41: 61–67.

Luftensteiner, H. W., 1980. Der Reproduktionsaufwand in vier mitteleuropäischen Pflanzengemeinschaften. Pl. Syst. Evol. 135: 235–251.

Luftensteiner, H. W., 1981. Myxochory in four different plant communities in Austria. Israel J. Bot. 30: 95–98.

Mayer, H., 1974. Wälder des Ostalpenraumes. Fischer, Stuttgart. 344 pp.

Meusel, H., Jäger, E. & Weinert, E., 1965. Vergleichende Chorologie der zentraleuropäischen Flora (Text, Karten). Fischer, Jena.

Meusel, H., Jäger, E., Rauschert, S. & Weinert, E., 1978. Vergleichende Chorologie der zentraleuropäischen Flora II (Text, Karten). Fischer, Jena.

Niklfeld, H., 1964. Zur xerothermen Vegetation im Osten Niederösterreichs. Verh. Zool.-Bot. Ges. Wien 103/104: 152–181.

Niklfeld, H., 1972. Der niederösterreichische Alpenostrand – ein Glazialrefugium montaner Pflanzensippen. Jahrb. Ver. Schutze Alpenpfl. -tiere 37: 42–94.

Niklfeld, H., 1973. Über Grundzüge der Pflanzenverbreitung in Österreich und einigen Nachbargebieten. Verh. Zool.-Bot. Ges. Wien 113: 53–69.

Niklfeld, H., 1974. Zur historischen Deutung von Pflanzenarealen am Ostrand der Alpen. Wiss. Arb. Burgenland 54: 46–52.

Niklfeld, H., 1979. Vegetationsmuster und Arealtypen der montanen Trockenflora in den nordöstlichen Alpen. Stapfia 4: 1–229.

Poelt, J., 1966. Bestimmungsschlüssel europäischer Flechten. J. Cramer, Lehre.

Poelt, J. & Vĕzda, A., 1977 (+ 1981). Bestimmungsschlüssel europäischer Flechten. Ergänzungsheft I (+ II). Cramer, Lehre.

Polatschek, A., 1966. Cytotaxonomische Beiträge zur Flora der Ostalpenländer, I; II. Österr. Bot. Z. 113: 1-46, 101-147.

Rabenhorst, L., 1884-1918. Kryptogamenflora von Deutschland, Österreich und der Schweiz, 2. Auflage. Leipzig.

Schmid, E., 1936. Die Reliktföhrenwälder der Alpen. Beitr. Geobot. Landesaufn. Schweiz 21: 1-190.

Straka, H., 1970. Arealkunde, 2. Auflage. In: H. Walter (ed.), Einführung in die Phytologie. Ulmer, Stuttgart.

Tüxen, R., 1937. Die Pflanzengesellschaften Nordwestdeutschlands. Mitt. flor.-soz. Arb. gem. Niedersachsen 3: 1-170.

Wagner, H., 1941. Die Trockenrasengesellschaften am Alpenostrand. Denkschr. Akad. Wiss. Wien math.-nat. Kl. 104(1): 1-78.

Wagner, H., 1956. Die Pflanzengeographische Stellung des Wiener Raumes. In Exkursionsführer für die XI. Internationale Pflanzengeographische Exkursion durch die Ostalpen 1956. Angew. Pflanzensoziol. (Wien) 16: 73-79.

Wendelberger, G., 1962. Das Reliktvorkommen der Schwarzföhre (Pinus nigra Arnold) am Alpenostrand. Ber. Dt. Bot. Ges. 75: 378-386.

Wendelberger, G., 1963. Die Relikt-Schwarzföhrenwälder des Alpenostrandes. Vegetatio 11: 265-288.

Zimmermann, A., 1972. Pflanzenareale am niederösterreichischen Alpenostrand und ihre florengeschichtliche Bedeutung. Diss. Bot. 18: 1-199.

Zimmermann, A., 1976. Montane Reliktföhrenwälder am Alpen-Ostrand im Rahmen einer gesamteuropäischen Übersicht. In: J. Gepp (ed.), Mitteleuropäische Trockenstandorte in pflanzen- und tierökologischer Sicht. Ludwig-Boltzmann-Institut für Umweltwissenschaften und Naturschutz, Graz.

Zoller, H., 1954. Die Typen der Bromus erectus-Wiesen des Schweizer Juras. Beitr. geobot. Landesaufn. Schweiz 33: 1-309.

Zólyomi, B., 1966. Neue Klassifikation der Felsen-Vegetation im pannonischen Raum und der angrenzenden Gebiete. Bot. Közlem. 53: 49-54.

Zukrigl, K., 1973. Montane und subalpine Waldgesellschaften am Alpenostrand. Mitt. Forstl. Bundes-Versuchsanst. Wien 101: 1-417-

Accepted 19.1.1984.

.

Some synchorological aspects of basiphilous pine forests in Fennoscandia*

Jørn Erik Bjørndalen
Sogn og Fjordane Regional College, P.O. Box 39, N-5801 Sogndal, Norway

Keywords: Basiphilous *Pinus sylvestris* forests, Classification, Fennoscandia, Geographical race, Phytogeography, Synchorology

Abstract

Basiphilous pine forests and related birch forests are herb- and grass-rich forests on calcareous substrate. These forests are complex communities with floristic/ecological elements from different vegetation types occurring in a subtle micromosaic. These elements are e.g. species from acidophilous conifer forests, thermophilous forest-rim communities, calcareous shallow-soil and steppe communities, eutrophic wet meadows and fens, and in northern Fennoscandia also species from alpine *Dryas* heaths. Four associations are recognized in Fennoscandia: *Convallario–Pinetum, Melico–Piceetum pinetosum, Peucedano–Pinetum* and *Epipacto atrorubentis–Betuletum*. The main association is the *Convallario–Pinetum*, a widespread community in Fennoscandia and Estonia with a considerable floristic variation between the different regions. Examples of the floristic variation along west–east profiles and south–north profiles in Fennoscandia are presented. The basiphilous pine forest complex can be divided into a number of ecological types along the moisture and nutritional gradients. A further subdivision into geographical types (races) is presented.

Introduction

Herb-rich forests dominated by *Pinus sylvestris* (occasionally also by *Betula pubescens* and *Picea abies*) occur on more or less calcareous substrate throughout Norway, Sweden and Finland. These forests are rich in calciphilous and eutrophic species, and the term 'basiphilous pine forests' will be used to designate this forest type. The basiphilous pine forests (and related birch forests) are complex communities which contain elements from a variety of floristic/ecological species groups (e.g., species from forest-rim communities, calcareous shallow-soil communities, meadows, steppe communities, acidophilous pine forests and alpine *Dryas* communities) in a subtle micro-mosaic. The floristic composition shows a considerable variation in Fennoscandia, and numerous subtypes can be recognized along ecological and phytogeographical gradients (Bjørndalen, 1980a). The classification into syntaxa of the Braun-Blanquet system is difficult, especially higher syntaxonomical units (cf. Fig. 1).

I have recently started a phytosociological investigation of these forests. A preliminary report introducing the project has been published (Bjørndalen, 1980a). A series of regional papers are under preparation by the author or together with co-workers. The basiphilous pine forests of an area in Telemark, SE Norway, have been described by Bjørndalen (1980b, 1981), including the formal description of the association *Convallario–Pinetum*.

The present contribution presents some preliminary results on the synchorology of the basiphilous pine and birch forests in Fennoscandia, a study of the synchorological aspects will be given in a concluding paper of the project. By Winter 1982 based

* Nomenclature follows Lid (1974) for vascular plants, Nyholm (1954–1969) for musci and Dahl & Krog (1973) for lichens.

Vegetatio 59, 211–224 (1985).
© Dr W. Junk Publishers, Dordrecht. Printed in the Netherlands.

212

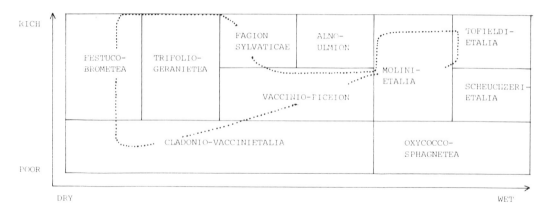

Fig. 1. Schematical representation of the syntaxonomical range (dotted line) of the basiphilous pine forest association *Convallario–Pinetum* in S Fennoscandia.

on almost 1 100 relevés from Fennoscandia and Estonia.

Syntaxonomy and distribution

Figure 2 shows the distribution of basiphilous pine forests and related communities in Fennoscandia state of knowledge (December 1982). The association *Convallario–Pinetum* seems to be the main association of basiphilous pine forests in Fennoscandia, extending from the Oslofjord area and southern Sweden north to the Bodø area in N Norway. Some herb-rich pine forests in S Finland (especially the Hämeenlinna area) probably belong to the *Convallario–Pinetum* as well. The basiphilous pine forests in Gotland and Estonia can probably be regarded as an own subassociation characterized by *Sesleria caerulea*. The more acidophilous pine forest association *Peucedano–Pinetum*, described by Matuszkiewics (1962), interferes with the true basiphilous pine forests, especially in S Finland. Drought seems to be the common ecological factor which links the basiphilous pine forests on calcareous soils with the acidophilous herb-rich pine forests on well-drained sandy soils. The *Peucedano–Pinetum* occurs in the extreme southeastern part of Fennoscandia (cf. Kielland-Lund, 1967). The *Melico–Piceetum pinetosum* (cf. Kielland-Lund, 1973, 1981) occurs throughout the whole distribution area of the basiphilous pine forest complex. Also the oceanic basiphilous pine forests at the Norwegian west coast ('*Saniculo–Pinetum*'

sensu Bjørndalen, 1980a) can be included in the *Melico–Piceetum*. Basiphilous birch forests on calcareous scree material in N Norway ('*Epipacto atrorubentis–Betuletum*' sensu Bjørndalen, 1980a) is related to the alpine *Dryas octopetala* heaths.

The *Convallario–Pinetum* is characterized by a significant amount of thermophilous forest-rim and calcareous shallow-soil species, especially in the southeastern part of Fennoscandia. This association can be regarded as a marginal community in the class *Trifolio–Geranietea sanguinei*. *Peucedano–Pinetum* and *Melico–Piceetum pinetosum* belong to the *Vaccinio–Piceetea*, in the alliances *Dicrano–Pinion* and *Vaccinio–Pinion*, respectively. (For further details, see Bjørndalen, 1980a, b.)

Species groups (ecological groups)

The basiphilous pine forests (and related communities) are characterized by a mixture of different floristical/ecological elements in a subtle micromosaic pattern within apparently homogeneous stands. Some of the important species groups are:

1. Species from more or less acidophilous conifer (*Vaccinio–Piceetea*) forests, e.g., *Arctostaphylos uva-ursi, Calluna vulgaris, Cladonia* spp., *Dicranum* spp., *Hylocomium splendens, Picea abies, Pinus sylvestris, Pleurozium schreberi, Vaccinium myrtillus* and *V. vitis-idaea*. This species group is prominent in all types of basiphilous pine forests in the whole distribution area.

2. Species belonging to rich spruce forests (e.g. low herb spruce forests), e.g., *Calamagrostis arundinacea, Carex digitata, Fragaria vesca, Hieracium sylvaticum* coll., *Melica nutans, Rhytidiadelphus triquetrus, Rubus saxatilis* and *Viola riv-*

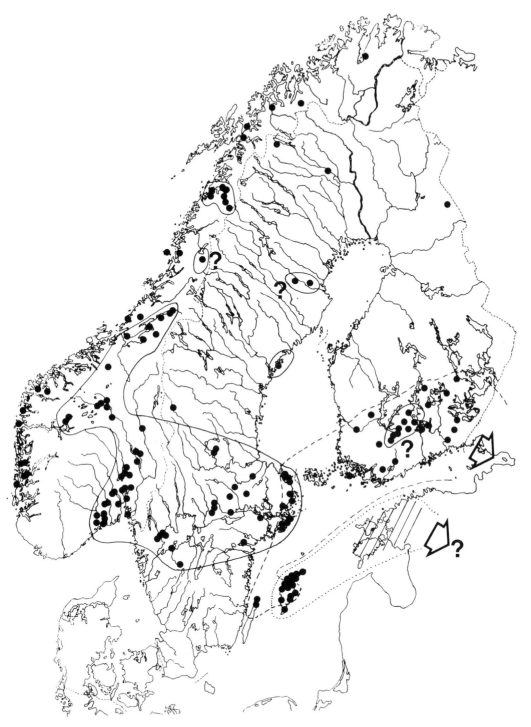

Fig. 2. Distribution of basiphilous pine forests and related forest types in Fennoscandia (pr. Winter 1982). Whole line marks the distribution area of *Convallario–Pinetum*, dotted line the area of the Gotlandic/ Estonian subtype of *Convallario–Pinetum* and dashed line the area of *Peucedano–Pinetum*. Other localities are mostly *Melico–Piceetum pinetosum* (this forest type occurs also within the distribution areas of the other associations). The occurrences in N Norway (north of Salten) are mostly different types of basiphilous birch forests (*Epipacto atrorubentis–Betuletum*). Data on Norway and Sweden mostly based on own research, the Finnish data mostly on Jalas (1950) and Mäkirinta (1968).

iniana. An important group in all parts of the Fennoscandian distribution area.

3. Characteristic species of thermophilous deciduous forests (*Querco–Fagetea*), e.g., *Brachypodium sylvaticum, Convallaria majalis, Corylus avellana, Fraxinus excelsior, Hepatica nobilis, Lonicera xylosteum, Quercus robur* and *Sanicula europaea.* This group is most important in the southern parts of Fennoscandia.

4. Thermophilous southern and southeastern species belonging to *Trifolio–Geranietea* communities, e.g., *Brachypodium pinnatum, Epipactis atrorubens, Filipendula vulgaris, Geranium sanguineum, Hypochoeris maculata, Inula salicina, Laserpitium latifolium, Origanum vulgare* and *Trifolium medium.* In addition, many species from the *Festuco–Brometea* and *Sedo–Scleranthetea* are also common, e.g., *Anthyllis vulneraria, Artemisia campestris, Carlina vulgaris, Festuca ovina, Galium verum, Pimpinella saxifraga, Polygonatum odoratum, Thymus serpyllum* and *Tortella tortuosa.* This species group is very important in SE Norway, S and central Sweden and S Finland. A marked decline occurs in the number of these species towards N and W in Fennoscandia.

5. Calciphilous/hygrophilous species confined to eutrophic wet meadows and rich fens (eutrophic groups within the *Molinietalia,* and *Tofieldietalia*), e.g., *Angelica sylvestris, Carex capillaris, C. flacca, C. panicea, C. pulicaris, C. vaginata, Molinia caerulea, Sesleria caerulea, Succisa pratensis* and mosses such as *Aulacomium palustre, Campylium stellatum, Ctenidium molluscum* and *Tomenthypnum nitens.* This group occurs in certain seasonal hygrophilous types of basiphilous pine forests, and is found scattered throughout the distribution area of the basiphilous pine forests.

6. Alpine, calciphilous species from *Dryas* heaths (*Kobresio–Dryadion*), e.g., *Antennaria alpina, Astragulus alpinus, Carex rupestris, Dryas octopetala, Rhytidium rugosum, Saxifraga oppositifolia* and *Silene acaulis.* Also more hygrophilous species occur, e.g., *Salix reticulata, S. myrsinites, Saxifraga aizoides, Saussurea alpina* and *Thalictrum alpinum.* This species group is most important in basiphilous pine and birch forests in N Norway.

Phytogeographical elements

The species which occur in the basiphilous pine and birch forests can be divided into different phytogeographical elements. A division of the 180 most important species into phytogeographical groups (based on the maps in Hultén, 1971) gives the following results: (1) ubiquitous species within the distribution area of the basiphilous pine forests: 40%; (2) species with southeastern (somewhat continental) distribution: 30%; (3) species with southern (somewhat oceanic) distribution: 19%; (4) species with eastern or northeastern distribution: 5%; (5) alpine species: 6%.

The ubiquitous species mostly belong to acidophilous *Vaccinio–Piceetea* communities (species group 1) and eutrophic spruce forests (group 2). The species with southern and southeastern, continental distribution are important constituents of *Convallario–Pinetum,* and these species are also important differential species in the formation of geographical races (regional types) of this association. The number of southeastern species decreases rapidly northwards and westwards in Fennoscandia, i.e. northwest of Østlandet/Sørlandet in SE Norway, north of the 'limes norrlandicus' in central Sweden and north of the line Pori–Savonlinna in Finland. Species with eastern distribution are most important in Finland and northern Sweden. The alpine species are mostly confined to N Norway and adjacent parts of Sweden and Finland, but are found also at higher elevation in SE Norway.

Floristic variation of basiphilous pine and birch forests: some examples

Profile 1: W–E in southern Norway and Sweden (Fig. 3)

Species groups with high constancy in all of the regions are the *Vaccinio–Piceetea* species (e.g., *Calluna vulgaris, Dicranum scoparium, Hylocomium splendens, Pinus sylvestris, Pleurozium schreberi, Vaccinium myrtillus* and *V. vitis-idaea*) and the characteristic species of *Melico–Piceetum* (e.g., *Carex digitata, Fragaria vesca, Melica nutans, Rhytidiadelphus triquetrus* and *Rubus saxatilis*).

The basiphilous pine forests in W Norway deviate from those of the other regions in different respects: (1) most of the southeastern species are missing; (2) southern, thermophilous species with affinity to *Querco–Fagetea* are more common (e.g., *Brachypodium sylvaticum, Fraxinus excelsior, Hedera helix* and *Sanicula europaea*); (3) hygrophilous species occur with higher constancy; (4) many oceanic species occur, which are missing or rare in SE parts of Scandinavia (e.g., *Blechnum spicant, Festuca vivipara, Hypericum pulchrum, Hypnum ericetorum, Ilex aquifolium, Plagiothecium undulatum, Polytrichum formosum* and *Primula vulgaris*). Most basiphilous pine forests in W Norway belong to the *Melico–Piceetum,* probably in a separate oceanic subassociation or variant.

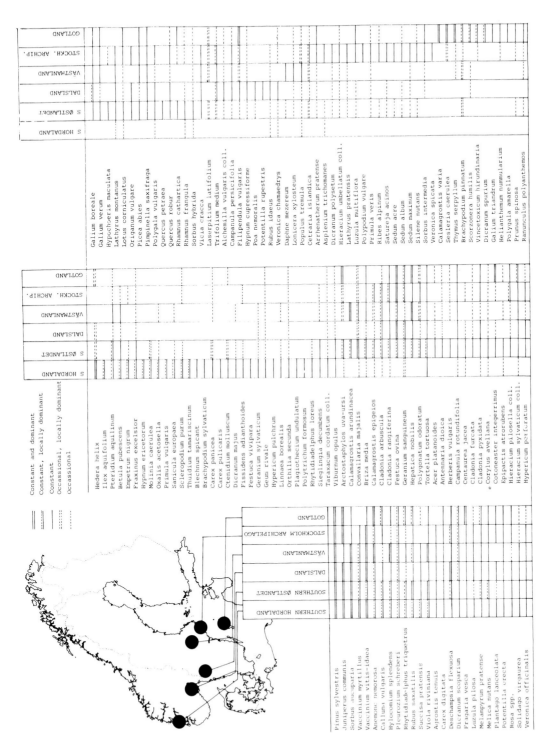

Fig. 3. W–E profile showing the variation in the floristic composition of basiphilous pine forests in S Scandinavia. The diagram is based on the following material: S Hordaland (Bjørndalen, 1980a; Bjørndalen & Vevle, in manuscript); Østlandet (Bjørndalen, 1980b, Bjørndalen & Brandrud, in prep.); Dalsland, Västmanland and the Stockholmarchipelago (Bjørndalen, in manuscript); Gotland (Bjørndalen, in prep.).

The southeastern, continental species are well represented in the other regions. The *Convallario-Pinetum* has its optimal development in the southern parts of the Cambro-Silurian districts of the Oslo Region, where a series of variants can be distinguished along different ecological gradients (cf. Bjørndalen, 1980b, 1981). The stands in central Sweden (Dalsland, Västmanland and the Stockholm area) belong to other geographical races of the *Convallario-Pinetum*. The basiphilous pine forests in the Stockholm archipelago deviate somewhat from the other regions because xerophilous pine forest types with xeric species are well developed in the area. The basiphilous pine forests in Gotland, which are closely related to those of Estonia (cf. Linkola, 1930), are characterized by species such as *Calamagrostis varia, Galium triandrum, Helianthemum nummularium, Scorzonera humilis, Sesleria caerulea, Thymus serpyllum* and many of the typical alvar species. The Gotlandic (and Estonian) basiphilous pine forests can probably be distinguished as a separate subassociation of the *Convallario-Pinetum* (I have earlier designated these forests as '*Seslerio-Pinetum*', cf. Bjørndalen, 1980a).

Profile 2: W–E in northern Fennoscandia (Fig. 4)

The *Vaccinio-Piceetea* species are shared by all regions. The low herb spruce forest group is reduced compared to further south, but species such as *Carex digitata, Hieracium sylvaticum* coll., *Melica nutans* and *Rubus saxatilis* are still common. The basiphilous pine and birch forests in the Bodø-Fauske area (Salten) are very rich and diverse communities, and larger, continuous stands are found. The number of calciphilous species is high, with examples such as *Antennaria alpina, Carex rupestris, Poa alpina, Rhytidium rugosum, Saxifraga aizoides* and *S. oppositifolia. Ophrys insectifera* can occur relatively numerously in some of the stands. The basiphilous birch forests of N Norway are often dominated by *Dryas octopetala* (not included in the diagram). Some oceanic species occur more commonly in outer Salten, e.g., *Festuca vivipara, Molinia caerulea, Rhacomitrium lanuginosum* and *Succisa pratensis*. The basiphilous pine forests in Salten must be regarded as a thermophilous community despite the occurrences of alpine species, and represents a northern development of *Convallario-Pinetum*.

The basiphilous spruce and pine forests of Masugnsbyn in Swedish Lappland contain some calciphilous species such as *Carex capillaris, Dryas octopetala, Salix starkeana, Selaginella selaginoides* and others, but are generally poor in species. The content of *Vaccinio-Piceetea* species is pronounced. The basiphilous pine forests in Oulanka, N Finland, contain *Equisetum scirpoides, Selaginella selaginoides* and other calciphilous species (e.g. *Cypripedium calceolus*). However, the most important species are *Vaccinio-Piceetea* species. *Betula verrucosa, Daphne mezereum* and *Viola rupestris* represent an eastern species element. The basiphilous pine forests of Oulanka belong probably to the *Melico-Piceetum*.

Profile 3: S–N in Norway (Fig. 5)

Several *Vaccinio-Piceetea* species and species from low herb spruce forests are shared by all regions. The basiphilous pine forests are optimally developed in SE Norway. Especially the southern and southeastern species become rare in the interior parts of SE Norway, e.g., *Briza media, Corylus avellana, Geranium sanguineum, Quercus petraea, Q. robur* and *Rhamnus cathartica*. Species such as *Abietinella abietina, Knautia arvensis* and *Rhytidium rugosum* are more common in the interior and more elevated parts of SE Norway (e.g. Gudbrandsdalen/Ottadalen). The southeastern element is less pronounced in the basiphilous pine forests in Nordmøre and the Trondheimsfjord area. *Empetrum hermaphroditum* is an important species north of the Trondheimsfjord. A transitional type between the SE Norway and the Salten type of *Convallario-Pinetum* is found in the Snåsa area in Nord-Trøndelag. Alpine species are important constituents in the basiphilous pine and birch forests in N Norway, e.g., *Antennaria alpina, Carex rupestris, Dryas octopetala, Saxifraga aizoides* and *S. oppositifolia*.

Profile 4: S–N in Sweden (Fig. 6)

The species shared by the regions are mostly species with affinity to xerophilous *Pinion* communities, to a lesser extent to the more eutrophic parts of *Vaccinio-Piceetea* (e.g. *Melico-Piceetum*). Southeastern, continental species are well represented in Gotland and the Stockholm archipelago, but the number of such species declines rapidly towards the north. Species such as *Hepatica nobilis*, *Thymus serpyllum* and *Trifolium medium* occur in the basiphilous pine forests north to the coast of Ångermanland (Höga kusten). The stands in Västerbotten (the Skellefteå area) are more hygrophilous types with several eutrophic species, e.g., *Alnus incana*, *Carex capillaris*, *C. vaginata*, *Cirsium heterophyllum*, *Listera ovata*, *Pyrola rotundifolia* and *Selaginella selaginoides*. *Cypripedium calceolus* is an important species in some of the stands. This pine forest represents a northern equivalent to the seasonal hygrophilous variant of *Convallario–*

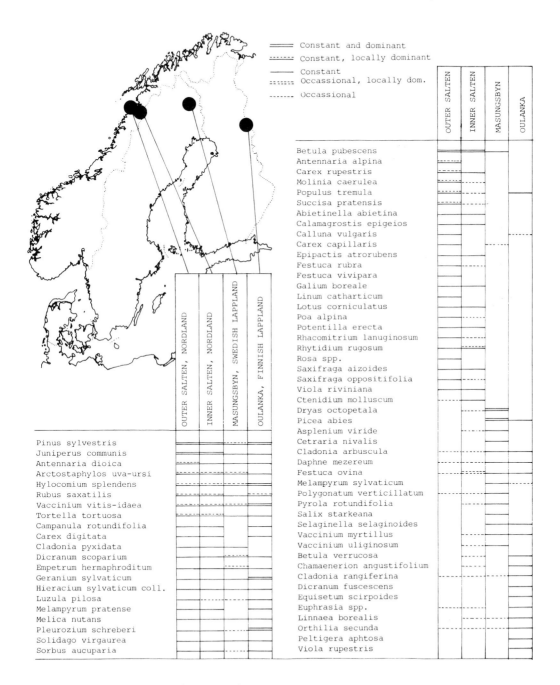

Fig. 4. W–E profile through N Fennoscandia. Salten (Bjørndalen, 1980a, in manuscript); Masugnsbyn (Bjørndalen, in manuscript); Oulanka (Söyrinki *et al.*, 1977).

218

Pinetum. The *Dryas*-dominated spruce forest at Masugnsbyn in Lapland is a very special forest type occurring on dolomite outcrops. Species such as *Asplenium viride, Cetraria nivalis, Dryas octopetala, Polygonatum verticillatum, Salix starkeana* and *Selaginella selaginoides* are important species in this community.

General phytogeographical patterns

The profiles show only examples of the floristic variation of the basiphilous pine forests in Norway and Sweden, but give an indication of some major trends. Further studies from more localities and numerical treatment of the material (e.g. by ordina-

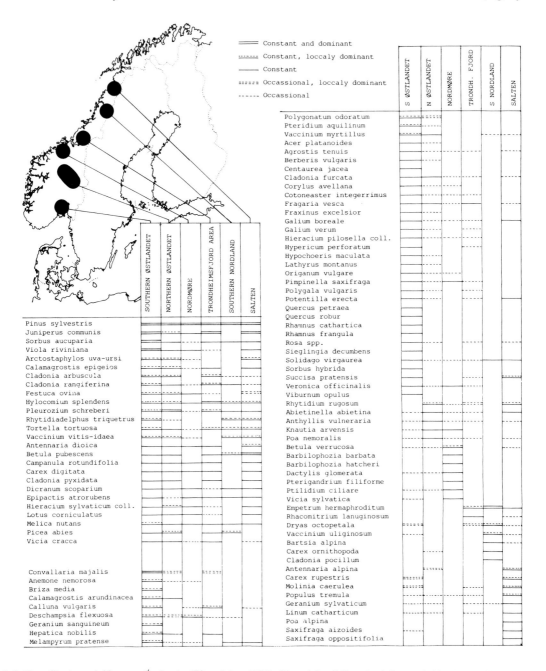

Fig. 5. S–N profile through Norway. Østlandet (Bjørndalen, 1980b; Bjørndalen & Brandrud, in prep.); Nordmøre (Holten, 1977); the Trondheimsfjord area (Aune, 1980; Bjørndalen, in manuscript); Salten (Bjørndalen, 1980a, in manuscript).

219

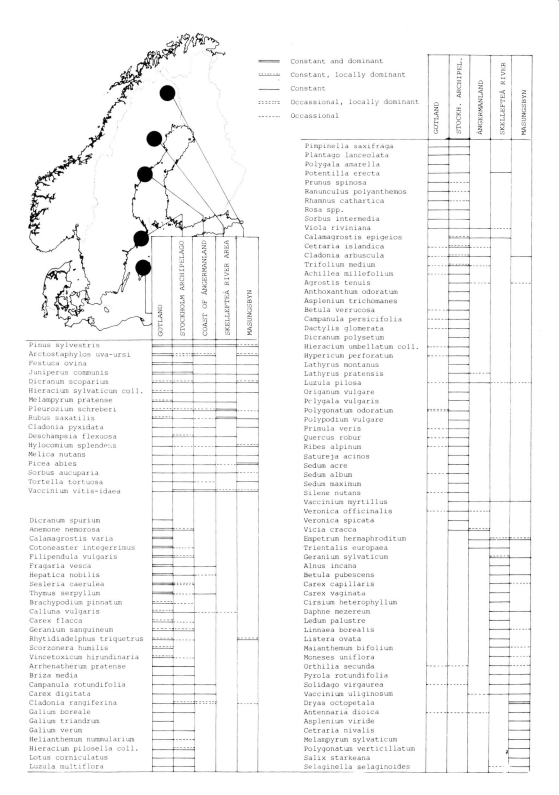

Fig. 6. S–N profile through Sweden. Gotland (Bjørndalen, in prep.); the other regions (Bjørndalen, in manuscript).

tion techniques) will give a more complete picture of the regional variation of the forest types. However, the differences between the profiles reflect general phytogeographical patterns in Norway and Sweden, especially with respect to the relative content of (1) southeastern, continental species; (2) oceanic species (heathland species and southern, thermophilous species); (3) eastern or northeastern species; and (4) alpine species.

The southeastern species (usually species from thermophilous forest-rim communities, calcareous shallow-soil communities and calcareous dry meadows) is the most important species group in SE Fennoscandia. The basiphilous pine forests in Gotland and along the Swedish Baltic Sea coast have a marked concentration of extreme southeastern species which are rare or absent in other regions, e.g., *Calamagrostis varia*, *Helianthemum nummularium*, *Galium triandrum*, *Melica ciliata*, *Sesleria caerulea* and *Vincetoxicum hirundinaria*. *Thymus serpyllum* is common in the basiphilous pine forests of eastern Sweden north to the coast of Ångermanland, but are almost absent from the basiphilous pine forests in Dalsland and the Oslo region. Species such as *Brachypodium pinnatum*, *Filipendula vulgaris*, *Laserpitium latifolium*, *Scorzonera humilis* and *Veronica spicata* reach SE Norway, but are most important in the basiphilous pine forests of central Sweden. The majority of the southeastern species are well represented in SE Norway, Sweden south of the 'limes norrlandicus' and S Finland, e.g., *Arrhenatherum pratense*, *Artemisia campestris*, *Calamagrostis arundinacea*, *Campanula persicifolia*, *Carlina vulgaris*, *Geranium sanguineum*, *Inula salicina*, *Rhamnus cathartica* and *Silene nutans*. The southeastern element is less pronounced west and north of the major areas with basiphilous pine forests in SE Fennoscandia. Some species reach the interior fjord districts of W Norway and the Trondheimsfjord area, e.g., *Berberis vulgaris*, *Hypericum perforatum*, *Knautia arvensis*, *Origanum vulgare*, *Pimpinella saxifraga*, *Polygonatum odoratum*, *Sedum album* (W Norway) and *Trifolium medium*. Most of these species are usually not represented in the basiphilous pine forests north of the 'limes norrlandicus' in Sweden.

Oceanic species are important constituents in the basiphilous pine forests of W Norway. Heathland and oligotrophic forest species such as *Blechnum spicant*, *Festuca vivipara*, *Hypericum pulchrum*,

Hypnum ericetorum, *Plagiothecium undulatum* and *Polytrichum formosum* do not occur in basiphilous pine forests in other regions in S Norway. Also *Ilex aquifolium* and *Primula vulgaris* are found only in W Norway. Some of these species occur in basiphilous pine forests along the coast of Møre og Romsdal and Trøndelag, *Festuca vivipara* also in the oceanic parts of the Salten district in N Norway. Many of the southern, thermophilous species have an oceanic distribution in Scandinavia. Species such as *Brachypodium sylvaticum*, *Hedera helix* and *Sanicula europaea* have their optimum in the basiphilous pine forests of W Norway, but occur also in other coastal areas (especially in the southern parts of the Oslo region and in Gotland and Öland).

Many southern, thermophilous species are also important in the basiphilous pine and birch forests in N Norway, e.g., *Calamagrostis epigeios*, *Carex digitata*, *Epipactis atrorubens*, *E. helleborine* and *Ophrys insectifera*. In fact, some of these species seem to be more or less confined to the basiphilous pine and birch forests on their northernmost localities in Norway. The southern, thermophilous species are of minor importance in Sweden north of the 'limes norrlandicus', except for Höga kusten in Ångermanland. The basiphilous pine forests there occur often on scree material on southfacing slopes.

Eastern/ northeastern species are abundant in the basiphilous pine forests in N Sweden and N Finland, e.g., *Ledum palustre*, *Salix starkeana* and *Viola rupestris* (*Actaea erythrocarpa* in the Oulu area). Species such as *Carex vaginata*, *Daphne mezereum*, *Goodyera repens*, *Pyrola chlorantha* and *Betula verrucosa* are widespread also in SE Fennoscandia.

The marked concentration of alpine species is characteristic for the basiphilous pine forests in N Norway. Species with affinity to *Dryas* heaths and other calciphilous alpine communities are important, e.g., *Antennaria alpina*, *Bartsia alpina*, *Carex rupestris*, *Dryas octopetala*, *Poa alpina*, *Rhytidium rugosum*, *Saussurea alpina*, *Saxifraga aizoides* and *S. oppositifolia*. Alpine species occur also in the basiphilous pine forests of N Sweden and N Finland, e.g., *Cetraria nivalis*, *Dryas octopetala*, *Equisetum scirpoides* and *Selaginella selaginoides*. *Empetrum hermaphroditum* is an important constituent of basiphilous pine forests in all parts of N Fennoscandia. Some relic localities of *Carex rupes-*

tris and *Dryas octopetala* are found at some places in lowland areas of SE Norway and Trøndelag, and these species occur abundantly in basiphilous pine forests at those localities.

Species with an affinity to more oligotrophic conifer forests are common throughout the distribution area of the basiphilous pine forests, e.g., *Arctostaphylos uva-ursi, Deschampsia flexuosa, Dicranum scoparium, Hylocomium splendens, Juniperus communis, Melampyrum pratense, Pleurozium schreberi, Solidago virgaurea, Vaccinium myrtillus* and *V. vitis-idaea.* Members of the 'low herb species group' (i.e. species with affinity to the low herb spruce forest association *Melico-Piceetum*) are also common in almost all regions, e.g., *Carex digitata, Hieracium sylvaticum* coll., *Melica nutans, Rhytidiadelphus triquetrus, Rubus saxatilis* and *Viola riviniana.*

Syntaxonomical remarks

I find a synchorological approach to the classification of these basiphilous pine forests and related communities useful. A comprehensive classification scheme based on floristic composition in combination with ecological and phytogeographical criteria will be proposed when more material is available. Even if common methods of the Braun-Blanquet system are used, I find it difficult to assign the basiphilous pine forests to higher syntaxonomical levels. These communities are complex, and elements from a variety of floristic/ecological species groups occur. I prefer a reticulate classification where the main types comprise series of communities along the dry-wet gradient (e.g. the series: *Convallario-Pinetum,* xerophilous variant – *Convallario-Pinetum,* herb-rich variant – *Convallario-Pinetum,* hygrophilous variant) and the poor-rich gradient, e.g. the series: poor ericaceous pine forests with only sporadic occurrences of eutrophic species – *Peucedano-Pinetum* – *Melico-Piceetum pinetosum* – *Convallario-Pinetum.* The main types can be subdivided into geographical types (races). For instance, the herb-rich variant of the *Convallario-Pinetum* can be separated in several geographical races: SE Norway race, W Norway race, Trondheimsfjord race, Salten race, central Sweden race, Gotland race, Öland race and Ångermanland race. Such a classification based on ecological and geo-

graphical aspects is useful also for practical purposes, e.g. the evaluation of the basiphilous pine forests as nature preservation objects (especially evaluation of representability). This system will now be adopted for national preservation plans for basiphilous pine forests and related communities in Norway and Sweden. The preliminary results from the investigation of the Fennoscandian basiphilous pine forests indicate that synchorological aspects can be a useful and important compliment to the ordinary syntaxonomical classification of these communities.

References

Aune, E. I., 1980. Vegetasjonen på Bergsåsen, Snåsa. K. Norske Vidensk. Selsk. Mus. Bot. Avd., Trondheim.

Aune, E. I. & Kjærem, O., 1977. Botaniske undersøkingar ved Vefsnavassdraget, med vegetasjonskart. K. Norske Vidensk. Selsk. Mus. Rapp. Bot. Ser. 1977, 1: 1–138.

Bjørndalen, J. E., 1980a. Kalktallskogar i Skandinavien – ett förslag till klassificering. Svensk Bot. Tidskr. 74: 103–122.

Bjørndalen, J. E., 1980b. Phytosociological studies of basiphilous pine forests in Grenland, Telemark, SE Norway. Norw. J. Bot. 27: 139–161.

Bjørndalen, J. E., 1981. Classification of basiphilous pine forests in Telemark, SE Norway: a numerical approach. Nord. J. Bot. 1: 665–670.

Bjørndalen, J. E., 1984a. Kalktallskogar som naturvårdsobjekt i Sverige. Statens naturvårdsverk.

Bjørndalen, J. E., 1984b. En naturvårdsinventering av kalktallskogar på Gotland. Naturvårdsenheten, Länsstyrelsen i Gotlands län, Visby (in press).

Bjørndalen, J. E. & Brandrud, T. E., 1984. Landsplan for verneverdige kalkfuruskoger og beslektede skogstyper i Norge. Miljøverndepartementet rapport (in press).

Bjørndalen, J. E. Classification of basiphilous pine forests in Gotland, Sweden (in prep.).

Bjørndalen, J. E. & Brandrud, T. E. Phytosociological studies of basiphilous pine forests in SE Norway (in prep.).

Dahl, E. & Krog, H., 1973. Macrolichens of Denmark, Finland, Norway and Sweden. Scandinavian University Books, Oslo – Bergen – Tromsø.

Holten, J. I., 1977. Floristiske og vegetasjonsøkologiske undersøkelser i sør- og nordeksponerte lier ved Gjøra i Sunndal. Thesis, Univ. of Trondheim (unpubl.).

Hultén, E., 1971. Atlas över växternas utbredning i Norden. Fanerogamer och kärlbunksväxter. Generalstabens litografiska anstalts förlag, Stockholm.

Jalas, J., 1950. Zur Kausalanalyse der Verbreitung einiger nordischen Os- und Sandpflanzen. Ann. Bot. Soc. Vanamo 24 (1): 1–345.

Kielland-Lund, J., 1967. Zur Systematik der Kiefernwälder Fennoskandiens. Mitt. Flor.-soz. Arbeitsgem. N.F. 11/12: 127–141.

222

Kielland-Lund, J., 1973. A classification of Scandinavian forest vegetation for mapping purposes. IBP i Norden 11: 173–206.

Kielland-Lund, J., 1981. Die Waldgesellschaften SO-Norwegens. Phytocoenologia 9: 53–250.

Lid, J., 1974. Norsk og svensk flora. 2. utg. Det norske samlaget, Oslo.

Linkola, K., 1930. Über die Halbhainwälder in Eesti. Acta Forest. Fenn. 36 (3): 1–30.

Mäkirinta, U., 1968. Haintypenuntersuchungen im mittleren Süd-Häme, Südfinnland. Ann. Bot. Fenn. 5: 34–64.

Matuszkiewics, W., 1962. Zur Systematik der natürlichen Kiefernwälder des mittel- und osteuropäischen Flachlandes. Mitt. Flor.-soz. Arbeitsgemein. N.F. 9: 145–186.

Nyholm, E., 1954–1969. Illustrated moss flora of Fennoscandia. II. Musci. Swedish Natural Science Research Counsil, Lund.

Söyrinki, N., Salmela, R. & Suvanto, J., 1977. Oulangan kansallipuiston metsä- ja suokasvillisuus. Acta Forest. Fenn. 154: 1–150.

Accepted 16.4.1984.

Addendum

Two years have passed since this contribution was presented and much more material on the basiphilous pine forests and related communities in Fennoscandia has become available. New localities were discovered in many parts of Norway and Sweden. The Swedish and Norwegian environmental departments recently initiated national nature preservation plans for basiphilous pine forests and related communities on the basis of this research. Intensive field work in all parts of Sweden and Norway was carried out in 1982 and 1983 (in Norway together with Tor Erik Brandrud), and reports will soon be published (Bjørndalen, 1984a, b; Bjørndalen & Brandrud, 1984). The classification scheme based on the combination of floristic, ecological and geographical aspects presented in the present paper proved to be useful for the work with nature preservation plans. The system was further developed and additions were made. The comprehensive classification scheme is presented in the reports (see Bjørndalen, 1984a; Bjørndalen & Brandrud, 1984).

The true basiphilous pine forests (i.e. relative open pine forests with a significant content of calciphilous, heliophilous and thermophilous species) are divided in three main types according to moisture conditions: xerophilous type – herb-rich (mesic) type – seasonal hygrophilous type. These types are assigned to the association *Convallario–Pine-tum* Bjørndalen 80. Some basiphilous pine forests are more closed forest communities on deeper brown forest soil. The stands are herb-rich, but most of the calciphilous, heliphilous and thermophilous species are absent. These forests resemble closely the low herb spruce forests, and are referred to as, low herb pine forests' (placed as subassociation *pinetosum* of *Melico–Piceetum* Kielland-Lund 62). The basiphilous birch forests (*Epipacto atrorubentis–Betuletum* Bjørndalen 80 prov.) are also included in the national nature preservation plans. *Convallario–Pinetum, Melico–Piceetum pinetosum* and *Epipacto–Betuletum* are divided into different geographical races (see below).

Some other communities may interfere with the main types of basiphilous pine and birch forests, e.g., the herb-rich pine forests on sandy soil (*Peucedano–Pinetum* Matuszkiewics 62) and species-poor *Festuca ovina*-dominated pine forests in continental areas. The *Peucedano–Pinetum* is optimally developed on dry, sandy eskers in S Finland, and is only occasionally represented in Sweden. Oligotrophic pine forests with dominance of ericaceous plants (e.g., *Calluna vulgaris, Empetrum hermaphroditum* and *Vaccinium* spp.) are found even on limestone, especially in W Norway and N Fennoscandia where thich humus layer often is developed. Some eutrophic species may occasionally occur in these forests, and such stands can form transitionary types towards *Melico–Piceetum pinetosum* or *Convallario–Pinetum*.

The addendum may now be concluded with an outline of the geographical races of the main associations of basiphilous pine forests and birch forests (*Convallario–Pinetum, Melico–Piceetum pinetosum* and *Epipacto atrorubentis–Betuletum*) in Fennoscandia:

I. True basiphilous pine forests (*Convallario–Pinetum*)
    A. Xerophilous type
        (1) *SE Norway race*. The Cambro-Silurian districts of the Oslo region. Often dominated by *Arctostaphylos uva-ursi* and *Festuca ovina*. Rich in calciphilous southeastern species. Lichens (*Cladonia* spp., *Cetraria islandica*) often dominant in the bottom layer.
        (2) *Baltic Sea race*. Mainly in the Stockholm archipelago and adjacent coastal areas. The floristic composition resembles that of the SE Norway race, but some more extreme southeastern species are differential e.g., *Melica ciliata, Thymus serpyllum, Veronica spicata* and *Vincetoxicum hirundinaria*.

(3) *Gotland race.* *Arctostaphylos uva-ursi* is the most important dominant. A very rich forest type with numerous southeastern species. Typical species: *Anemone pratensis, Calamagrostis varia, Galium triandrum, Helianthemum nummularium, Sesleria caerulea* and *Vincetoxicum hirundinaria.*

(4) *Ottadal race.* *Festuca ovina*-dominated pine forests in the northern Gudbrandsdal area. Alpine species such as *Astragalus alpinus, Poa glauca* and *Rhytidium rugosum* occur together with thermophilous species. Ottadalen is one of the most continental areas in Fennoscandia.

(5) *Sognefjord/Sunndal race.* Interior fjord and valley districts with continental climate in W Norway. Some southeastern species occur, e.g., *Berberis vulgaris, Origanum vulgare, Sedum album* and *Woodsia ilvensis.*

(6) *Trondheimsfjord race.* Trondheimsfjord area in C Norway. The number of southeastern species is strongly reduced. Typical species are e.g., *Calamagrostis epigeios, Epipactis atrorubens, Sedum acre* and *Woodsia ilvensis.*

(7) *Alnö race.* Sporadically found in the geological interesting island Alnön outside Sundsvall in N Sweden. Many of the southeastern species are absent. Typical species are e.g., *Hypochoeris maculata, Potentilla argentea, Sedum acre, Silene nutans, Viola tricolor* and *Viscaria vulgaris.*

B. Herb-rich (mesic) type

(1) *SE Norway race.* Optimally developed in the southern Cambro-Silurian districts of the Oslo region. *Convallaria majalis* and *Calamagrostis epigeios* are often dominating. A very species-rich community.

(2) *Kragerø/C Telemark race.* Basiphilous pine forests occur occasionally on Precambrium gneisses in SE Norway. This type is poorer in species than the ordinary SE Norway race. *Convallaria majalis* and *Geranium sanguineum* are often dominant.

(3) *C Sweden race.* Best developed in the provinces of Västmanland and Närke, but occurs also in other parts of C Sweden. Often grass-rich, dominated by *Brachypodium pinnatum* and *Calamagrostis arundinacea.* Generally poorer in species than in SE Norway and the Baltic Sea area.

(4) *Baltic Sea race.* Coastal areas between Östergötland and Billudden in Uppland. Resembles closely the SE Norway race, but species such as *Filipendula vulgaris, Laserpitium latifolium* and *Primula veris* are more important in the Baltic Sea race.

(5) *Öland race.* Basiphilous pine forests with an understory vegetation which resembles closely that of thermophilous deciduous forests occur in northern parts of Öland. Grasses such as *Brachypodium sylvaticum, Bromus benekeni, Melica uniflora* and *Milium effusum* are important constituents together with *Cephalanthera longifolia, Convallaria majalis* and *Hedera helix.*

(6) *Gotland race.* Extremely rich in species. Many species can dominate, e.g., *Brachypodium pinnatum, B. sylvaticum, Calamagrostis varia, Convallaria majalis, Filipendula vulgaris, Geranium sanguineum, Sesleria caerulea* and *Vincetoxicum hirundinaria.*

(7) *Sunnhordland race.* Occurs on some limestone islands south of Bergen. *Calamagrostis epigeios* and *Geranium sanguineum* can dominate. Many oceanic species occur, and this forest type forms a transitionary community towards the *Sanicula europaea* dominated low herb pine forests of W Norway.

(8) *Trøndelag race.* Optimal developed in the Snåsa area in Nord-Trøndelag, C Norway. *Convallaria majalis* and *Calamagrostis epigeios* are common dominants. This forest type is rich in orchids, especially *Cypripedium calceolus, Epipactis atrorubens* and *Ophrys insectifera.*

(9) *Salten race.* Many well-developed stands in the Bodø-Fauske area in N Norway (north of the Arctic Circle). Mixture of alpine calciphilous species and more thermophilous species. However, most of the southern and southeastern species are absent. Typical species are e.g., *Antennaria alpina, A. dioica, Calamagrostis epigeios, Carex rupestris, Epipactis atrorubens, Molinia caerulea* and *Succisa pratensis.* Often very rich in orchids, e.g., *Ophrys insectifera.*

(10) *Ångermanland race.* Found on steep hillsides at Höga kusten in N Sweden. Dominant species are *Arctostaphylos uva-ursi, Calamagrostis epigeios, Convallaria majalis, Festuca ovina* and *Thymus serpyllum.* Most of the southeastern species are absent.

C. Seasonal hygrophilous type

(1) *SE Norway race.* Occurs scattered throughout the Cambro-Silurian districts of the Oslo region, occasionally also in other parts of SE Norway. One type is dominated by *Inula salicina* and *Trifolium medium,* but most stands are characterized by *Molinia caerulea* and different calciphilous/hygrophilous *Carex* spp. Also more xerophilous species are common. This forest type is often rich in orchids.

(2) *Kristiansand race.* Known from one area near Kristiansand in southernmost Norway. Dominated by *Convallaria majalis, Geranium sanguineum* and *Molinia caerulea.*

(3) *Sunnhordland race.* Occurs in the island Tysnes south of Bergen. Typical species are e.g., *Carex flacca, C. panicea, C. pulicaris, Molinia caerulea* and *Succisa pratensis.* Many oceanic species occur, but southeastern species are almost absent.

(4) *Østerdal/Älvdal race.* Found near spring horizons on calcareous morain deposits in the Røros/Østerdalen/Atnadalen area in Norway and northern Dalarna in Sweden. Many of the stands are rich in *Cypripedium calceolus.* Southeastern species are absent, but many rich fen species and alpine species occur.

(5) *Gotland race*. Known from the southernmost part of Gotland. *Inula salicina* and calciphilous/hygrophilous species are prominent, but many of the southeastern species which characterize the basiphilous pine forests in Gotland are found also in this type.

(6) *Stockholm archipelago race*. Fragments of hygrophilous types of basiphilous pine forests are found in the island Utö. Dominants are *Sesleria caerulea* and calciphilous/hygrophilous *Carex* spp.

(7) *Västmanland race*. Resembles the grass-rich basiphilous pine forests of C Sweden, but species such as *Briza media, Carex flacca, Equisetum hyemale, Gymnadenia conopsea* and *Listera ovata* are important constituents.

(8) *Skellefteå race*. Occurs at Brännberget near Skellefteå in N Sweden. Very rich in orchids, especially *Cypripedium calceolus. Convallaria majalis* can dominate, but hygrophilous species are also important. A special feature is the strong position of *Ledum palustre* in this community.

II. Low herb pine forests (*Melico-Piceetum pinetosum*)

(1) *SE Scandinavian race*. Occurs throughout S/C Sweden and SE Norway, with fringes north to Trøndelag, Jämtland and Medelpad. The typical species of *Melico–Piceetum* (the low herb species group) are dominating in the forest type. Other characteristic species are *Goodyera repens, Pyrola chlorantha* and in some areas also *Carex pediformis*. Relative constant floristic composition, even if some minor regional variation are found.

(2) *W Norway race*. In coastal areas in Hordaland and Sogn og Fjordane. Often extreme lush understory vegetation. Common dominants are *Hedera helix, Primula vulgaris* and *Sanicula europaea*. Numerous oceanic species occur.

(3) *Sunnmøre race*. A special type which occurs on olivine bedrocks in Møre og Romsdal, W Norway. Poorer in species than the ordinary W Norway race, but many oceanic species are common. Species with affinity to serpentine conditions occur, e.g., *Asplenium adulterinum, A. adianthumnigrum, Cardaminopsis petraea* and *Cerastium alpinum* var. *nordhageniana*.

(4) *Salten race*. A northern equivalent to the SE Scandinavian race. Often a lush understory vegetation. *Geranium sylvaticum* is often a prominent species.

(5) *Österbotten race*. Occurs in the area between Oulu and the Torne Valley in N Finland. More extreme eastern species are found, e.g. *Actaea erythrocarpa*.

(6) *Oulanka race*. Found on dolomite in the Kuusamo area in N Finland. Typical species are e.g., *Cypripedium calceolus, Equisetum scirpoides, Geranium sylvaticum, Polygala amarella, Thymus serpyllum* and *Viola rupestris*.

(7) *Skoganvarre race*. Some stands of herb and fern dominated pine forests are reported from the area between Lakselv and Karasjok in Finnmark, N Norway. No information of the floristic composition is available at the time.

III. Basiphilous birch forests (*Epipacto atrorubentis–Betuletum*)

(1) *Salten/S Troms race*. The basiphilous birch forests are optimal developed in interior parts of Salten and in southern parts of Troms, N Norway. Dominant species are *Arctostaphylos uva-ursi, Carex rupestris* and *Dryas octopetala*. More thermophilous species are common, e.g., *Calamagrostis epigeios, Carex digitata* and *Epipactis helleborine*.

(2) *Hattfjelldal race*. A special type dominated by spruce is found in Hattfjelldal in Nordland. *Dryas octopetala* is a dominant species in the field layer. The community is rich in orchids, especially *Cypripedium calceolus* and *Gymnadenia conopsea*.

(3) *S Norway subalpine race*. *Dryas octopetala* dominated birch forests occur occassionally in subalpine areas of S Norway. These forests occur at higher altitudes than the corresponding basiphilous birch forests in Salten, and thermophilous species are of minor importance.

(4) *N Troms/Porsanger race*. Occurs occasionally in N Troms and Finnmark, especially on dolomite in Kvænangen and Porsanger. The floristic composition is made up almost entirely of alpine species. *Dryas octopetala* is often the most important dominant.

(5) *Masugnsbyn race*. A special type dominated by spruce occurs near a dolomite quarry in Masugnsbyn, Swedish Lapland. *Dryas octopetala* is the dominant species in the field layer. Typical species are e.g., *Asplenium viride, Betula nana, Carex capillaris, Daphne mezereum, Empetrum hermaphroditum* and *Salix starkeana*.

Floristic changes in the *Castanopsis cuspidata* var. *sieboldii*-forest communities along the Pacific Ocean coast of the Japanese Islands*

A. Miyawaki[1] & Y. Sasaki[2]
[1] *Department of Vegetation Science, Institute of Environmental Science and Technology, Yokohama National University, 196 Tokiwadai, Hodogaya-ku, Yokohama, Kanagawa, 240, Japan*
[2] *Junior College of Economics, Saitama University. 255 Shimo-Ohkubo, Urawa, Saitama 338, Japan*

Keywords: *Castanopsis cuspidata* var. *sieboldii*, Community dynamics, Distribution limit, Eg(t) quotient, Forest vegetation, Laurel forests, (Sg(t) quotient

Abstract

The forest vegetation of Japan can be classified into three major regions: (1) the *Camellietea japonicae* evergreen broad-leaved forest region, e.g. Laurel forest which can be compared with the sclerophyllous gorest (durilignosa sensu Rübel, 1930), in the Mediterranean region, (2) the *Fagetea crenatae* summergreen broad-leaved forest region, (3) the *Vaccinio-Piceetea japonicae* subalpine and subboreal conifer forest region. The distribution of these forest types on the Japanese Islands is related to both the warmth index, *WI*, and to the coldness index, *CI*, after Kira (1945). The borderline between the evergreen *Camellietea japonicae* and the summergreen *Fagetea crenatae* in Japan almost coincides with the 85 °C line of *WI*. The chorological variation of the forest vegetation in Japan and Korea shows a close correlation with the amount of warmth in the actual vegetation season.

The evergreen broad-leaved *Castanopsis cuspidata* var. *sieboldii* forests can be classified into three major alliances; *Quercion acuto-myrsinaefoliae*, *Maeso japonicae-Castanopsion sieboldii*, and *Psychotrio-Castanopsion sieboldii*.

The distribution limit of these three alliances on the Japanese Islands is again related to temperature. Changes in temperature and latitude correspond closely to changes in the Evergreen Broad-leaved Forests along the Pacific Ocean coast of the Japanese Islands. The three forest alliances differ in their number of evergreen and deciduous broad-leaved woody species, secondary forest types and syndynamic processes.

Introduction

The forest communities on the Japanese islands, their syntaxonomy, structure, natural distribution and changes due to human influence, have been repeatedly dealt with elsewhere (e.g. Miyawaki, 1967, 1975, 1977, 1979; Miyawaki & Itow, 1966; Miyawaki *et al.*, 1980, 1981).

The evergreen broad-leaved forests of Japan are Laurel forests forming the class of *Camellietea ja-ponicae*. They were repeatedly driven back south and advanced again to the north by climatic fluctuations. The evergreen broad-leaved forests which are actually present on the Japanese islands today have migrated to the north after the last glacial period and form plant communities of their own.

During the Upper Würm 25 000–17 000 years ago, the Japanese evergreen broad-leaved forests were driven back up to the southernmost part of Kyushu and the plains in most parts of Japan were occupied by deciduous broad-leaved forests. However, today they reach as far as Yamada-cho in the prefecture of Iwate on the Pacific coast up to latitude 39° 30′N. This means that from the past until

* Contribution from the Department of Vegetation Science, Institute of Environmental Science and Technology, Yokohama National University, No. 154.

today evergreen broad-leaved forests on the Japanese islands have repeatedly migrated over a distance of up to 850 km.

Evergreen broad-leaved forests with *Castanopsis cuspidata* var. *sieboldii* belong to the most important ones in Japan. Today they range over almost all parts of the *Camellietea japonicae* region from 37° N to 24° N, the southernmost part of Japan. All 26 important towns with a number of inhabitants exceeding 500 000 (except Sapporo) are situated in the region of the *Camellietea japonicae* (Miyawaki, 1979). This indicates the human pressure on these forests.

The *Castanopsis cuspidata* var. *sieboldii* forests were studied as an example demonstrating the distribution boundaries of evergreen broad-leaved forests on the Japanese islands with respect to the gradient of climatic conditions, as well as their structure, notably the ratio of evergreen and deciduous woody species in the associations, and their dynamics, including the changes from natural woods to substitute communities under human influence. At higher latitudes devoid of *Castanopsis cuspidata* var. *sieboldii* forest surveys of *Persea thunbergii* and *Camellia japonica* forests were performed for comparison.

Investigated regions

The present report is concerned with the evergreen broad-leaved forests of the Pacific climatic district. In order to ensure a comparison under otherwise identical topographical and other geographical conditions, Pacific coastal areas situated below 200 m altitude were chosen as the main area of research. The evergreen broad-leaved forests of Japan appear to be differentiated within this coastal region according to temperature, but at the same time wood associations may develop under the same temperature conditions in relation to edaphic differences.

The sites of the climatic stations (Fig. 1) and the localities of the vegetation relevés (Table 1) are not completely identical.

Phytosociological classification

The Evergreen Broad-leaved, *Castanopsis cuspidata* var. *sieboldii* forests can be classified into three alliances: (a) the *Quercion acuto-myrsinaefoliae* (Qam), (b) the *Maeso japonicae-Castanopsion sieboldii* (MCs), (c) the *Psychotrio-Castanopsion sieboldii* (PCs); and 7 associations occupying the Pacific Ocean coastal areas (see Table 1).

The climatic gradient from south to north on the Japanese islands

The annual precipitation on the Japanese islands amounts to 1 500–2 500 mm (with the exception of the Setouchi region, the region of the so-called Japanese Central Sea and small parts of northern Japan). The amount of precipitation and its season-

Table 1. Localities of vegetation releves of evergreen broad-leaved forests along the Pacific coast.

No.	Locality	Prefecture	Latitude	Natural forest community
1	Iwaki	Fukushima	35°58′	Ardisio–Castanopsietum sieboldii (Qam)
2	Kashima	Ibaragi	35°58′	Ardisio–Castanopsietum sieboldii
3	Yokohama	Kanagawa	35°25′	Ardisio–Castanopsietum sieboldii
4	Kamakura	Kanagawa	35°20′	Rumohro–Castanopsietum sieboldii (MCs)
5	Hamaoka	Shizuoka	34°40′	Symploco–Castanopsietum sieboldii
6	Ise	Mie	34°20′	Symploco–Castanopsietum sieboldii
7	Susami	Wakayama	33°30′	Symploco–Castanopsietum sieboldii
8	Anan	Tokushima	33°40′	Symploco–Castanopsietum sieboldii
9	Miyazaki	Miyazaki	31°50′	Symploco–Castanopsietum sieboldii
10	Island Yaku	Kagoshima	30°20′	Hydrangeo–Castanopsietum sieboldii (PCs)
11	Island Amami	Kagoshima	28°25′	Symploco liukiuensis–Castanopsietum sieboldii
12	Okinawa	Okinawa	26°40′	Illicio anisati–Castanopsietum sieboldii
13	Island Iriomote	Okinawa	24°20′	Adinandro–Castanopsietum sieboldii

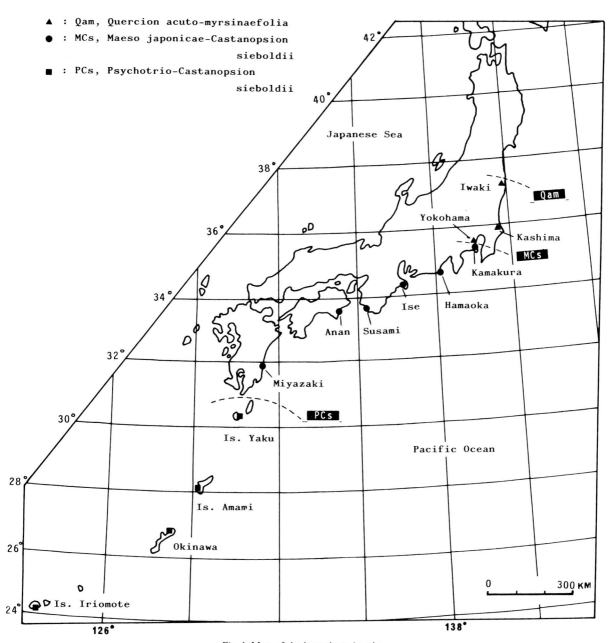

Fig. 1. Map of the investigated region.

al distribution do not impose limitations on the formation of forests in Japan. On the other hand, temperature plays a decisive part as limiting factor for the development of evergreen broad-leaved forests.

Kira (1945) paid attention to the relation of temperature to the boundaries between Japanese evergreen broad-leaved forests (Laurel forests) and the deciduous ones and, accordingly, proposed a warmth index, *WI*, and a coldness index, *CI*. However, re-examination of the real conditions proved the distribution of evergreeen broad-leaved forests to show a clear-cut relation only to the *CI*. The distribution of the forests is in better agreement with the *CI* isolines than with the *WI* lines. This is to say that in Japan the severest conditions for plants of the evergreen vegetation are encountered at the boundary of their distribution and that low temperatures in winter play a decisive part, the recorded minimal temperature, *EMT*, being most important.

On the Japanese islands the temperature gradient from south to north along the coasts approximately parallels the increase in latitude (Miyawaki, 1977). It is therefore possible to determine the northern boundaries of a community according to the temperature conditions it requires. On the Japanese islands the temperature gradient from south to north on the Japanese sea coast somewhat differs from that on the Pacific coast. It is less marked on the Japanese sea coast than on the Pacific coast (Fig. 2). This is caused by the difference in snow

cover, which is thicker in the west, and by the warm Tushima maritime stream, which exerts a moderating effect on the whole climate along the west coast and especially on the winter temperatures.

Distribution boundaries of evergreen broad-leaved forests on the Japanese islands

The evergreen species gradually decrease in number from south to north, which leads to an impoverishment of the evergreen broad-leaved forests. Within the *Castanopsis cuspidata* forests, Japan's most important evergreen broad-leaved forests, the *Psychotrio–Castanopsion sieboldii* has its

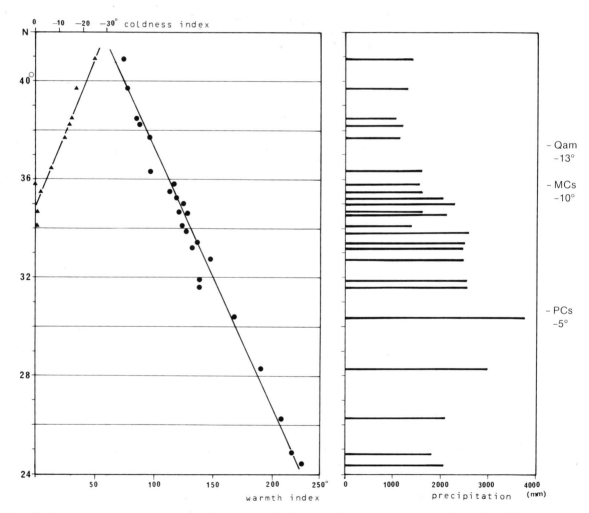

Fig. 2. Coldness index, warmth index and precipitation on the Pacific coast. Boundaries of the three alliances Qam, MCs, and PCs are also indicated.

northern boundary at 30°30′N. This latitude represents the first boundary of Japan's evergreen broad-leaved forests from south to north at an average annual temperature of 18 °C, a *WI* of 164 °C and an *EMT* of –5 °C. The second distribution boundary of the evergreen broad-leaved forests to the north is situated at 35°35′N. This is the boundary of the *Maeso japonicae-Castanopsion sieboldii*. In this region the average annual temperature amounts to 15 °C, *WI* to 116 °C, *CI* to 0 °C and *EMT* to –10 °C. The northern distribution boundary of the species *Castanopsis cuspidata* var. *sieboldii* itself lies still farther north at 38°00′N with an average annual temperature of about 13 °C, *WI* = 95 °C, *CI* = –10 °C and *EMT* = –13 °C. At still higher latitudes the evergreen broad-leaved forests are confined to the coast and occur only sparsely and locally. The distribution boundary of *Persea thunbergii* forests which should be considered to represent an impoverished *Polysticho–Perseetum thunbergii* lies at 39°50′N. The same is true for the *Quercion acuto-myrsinae-foliae* and the evergreen broad-leaved forest in general. The average annual temperature is 10 °C, *WI* = 80 °C, *CI* = 20 °C and *EMT* = –15 °C.

North of 39°50′N evergreen coppice woods with *Camellia japonica* develop locally on southern slopes along the coast. These shrub-like communities attain their northern distribution limit at 40°55′N where the average annual temperature amounts to 8 °C, the *WI* to 75 °C, the *CI* to –25 °C and the *EMT* to –18 °C.

Changes in number of evergreen and deciduous broad-leaved woody species in forests communities from south to north

On the Japanese islands the *Castanopsis cuspidata* var. *sieboldii* forests are important evergreen woods. Due to the dominating species *Castanopsis cuspidata* var. *sieboldii* they show almost the same physiognomy, however they become poorer in evergreen woody species in a northward direction.

The total number of evergreen woody and herbaceous species in these communities is more than 80 in the southernmost parts of Japan, as for example on the island of Iriomote. Two quotients are used to express these relationships, the *Eg(t)* and the *Sg(t)* quotient (Miyawaki, 1979): *Eg(t)* quotient (quotient of evergreen woody species) = number of evergreen woody species/total number of species

Table 2. Temperature gradient and floristic changes in the *Castanopsis cuspidata* var. *sieboldii* forests.

Locality	Annual temp.	WI (°C)	CI (°C)	Average number of species	Absolute number of evergreen woody species**	Absolute number of deciduous woody species**	Eg(t) (%)	Sg(t) (%)
(a) Qam region								
Iwaki*	12.0	95	–10	38	22	31	22	31
Kashima*	13.5	112	–5	30	25	15	43	26
Yokohama	15.0	114	–3	25	21	13	46	28
(b) MCs region								
Kamakura	15.2	123	0	42	28	15	40	22
Hamaoka	15.7	128	0	44	47	14	51	16
Ise*	15.0	127	0	51	50	14	57	16
Susami*	15.5	134	0	32	40	8	62	11
Anan*	16.0	137	0	37	40	7	59	12
Miyazaki	16.6	139	0	43	46	5	63	31
(c) PCs region								
Is. Yaku	19.0	168	0	53	73	9	74	10
Is. Amami	20.9	191	0	45	59	2	78	4
Okinawa	22.3	207	0	46	85	5	63	5
Is. Iriomote	24.0	224	0	44	68	7	73	7

* Temperature gradient is calculated using Figure 2.
** Woody species have been counted with a constancy of over 10%.

× 100. *Sg(t)* quotient (quotient of deciduous woody species) = number of deciduous woody species/total number of species = 100. Only woody species with a constancy ≥ 10% were accounted for. The *Eg(t)* quotient on Iriomote is 80%, while in the northernmost border regions only 20 of all evergreen species are left and the *Eg(t)* quotient does not exceed 30%, see Table 2.

Only 2 to 3 evergreen woody species occur in the *Persea thunbergii* forests, (*Polysticho–Perseetum thunbergii*) and in the *Camillia japonica* coppice woods (*Camellia japonica* community) growing farther to the north. On the other hand, in southern Japan at 35°N the absolute number of deciduous species is lower than 10 and the *Sg(t)* quotient is also less than 10%.

Farther to the north the absolute number of deciduous woody species rises to 20–40 and the *Sg(t)* quotient is 10–30%.

North of 36°N the absolute number of deciduous woody species exceeds that of the evergreen ones and the *Sg(t)* values are higher than the *Eg(t)* values. Thus, the curves of the *Eg(t)* and *Sg(t)* quotient are parabola-shaped (Fig. 3). The values of the *Eg(t)* quotient and the corresponding *Sg(t)* quotient are reciprocal, however, the *Sg(t)* quotient rises somewhat more slowly.

Community dynamics

The syndynamic processes of the Japanese evergreen forests show a specific course in each area and correspond to the following 3 series:

(a) *Quercion acuto-myrsinaefoliae* series, *Qam*

Evergreen broad-leaved forests of *Ardisio-Castanopsietum sieboldii* and *Polysticho-Perseetum thunbergii*/*Quercion acuto-myrsinaefoliae*; the *Qam* regions are substituted by secondary woods of *Carpino-Quercion serratae* following cut. Under further and stronger human influence they turn into shrubs of *Actinidio-Vition coignetiae* or into *Miscanthion sinensis* meadows and eventually are converted to annual communities of *Cypero-Molluginion strictae*. These changes of the communities occur between 35°35′ and 39°N, i.e. from the northern borderline of *Rumohro-Castanopsietum sieboldii* to the northern boundary of *Polysticho-Perseetum thunbergii*.

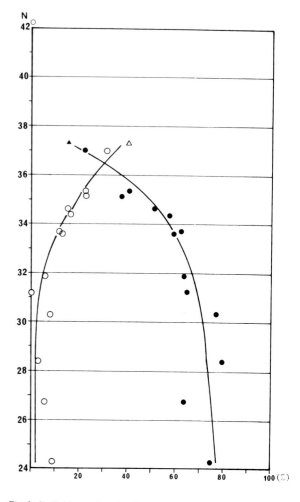

Fig. 3. ○: *Sg(t)* quotient in *Castanopsis sieboldii* forests; △: in other evergreen broad-leaved forests; ●: *Eg(t)* quotient in *Castanopsis sieboldii* forests; ▲: in other evergreen broad-leaved forests.

(b) *Maeso japonicae-Castanopsion sieboldii* series, *MCs*

In the region of the natural forests of *Symploco-Castanopsietum sieboldii* and *Rumohro-Castanopsietum sieboldii* from the alliance of *Maeso-japonicae-Castanopsion sieboldii* evergreen woods also develop as rejuvenated secondary woods following a cut. After repeated cuts and other strong human influence they turn into deciduous coppice wood of *Clerodendro-Mallotion japonicae* and further to *Miscanthion sinensis* and, eventually, e.g. on arable soil, to annual field weed communities of *Cypero-Molluginion strictae*.

This sere of community dynamics is confined to the region between 35°35′ and 30°30′N, i.e. from the northern boundary of *Rumohro–Castanopsietum sieboldii* to the southern border of *Symploco–Castanopsietum sieboldii* occurring in the southern parts of Kyushu.

(c) *Psychotrio–Castanopsion sieboldii* series, *PCs*

In the area of the *Adinandro–Castanopsietum sieboldii* (island of Okinawa), the *Symploco liukiuensis–Castanopsietum sieboldii* (island of Amami) and the *Hydrangeo–Castanopsietum sieboldii* (island of Yaku), all belonging to the alliance of *Psychotrio–Castanopsion sieboldii*, evergreen woods mainly of *Lasiantho–Castanopsietum sieboldii* also occur as secondary woods following a cut.

When human influence increases they change into deciduous coppice woods of the *Villebruno-Tremion orientalis* showing a simpler structure, to *Miscanthion sinensis* meadow communities and, eventually, down to annual weed communities of the *Siegesbeckion orientalis*.

A consideration of community dynamics according to the potential natural forest regions as shown above proves the rejuvenation ability of evergreen woods to be very different.

To the north of the *Rumohro–Castanopsietum sieboldii* area at 35°35′N deciduous broad-leaved woods mostly occur as secondary communities; the physiognomy, species composition and the processes of vegetation transformation differ here from all the other potential natural forest regions. The changes in the regions of natural evergreen forest associations, such as *Qam*, *MCs* and *PCs*, are schematically shown in Figure 4. Evergreen woods appear in the area of both the *PCs* and *MCs* association following a cut of the natural forests. However, in the *Qam* area, beyond 35°35′N, secondary deciduous woods of *Quercetum acutissimo-serratae* occur (Fig. 4).

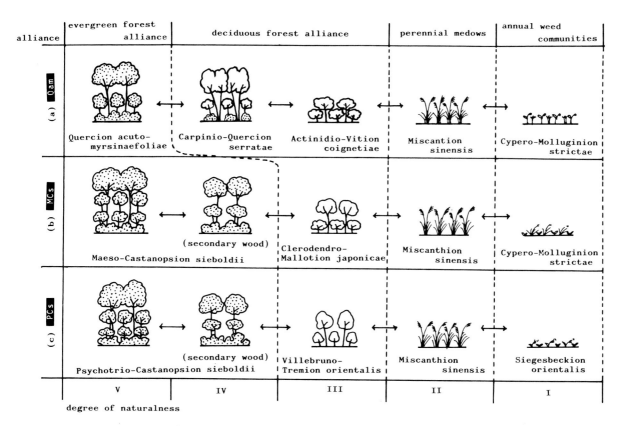

Fig. 4. Scheme of the community dynamics of the three *Castanopsis cuspidata* var. *sieboldii* forest alliances.

232

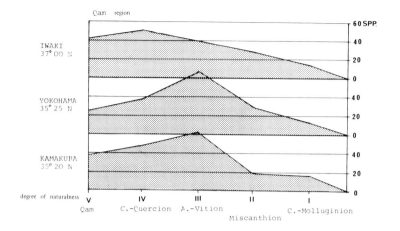

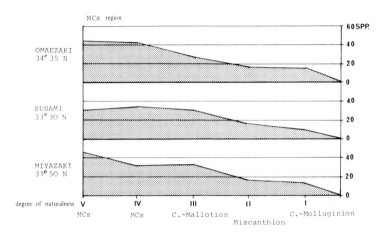

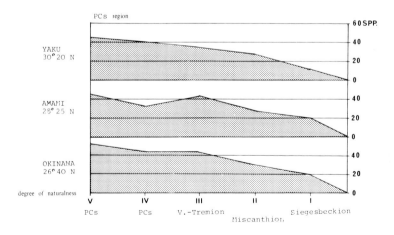

Fig. 5. Changes in the number of species of the community series within the three alliances from north to south.

Changes in the number of species in the community series of the evergreen forest regions

The succession of communities on bare soils is initiated by annual plant communities. In the course of time and with soil development perennial meadow or herb communities develop, which are followed by shrubs and deciduous woods, eventually leading to the stage of the evergreen final community, the so-called climax community. Physiognomy, combination of species, as well as morphology of the community undergo changes during succession.

If the natural forests are cut and destroyed the communities undergo changes leading to impoverishment. As a consequence these progressive and regressive successions lead to different potential natural communities. This transformation of communities may also be demonstrated by an analysis of the total number of species in the single community series (Fig. 5).

With exception of *Qam* the number of species in the communities of the 'community cycle' (Schwickerath, 1954) or of the series show maximal values in the potential natural evergreen broad-leaved forest. However, in the *Quercion acuto-myrsinaefolium* region in the north where the *Ardisio-Castanopsietum sieboldii* occurs, the highest number of species is shown by the *Actinidio-Vition*.

(a) *Quercion acuto-myrsinaefoliae* (*Qam*) series

The average number of species increases from the annual plant communities to the perennial herb or meadow communities. The largest number of species is met with in the deciduous coppice woods and other secondary woods. In stabilized natural forests corresponding to their local habitat, the so-called final communities, the total number of species decreases again. The number of evergreen woody species in the natural stands amounts to about 20, which is 10–20% of all occurring species. The species number curve culminates in the secondary shrub-like communities of the *Actinidio-Vition coignetiae* with degree III of vegetation naturalness (Miyawaki & Fujiwara, 1971). This phenomenon becomes apparent in the northern-border communities of the *Camellietea japonicae* which are affected by *Fagetea crenatae*.

(b) *Maeso japonicae-Castanopsion sieboldii* (*MCs*) series

Natural forests formed as a result of the climatic conditions and corresponding to a potential natural vegetation of evergreen broad-leaved forests, namely as the *Symploco liukiuensis-Castanopsietum sieboldii* show the largest number of species in the community cycle of the *Maeso japonicae-Castanopsion sieboldii*, viz. 40–50. Thus, the evergreen woody species make up about 50–60% of all species occurring in forest stands. The final number of species characteristic of evergreen broad-leaved forests is rapidly attained during succession.

(c) *Psychotrio-Castanopsion sieboldii* (*PCs*) series

Likewise, the *Psychotrio-Castanopsion sieboldii*, as well as the *Casiantho-Castanopsietum sieboldii*, the *Symploco liukiuensis-Castanopsietum sieboldii* on the islands of Yaku and Amami, and the *Illicio yaeyamenesis-Castanopsietum sieboldii* on the southern islands of Japan show distinct species number curves according to the secondary succession of forest communities. As shown for other community cycles (Miyawaki, 1981) the species number curve rises regularly with development of the vegetation. The terminal communities of the above-mentioned associations of *Castanopsis cuspidata* var. *sieboldii* forests are richest in species.

A temporary short-term decrease in the number of species occurs only in rejuvenated evergreen secondary woods after a cut (degree III or IV of naturalness). In the natural terminal forest communities the number of evergreen woody species amounts to 60–85, which is as much as 60–80% of the total number of occurring species. The evergreen woody species are supplemented with ferns and other herbaceous plants. Thus, transformation of the evergreen forest communities from south to north on the Japanese islands is a clear-cut and specific phenomenon with respect to spatial order and number of species.

References

Kira, T., 1945. Neue Klimagliederung Ostasiens als Grund für Agrogeographie (Agrogeographische Studie auf den Ostasien 1). Kyoto.

Miyawaki, A., 1975. Entwicklung der Umweltschutz-Pflanzungen und Ansaaten in Japan. In: Sukzessionsforschung, Ber. Internat. Sympos. IVV, pp. 273–254. Cramer, Vaduz.

Miyawaki, A., 1977. Klimabedingte Unterschiede und Gemeinsamkeiten der Vegetation an der japanischen und der pazifischen Meeresseite Japans. In: Vegetation und Klima, Ber. Internat. Sympos. IVV, pp. 235–247. Cramer, Vaduz.

234

Miyawaki, A., 1979. Die Umwandlung immergrüner in sommergrüne Laubwälder in Japan. In: Werden und Vergehen von Pflanzengesellschaften, Ber. Internat. Sympos. IVV, pp. 367–381. Cramer, Vaduz.

Miyawaki, A., 1981. Das System der Lorbeerwälder (Camellietea japonicae) Japans. In: Syntaxonomie, Ber. Internat. Sympos. IVV, pp. 589–598. Cramer, Vaduz.

Miyawaki, A., 1967, 1977. Vegetation of Japan compared with other regions of the world. Gakken, Tokyo.

Miyawaki, A. (ed.) et al., 1980. Vegetation of Japan. Vol. 1. Yakushima. Shibundo, Tokyo.

Miyawaki, A. (ed.) et al., 1981. Vegetation of Japan. Vol. 2. Kyushu. Shibundo, Tokyo.

Miyawaki, A. & Itow, S., 1966. Phytosociological approach to the conservation of natural resources in Japan. Pacific Sci. Congress. Tokyo.

Miyawaki, A. & Fujiwara, K., 1971. Vegetationskundliche Untersuchungen in der Stadt Zushi bei Yokohama – Besondere Betrachtung mit Camellietea japonicae-Wald (immergrüner Laubwald) Japans. Zushi.

Miyawaki, A., Okuda, S. & Mochizuki, R., 1978. Handbook of Japanese vegetation. Shibundo, Tokyo.

Schwickerath, M., 1954. Die Landschaft und ihre Wandlung auf geobotanischer und geographischer Grundlage entwickelt und erläutert im Bereich des Messtischblattes Stolberg. Aachen.

Accepted 16.4.1984.

The occurrence of communities with species of *Ranunculus* subgenus *Batrachium* in central Europe – preliminary remarks*, **

Gerhard Wiegleb & Wolfgang Herr***
Fachbereich 7 (Biologie), Universität Oldenburg, Ammerländer Heerstr. 67-99, D-2900 Oldenburg, F.R.G.

Keywords: *Batrachium,* Community, Distribution, Running waters

Abstract

A short outline of the history of the description of *Batrachium* communities with special regard to taxonomic reliability is given. Own investigations concern firstly the distribution of *Batrachium* taxa in Lower Saxony (north Germany). Some taxonomical problems are discussed. The *Batrachium* species are classified with regard to growth form into two structural groups and the different distribution of these groups within a drainage area is mentioned. The problems arising from the classification of the communities dominated by small *Batrachium* species with floating leaves are discussed in more detail using a distribution map of characteristic and associated species. With regard to a wider geographical area (central Europe), 3 types of *Batrachium* communities are distinguished and characterized floristically. The great complexity of the problem is pointed out. Finally, the phytosociological behaviour of the 7 most frequent *Batrachium* species is discussed with special reference to the different types of waters that are colonized.

Introduction

Many plant communities referring to *Batrachium* species have been described. In Tüxen & Schwabe (1972) 20 communities named after *Batrachium* species are recognized within the Potametea, obvious synonyms and nomina inversa already excluded! There were 5 communities from running waters, all of them named after '*Ranunculus fluitans*', and about 15 from stagnant waters most of them named after '*Ranunculus aquatilis*'. As to run-

ning waters, many *Batrachium* communities have been described under other names. One example is the *Callitricho-Myriophylletum*, which in Weber-Oldecop (1969), and especially in Wiegleb (1979) comprises *Ranunculus peltatus*-dominated relevés partly without *Callitriche* or *Myriophyllum*.

Regarding higher units, Neuhäusl (1959) proposed the name *Ranunculion fluitantis* for running water communities. This unit comprised plant communities, which are not exclusively built up by *Batrachium* species, not even batrachid species at all. It has to be looked upon as a heterogeneous mixture reflecting the very different ecological conditions found in different types of running waters.

For *Batrachium* communities in stagnant waters, Passarge (1964) proposed the name *Ranunculion aquatilis*. In his tables, he only displayed relevés from small eutrophic water bodies which are closely allied to the Lemnetea. Thus his alliance did not find very much recognition in the phytosociological literature. At the same time, den Hartog & Segal

* This paper is dedicated to the memory of D. W. Weber-Oldecop (1933–1982), a pioneer in macrophyte research in Germany.
** Nomenclature follows Cook (1966).
*** We thank Dr N. T. H. Holmes (Huntingdon, U. K.) and Prof. Dr C. D. K. Cook (Zürich, Switzerland) for their help with taxonomical problems of *Ranunculus* subgenus *Batrachium*. The research is granted by financial support of the Niedersächsisches Landesverwaltungsamt and the Landesamt für Natur- und Umweltschutz Schleswig-Holstein.

Vegetatio 59, 235–241 (1985).

(1964) combined all *Batrachium-* and *Callitriche-* rich communities (including such ones formerly included in the Littorelletea) into their *Callitricho-Batrachion*. This approach was based on the similar growth form of most species of the two genera. Passarge (1978) ranked the *Callitricho–Batrachion* as an order (*Callitricho–Batrachetalia*) with the two alliances *Ranunculion aquatilis* and *Ranunculion fluitantis*.

Wiegleb (1982) adopted Passarge's proposal in a modified form (excluding the *Sparganium-* and magnopotamid-rich communities) adding that the *Callitricho–Batrachetalia* still seems to be the most heterogeneous order of the Potametea with regard to floristic composition. Also with regard to growth form there are great differences, because *Ranunculus fluitans, R. calcareus, R. circinatus* and *R. trichophyllus* are no batrachid species.

Because of the taxonomical problems, the number of reliable phytosociological relevés is very low. Most of the taxa occurring in running waters are difficult to identify (Wiegleb, 1982). In older tables names like '*Ranunculus fluitans*' can be found, which could also mean one of the variants of *Ranunculus penicillatus* or one of the *fluitans* hybrids. Even more frequently '*Ranunculus aquatilis*' is quoted, a name that could refer to almost any *Batrachium* species. A valid description of *Batrachium* communities is only possible since the appearance of the monograph of Cook (1966). This view is supported by the discussions of Weber-Oldecop (1969) on the typification of the '*Ranunculetum peltati*', which proved to be impossible because of taxonomical uncertainty.

The search for material dealing with *Batrachium* vegetation need not be restricted to pure phytosociological publications. Reliable information can be found, for example, in Kohler *et al.* (1971), Kutscher & Kohler (1976), Holmes & Whitton (1977), Holmes (1980a), Mériaux (1981, 1982), Mériaux & Géhu (1980), Déthioux (1982), and Herr (1983), even though there are great differences in the relevéing procedure.

Own floristical and taxonomical investigations

To obtain a reliable new survey of *Batrachium* communities we started in a limited area (northern Germany). This project is part of an overall inventory of the river vegetation of northern Germany. Nearly all rivers in Lower Saxony and selected systems in Schleswig-Holstein have been investigated during the last 4 years with a total of about 1 200 relevés (see Herr, 1983; Wiegleb & Herr, 1982). Furthermore, several journeys within the area in concern have been undertaken, so far to Denmark, Ireland, The Netherlands, France and Italy. In every case herbarium material was collected and relevés in water courses were made.

Distribution maps of the *Batrachium* species given by Cook (1966) are relatively incomplete for some species. For example *Ranunculus penicillatus* (according to Cook's map) shows a great gap of distribution in the northern parts of Germany. This results from Cook's method, who only used such data for his maps which were absolutely sure. That means that he had either collected the plant himself or seen a specimen on a herbarium sheet.

The most frequent taxa in Lower Saxonian water courses are:
1. *Ranunculus peltatus* Schrank
   This is the most frequent species with a general distribution in the lowlands, also to be found in the Harz Mountains.
2. *Ranunculus fluitans* Lamarck
   This species occurs in the highlands and in rhitral river sections of the lowlands, especially the Lüneburger Heide region.
3. *Ranunculus penicillatus* s.l.
   This taxon is restricted to the lowland area, where it occurs in almost every region, where *R. peltatus* also occurs. *Ranunculus penicillatus*-like plants have been recognized for the first time in north German waters by Weber-Oldecop (1969). However, he considered them together with *R. peltatus,* which was also done by Wiegleb (1979). In the meantime, Weber-Oldecop (1977) had emphasized the occurrence of *R. penicillatus* with regard to the River Oertze.

Ranunculus penicillatus is a polymorphic taxon comprising fertile and non-fertile forms of different cytological status. In Lower Saxony the morphological range varies from forms that are close to *R. peltatus* to forms that are intermediate between *R. peltatus* and *R. fluitans.* There is a significant gap, between *R. penicillatus* and *R. fluitans* with regard to the morphological characters. All these plants have to be referred to *Ranunculus penicillatus* (Dum.) Bab.

var. *penicillatus*.

Ranunculus penicillatus (Dum.) Bab. var. *calcareus* (Butcher) C. D. K. Cook seems to occur only in one river system in Schleswig-Holstein, even though elsewhere strains of *R. penicillatus* can be found that produce floating leaves only occasionally, moreover, resembling *R. calcareus* morphologically a great deal.

4. *Ranunculus trichophyllus* Chaix

This species can be frequently found in calcareous rivulets in the highlands. It is more frequent in stagnant waters throughout the whole region. Especially in the lowlands it can be found in non-calcareous ponds and ditches.

5. *Ranunculus circinatus* Sibth.

This species is restricted to a few rivers with slow current velocity in the lowlands, which is a remarkable distribution, too, because the species is very frequent in stagnant waters throughout the region especially in the highlands, where calcareous waters are more common. It has never been found in fast running waters as found in southern Germany, France and Denmark.

Detailed distribution maps have been constructed. They will be published elsewhere.

Phytosociological problems

The first step in ordering the multiformity of *Batrachium* taxa is to classify them with regard to growth form:

1. Long species, normally > 2m (up to 6 or 8 m), with rigid or succulent submersed leaves, usually without floating leaves: *Ranunculus fluitans, R. calcareus* and certain *fluitans*-like hybrids like *R. fluitans* × *trichophyllus* and *R. fluitans* × *circinatus*.

2. Short species, usually about 1 m long (up to 2 m) with flaccid submersed leaves, usually producing floating leaves at the fertile stems: *Ranunculus peltatus* and *R. aquatilis* and the hybrids of these species with all small species. *Ranunculus trichophyllus* and *R. circinatus* do not fit so well in this group, because they do not produce floating leaves.

Of the taxa producing floating leaves, *Ranunculus penicillatus* including *R. fluitans* × *peltatus* is somewhat intermediate between the two groups, because it normally reaches a length of more than 2 m.

There are many examples for a regular distribution of these taxa in north Germany rivers (see also Herr, 1983; Worbes, 1979). In most cases *Ranunculus peltatus* is restricted to the upper parts of the main course and to some small tributaries. *Ranunculus fluitans* can only be found in the lower courses, where the rivers reach a certain depth and discharge. In the middle courses and some tributaries, mainly *Ranunculus penicillatus* and *Ranunculus fluitans* × *peltatus* can be found.

Special problems arise from the taxonomic uncertainty within the floating-leaved taxa. It is not possible to assign each specimen to either *R. peltatus* or *R. penicillatus* or a well-defined hybrid. Phytosociological relevés containing such forms of *Batrachium* cannot be treated by means of a conventional approach. When classifying them as to the *Batrachium* taxa, one would atomize the complex into different 'associations' (better called dominance communities). But there are also difficulties classifying the relevés according to the accompanying species. There are several species groups of approximately equal value.

The problem is closely related to the classification of the *Callitricho–Myriophylletum*. The affinities of the so-called 'character species': *Ranunculus peltatus, Ranunculus penicillatus, Callitriche hamulata, Myriophyllum alterniflorum, Potamogeton alpinus* and *Nitella flexilis* to this vegetation type are quite different and not so easy to interprete as publications might suggest. Closely allied to this species group are *Callitriche platycarpa* and *Elodea canadensis*.

In the central part of the north German pleistocene there is only a small region, where all these species may be found together, and there are even only two river systems known, where they actually occur together. Steusloff (1939) described this type in a region, where by chance most of the characteristic species really co-exist. There are many possible combinations of the species, and indeed, all of these combinations can be found in different parts of the region. In some places, the adequate water types will only be colonized by *Callitriche platycarpa* and *Elodea canadensis*.

Of the other species groups in question, the *Callitriche's* show a strong tendency to colonize equivalent niches. In some regions *Callitriche hamulata* can be totally replaced by *Callitriche obtusangula*, in other drainage areas one can find 3 or 4 *Callitriche* species within a relatively short section of the river. Regarding the potamid species some of them may be substituted by rare hybrids of the same

Table 1. Floristic composition of *Batrachium* communities in Central Europe based on own observations (LS – Lower Saxony, SH – Schleswig-Holstein, DK – Denmark, F – France, I – Italy, Ba – Bavaria).

Consecutive number	1	2	3	4	5	6	7	8	9	10	11	12	13	14	15
Country of origin	SH	DK	Ba	F	LS	LS	DK	Ba	LS	I	F	Ba	LS	DK	SH
Number of relevées	24	22	5	2	47	36	5	6	3	3	10	8	12	2	23
Ranunculus peltatus	3	3	1	3	3	3	1	.	.	.	.	.	1	.	1
Ranunculus penicillatus[a]	3	3	3	.	2	2	.	.	.	.	.	.	.	.	1
Ranunculus trichophyllus	.	.	.	.	.	.	.	2	2	.	.	.	.	.	.
Ranunculus circinatus	.	.	.	1	.	.	2	3	.	1	.	.	.	1	.
Ranunculus fluitans	.	.	.	.	.	.	.	1	.	1	3	3	3	1	.
Ranunculus calcareus[b]	.	.	.	.	.	.	.	.	.	3	1	.	1	3	3
Callitriche hamulata	1	.	2	.	1	1	.	.	.	.	.	1	1	.	.
Myriophyllum alterniflorum	.	1	1	.	1	.	.	1	.	.	.	1	1	.	.
Potamogeton alpinus	1	1	.	.	1	.	.	.	.	.	.	.	.	.	.
Potamogeton pectinatus	.	.	.	.	.	.	1	.	.	1	2	1	1	2	1
Butomus umbellatus	.	.	.	.	.	.	1	.	.	.	.	.	1	1	1
Hippuris vulgaris	.	.	.	.	.	.	1	.	1	.	.	.	1	.	.
Ceratophyllum demersum	.	.	.	.	.	.	.	.	.	1	1	1	1	.	.
Groenlandia densa	.	.	.	.	.	.	1	2	1	1	2	1	.	.	.
Myriophyllum spicatum	.	.	.	.	.	.	.	1	.	1	2	2	.	.	.
Zannichellia palustris s.l.	.	.	.	.	.	.	.	.	1	1	1	.	.	.	.
Potamogeton nodosus	.	.	.	.	.	.	.	.	.	1	2	1	.	.	.
Berula erecta	2	1	.	2	1	1	1	1	1	1	1	2	.	1	1
Sparganium emersum	1	2	1	1	2	1	1	1	.	.	1	1	1	2	2
Elodea canadensis	1	1	.	.	2	2	1	2	1	.	.	1	2	1	1
Potamogeton crispus	1	1	.	.	1	1	1	1	1	1	.	1	1	1	.
Callitriche platycarpa	2	2	.	2	2	2	2	.	2	.	.	.	1	1	1
Potamogeton perfoliatus	.	1	.	.	1	.	1	1	.	1	1	1	1	.	1
Fontinalis antipyretica	.	1	2	1	1	.	.	1	1	2	1	.	.	.	.
Lemna minor	.	1	.	.	1	1	1	1	.	.	1	1	.	1	.
Nuphar lutea	1	.	.	1	1	1	.	.	.	.	1	1	1	.	1
Potamogeton natans	2	1	.	.	1	.	.	.	.	.	.	1	1	.	1
Sagittaria sagittifolia	1	.	.	.	1	.	.	.	.	.	1	.	1	.	1
Potamogeton pusillus s.l.	.	.	.	1	1	1	.	.	.	1	.	,	1	.	.
Callitriche obtusangula	.	.	.	.	1	1	.	.	.	.	.	1	.	.	.
Callitriche stagnalis	.	.	.	.	.	1	.	.	.	.	.	.	1	.	.
Myriophyllum verticillatum	.	.	.	1	.	.	.	.	.	.	.	.	1	.	.
Potamogeton trichoides	.	.	.	1	.	.	.	.	.	.	.	.	.	.	.
Callitriche cf. cophocarpa	.	.	.	.	.	.	.	1	.	.	.	1	.	.	.
Potamogeton lucens	.	.	.	.	.	.	.	.	.	1	.	.	.	.	1
Lemna trisulca	.	.	.	.	.	.	1	.	.	.	.	.	.	1	.

Additionally in 2: *Potamogeton polygonifolius, Lemna gibba;* in 5: *Potamogeton × undulatus, Potamogeton × spathulatus, Potamogeton gramineus, Luronium natans, Nitella flexilis;* in 10: *Spirodela polyrhiza;* in 14: *Elodea nuttallii, Vallisneria spiralis, Lagarosiphon major.*

[a] *Ranunculus penicillatus: R. penicillatus* var. *penicillatus* incl. *R. fluitans × peltatus* and similar hybrids with floating leaves.

[b] *Ranunculus calcareus: R. penicillatus* var. *calcareus* incl. *R. fluitans × trichophyllus* and similar hybrids without floating leaves.

3 = species frequent and abundant, 2 = species characteristic, 1 = species present.

growth form (*Potamogeton crispus* by *P. × undulatus; P. alpinus* by *P. × spathulatus*). In addition, the occurrence of the species of the *Sparganium* group do not provide a good basis for classification either.

Obviously, this complex of communities fuzzes into several types which are similar with regard to structure but different with regard to floristic composition. This effect can be recognized in smallscale areas, because neighbouring drainage areas may have a totally different stock of *Batrachium* taxa and accompanying species though being similar with regard to physico-chemical conditions. In this way, drainage areas behave biogeographically like islands.

We will now discuss a larger area (Central Europe). Here, a scheme for classification of the communities is proposed, which is based on own observations (Table 1).

Three vegetation types can be recognized with regard to abundant *Batrachium* species. On the left side of the table, the communities built up by species with floating leaves (*Ranunculus peltatus* and *Ranunculus penicillatus* var. *penicillatus*, the latter including here *Ranunculus fluitans* x *peltatus* and other hybrids with floating leaves) are shown. Characteristic species of this vegetation type are *Myriophyllum alterniflorum, Callitriche hamulata* and *Potamogeton alpinus* (also very frequent in *Sparganium emersum*-dominated systems). *Callitriche platycarpa, Elodea canadensis, Sparganium emersum, Potamogeton natans, Berula erecta* and *Fontinalis antipyretica* are frequent and may occur as co-dominants. Species like *Potamogeton pectinatus, Groenlandia densa, Hippuris vulgaris, Butomus umbellatus, Myriophyllum spicatum, Zannichellia palustris* s.l., *Potamogeton nodosus* and *Ceratophyllum demersum* are regularly absent. Of the other *Batrachium* species only *Ranunculus circinatus* occurs (only rarely). This vegetation type colonizes waters poor in hydrogen carbonate, where floating leaves are of a certain advantage to get additional CO_2 from the air.

The second type is characterized by *Ranunculus circinatus* and *Ranunculus trichophyllus*, two small species without floating leaves. These communities are normally considered as part of the *Ranunculetum fluitantis* s.l. (Wiegleb 1979) or the *Ranunculo-Sietum* (Kohler et al., 1971; Oberdorfer, 1977). But they colonize fast-flowing small rivulets with instable sediment conditions, where *Ranunculus fluitans* cannot grow (see also Schuster, 1980). These communities are the counterpart to the preceding ones in calcareous waters, and *Zannichellia palustris* and *Groenlandia densa* are the most characteristic companions under meso- to eutrophic conditions.

A third type is characterized by *Ranunculus fluitans* and *R. calcareus*, the latter one including hybrids of *R. fluitans* without floating leaves. These kinds of communities have often been described by various authors under various names, so only some additional comment is given. They can be separated into two subunits:

- One with *Groenlandia densa, Zannichellia palustris* and *Miriophyllum spicatum*, calciphilous species which this type has in common with the *Ranunculus circinatus/trichophyllus* type. Within this type, *Potamogeton nodosus* characterizes a southern race.
- Another one lacking those species but sometimes containing species of type 1 (*Callitriche hamulata, Ranunculus peltatus, Myriophyllum alterniflorum*). Oberdorfer's (1977) *Ranunculo-Callitrichetum* is part of this type. It occurs in non-calcareous waters, or more precisely, in waters which are poor in hydrogen carbonate, a fact which has already been pointed out by Siefert (1976) and Wiegleb (1981).

Since the trophic status of this type of rivers (lower courses, broad streams) is nowadays relatively similar, there are not so marked floristical differences as in the case of the vegetation types of the upper courses. This floristical convergence of constitutionally calcareous and non-calcareous rivers subject to anthropogeneous influence has led to distortion in other investigations, too (Wiegleb, 1981).

Conclusions

The following aspects may be summarized as new:
1. Two taxa of the *Ranunculus penicillatus* complex have been described phytosociologically.
2. Vegetation dominated by the small *Batrachium* species *Ranunculus circinatus* and *R. trichophyllus* has been described as an independent type.
3. Well-documented types (*Callitricho-Myriophylletum* s.l. and *Ranunculetum fluitantis* s.l.) have been analyzed in more detail. The floristic variability of these types has become more evident. The same is true for the interrelations between the two types and their relations to associated *Callitriche* communities.

Furthermore, some conclusions from textbooks can be corrected. Ellenberg (1978) considered aquatic vegetation as an 'azonal vegetation type', which means that large-scale geographical variation is not reflected in the composition of the vegetation a great deal. This is obviously not true, as Felzines (1979) and Wiegleb (1980) have shown.

Oberdorfer (1977) stated 'that the *Ranunculetum fluitantis* occurs without considerable differences from Ireland to the Balkanese'. Though his approach to this community is not directly comparable to ours, this statement has to be refused. In many regions, for example Ireland, great parts of Great Britain, Denmark, also Schleswig-Holstein (in northern Germany), *Ranunculus fluitans* is replaced by *R. fluitans*-like hybrids or forms of the *R. penicillatus* complex.

The remaining *R. fluitans* has to be divided into at least 3 morphologically distinct types the phytosociological behaviour of which is not yet evident. The community comprises an oceanic race with species like *Oenanthe fluviatilis* (Carbiener, 1983), a mediterranean one with *Vallisneria spiralis*, and is deeply differentiated according to water chemistry

Table 2. Phytosociological behaviour of the most frequent *Batrachium* taxa in Central Europe with reference to water types.

	Ditches	Ponds	Lakes	Small rivulets	Rivers and streams
Ranunculus fluitans	–	–	–	rare	from mono-dominant stands to extremely species-rich communities
Ranunculus calcareus	–	–	rare	rare	similar to the above
Ranunculus penicillatus	–	–	–	non-calcareous waters of different trophic degree	
				assoc. to the *Call. ham group*	not frequent with the *P. pectinatus* group
Ranunculus peltatus	*Callitriche*- and parvopotamid-rich communities		not frequent	assoc to the *Call. ham* group	only marginal
Ranunculus circinatus	parvopotamid communities	parvo- and magnopotamid communities		calcareous waters with the *Groenlandia* group	only marginal in calcareous waters
Ranunculus trichophyllus	batrachid communities	parvopotamid communities		similar to the above	only marginal
Ranunculus aquatilis	batrachid communities	parvopotamid communities	–	–	–

(especially calcium hydrogen carbonate content and trophic status) and physical conditions. Changing accompaniment of different *Callitriche* species can be found, the pattern of which has to be deciphered; the status of the *Potamogeton nodosus*-like forms has to be made clear etc., just to mention a few problems of this complex community.

Finally, a short survey of the phytosociological behaviour of the most frequent *Batrachium* taxa is given as far as it has become evident from different types of water bodies in Central Europe. Here, 7 taxa, also including *Ranunculus aquatilis,* are considered (Table 2).

Ranunculus fluitans, R. calcareus and *R. penicillatus* are normally restricted to running waters. *Ranunculus calcareus* also occurs in subalpine Italian lakes. This is frequently the case in the Lago di Garda (not *R. fluitans,* as has been stated by Bianchini *et al.,* 1974), but it might be a local phenomenon. On the other hand, *Ranunculus aquatilis* is restricted to stagnant waters, which coincides with the observations of Mériaux & Géhu (1980). This species has not been found in running waters on the continent, but Kohler (1979) mentions its occur-

rence from a small rivulet in southern England. The species is much rarer than *R. peltatus,* a fact which has also been proven by herbarium experience. In Lower Saxony it shows a distinct oceanical distribution with only about ten points of recently confirmed occurrence.

The composition of *Batrachium* species of small rivulets is quite similar to that of ditches and ponds, and it is often difficult to distinguish these vegetations by means of floristical composition. But in many cases there are marked differences in life form and syndynamical relations. As an example, *Ranunculus peltatus* is discussed. In running waters the species is mainly perennial, and though showing seasonal fluctuations, the community occurs continuously. In small stagnant waters on the other hand, *R. peltatus* is annual. According to the hydrological conditions it can be replaced by small reed swamp communities with *Oenanthe aquatica* or *Glyceria fluitans.* In many cases one can find a kind of equilibrium between these two kinds of structurally different vegetation types. In this way, other criteria than floristical ones can be used to distinguish a *R. peltatus* community from stagnant waters.

The concept of similarity in vegetation research should not be restricted to floristical composition. Structural and syndynamical approaches can solve phytosociological problems. Structural analysis in combination with both small-scale and large-scale biogeographical analysis may reveal further ecological insight in *Batrachium* communities.

References

Bianchini, F., Bertoldo, G. & Tessari, M., 1974. Floristica e fitosociologia delle macrofite. In: Consiglio nazionale delle ricerche, indagini sul Lago di Garda. Quaderni dell'instituto di ricerca sulle acque 18: 225–240.

Carbiener, R., 1983. Modifications des écosystèmes des rivières phréatiques d'Alsace par l'eutrophisation de la nappe des graviers du Rhin. Xème Coll. Intern. Phytosoc., Lille, Septembre 1981.

Cook, C. D. K., 1966. A monographic study of Ranunculus subgenus Batrachium (DC.) A. Gray. Mitt. Bot. Staatssammlung München 6: 47–237.

Déthioux, M., 1982. Données sur l'écologie de Ranunculus penicillatus (Dum.) Bab. et R. fluitans Lam. en Belgique. In: Studies on Aquatic Vascular Plants, pp. 187–191. Brussels.

Ellenberg, H., 1978. Die Vegetation Mitteleuropas mit den Alpen. 2. Aufl. Stuttgart.

Felzines, J. C., 1979. L'analyse factorielle des correspondances et l'information mutuelle entre les espèces et les factures du milieu: Application à l'écologie des macrophytes aquatiques et palustres. Bull. Soc. Bot. N. France 32: 39–63.

Hartog, C. den & Segal, S., 1964. A new classification of water-plant communities. Acta Bot. Neerl. 13: 367–393.

Herr, W., 1983. Die Fließgewässervegetation im Einzugsgebiet von Treene und Sorge. Gedenkschrift für Prof. Dr. E. W. Raabe (ed. K. Dierßen) (in press).

Holmes, N. T. H., 1980. Preliminary results from river macrophyte survey and implications for conservation. Nature conservancy council, Chief scientist's team notes 24. 68 pp. London.

Holmes, N. T. H. & Whitton, B. A., 1977. The macrophytic vegetation of the River Tees in 1975: observed and predicted changes. Freshwater Biol. 7: 43–60.

Kohler, A., 1979. Bericht über die Forschungsreisen nach Südengland und Südschweden im Sommer 1978. Mskr. Hohenheim.

Kohler, A., Vollrath, H. & Beisl, E., 1971. Zur Verbreitung, Vergesellschaftung und Ökologie der Gefäßmakrophyten im Fließwassersystem Moosach. Arch. Hydrobiol. 69: 33–365.

Kutscher, G. & Kohler, A., 1976. Verbreitung und Ökologie submerser Makrophyten in Fließgewässern des Erdinger Mooses (Münchener Ebene). Ber. Bayer. Bot. Ges. 47: 175–228.

Mériaux, J. L., 1981. La classe Potametea dans le Nord de la France. Xème Coll. Inter Phytosoc., Lille, Septembre 1981.

Mériaux, J. L., 1982. Inventaire et distribution des espèces des genres Callitriche, Elodea et Ranunculus (sous-genre Batrachium) dans le Nord de la France. In: Studies on Aquatic Vascular Plants, pp. 311–312. Brussels.

Mériaux, J. L. & Géhu, J. M., 1980. Réaction des groupements aquatiques et subaquatiques aux changements de l'environnement: In: O. Wilmanns & R. Tüxen (eds.); Epharmonie, pp. 121–142. Vaduz.

Neuhäusl, R., 1959. Die Pflanzengesellschaften des südöstlichen Teiles des Wittingauer Beckens. Preslia (Praha) 31: 115–147.

Oberdorfer, E. (ed.), 1977. Süddeutsche Pflanzengesellschaften, Teil I. 2. Aufl. Stuttgart.

Passarge, H., 1964. Pflanzengesellschaften des nordostdeutschen Flachlandes I. Jena.

Passarge, H., 1978. Übersicht über mitteleuropäische Gefäßpflanzengesellschaften, Feddes Repert. 89: 133–195.

Schuster, H. J., 1980. Analyse und Bewertung von Pflanzengesellschaften im Nördlichen Frankenjura. Diss. Botanicae 53.

Siefert, A., 1976. Über die Verschmutzung von Fließgewässern im süd-niedersächsischen Raum und ihr Einfluß auf Vorkommen und Verbreitung einiger Makrophyten, Diatomeen und Bakterien. Diss. Göttingen.

Steusloff, U., 1939. Zusammenhänge zwischen Boden, Chemismus des Wassers und Phanerogamenflora in fließenden Gewässern der Lüneburger Heide um Celle und Uelzen. Arch. Hydrobiol. 35: 70–106.

Tüxen, R. & Schwabe, A., 1972. Potamogetonetea. Bibl. Phytosoc. Syntaxon. 14: 1–124.

Weber-Oldecop, D. W., 1969. Wasserpflanzengesellschaften im östlichen Niedersachsen. Diss. TU Hannover.

Weber-Oldecop, D. W., 1977. Fließgewässertypologie in Niedersachsen auf floristisch-soziologischer Grundlage. Göttinger Florist. Rundbr. 10: 73–80.

Wiegleb, G., 1979. Vorläufige Übersicht über die Pflanzengesellschaften der niedersächsischen Fließgewässer. Naturschutz und Landschaftspflege in Niedersachsen 10: 85–116.

Wiegleb, G., 1980. Some applications of principal components analysis in vegetation ecological research of aquatic communities. Vegetatio 42: 67–73.

Wiegleb, G., 1981. Application of multiple discriminant analysis on the analysis of the correlation between macrophyte vegetation and water quality in running waters in Central Europe. Hydrobiologia 79: 91–100.

Wiegleb, G., 1982. Probleme der syntaxonomischen Gliederung der Potametea. In: H. Dierschke (ed.); Syntaxonomie, pp. 207–249. Vaduz.

Wiegleb, G. & Herr, W., 1982. Übersicht über Flora und Vegetation niedersächsischer Fließgewässer und deren Bedeutung für Naturschutz und Landschaftspflege. 3 Bd. Mskr. Oldenburg.

Worbes, M., 1979. Die Makrophytenvegetation der Fulda. Diplomarbeit. Göttingen. Mskr.

Accepted 23.9.1983.

Expansion and retreat of aquatic macrophyte communities in south Bohemian fishponds during 35 years (1941–1976)

S. Hejný
Botanical Institute, Czechoslovak Academy of Sciences, 252 43 Průhonice, Czechoslovakia

Keywords: Aquatic macrophyte, Dynamic changes, South Bohemian fishponds

Abstract

The aquatic macrophyte communities of southern Bohemian fishponds were analyzed during 35 years (1941–1976). From 61 vegetational units 26 communities may be characterized as being well adapted to modern fishpond management, 4 units are on their way to extinction and 18 units are in regression. Our long-term observations evaluated the following types of destruction: destruction of the community, in which one dominant species retreats and another regenerates (*Potameto natantis–Nymphaeetum candidae* → *Potamogeton natans* comm.); destruction of the community, in which only one stratum regenerates (synusia) (*Glycerietum aquaticae utricularietosum australis* → *Utricularia australis* comm.). The development of the pleustophytic communities followed two pathways: transformation of more complicated forms into simpler ones (*Utricularietum australis* → *Lemno–Spirodeletum*); development of complicated forms from simpler ones (*Lemnetum minoris* → *Lemno–Spirodeletum*).

Introduction

After 35 years, the results of studies on macrophyte communities made in south Bohemia show rather great changes in the distribution of the communities. The components of the synantropization process were the following: (a) removal of the whole littoral zone by bulldozing and scraping; (b) intensive duck-farming; (c) intensive application of fertilizers in ponds. 120 ponds forming 11 groups appear as a sufficiently representative set to describe the evaluation with respect to both the number of objects and their mutual geographical proximity. In total, 61 vegetational units were analyzed; 30 units (21 associations and 9 communities) colonized the inner aquatic parts of the ponds and 31 units (23 associations and 8 communities) colonized their littoral parts. After 35 years 2 associations and one community were in extinction *(Sparganietum minimi, Hottonietum palustris, Nymphoides peltata* comm.); 6 associations and 3 communities

were in retreat (*Lemnetum minoris, Lemno–Utricularietum bremii, Hydrocharietum morsus-ranae, Potameto natantis–Nymphaetum candidae, Potametum lucentis, Potametum crispi, Batrachietum aquatile, Potamogeton natans* comm., *Trapa natans* comm.) in the group of water plant units (30), but 16 units (11 ass., 5 comm.) have been found developing (*Riccietum fluitantis, Lemno–Spirodeletum, Lemnetum gibbae, Lemnetum trisulcae, Utricularietum australis, Myriophyllo–Potametum, Elodeetum canadensis, Potamoeto–Zannichellietum, Batrachio trichophylli–Callitrichetum, Polygonum amphibium* comm., *Ceratophyllum demersum* comm., *Zannichellia palustris* comm., *Potamogeton acutifolius, Potamogeton obtusifolius* comm., *Batrachium circinnatum* comm.). The situation was very similar as regards the dynamics of vegetation units in the littoral belts (1 ass. in extinction, 9 in regression, 17 units in progression). In both biotopes, the 26 comm. in progress may be characterized as being well adapted to modern fishpond management.

Vegetatio 59, 243–245 (1985).

244

The empirical data collected in the past have been summarized in a number of tables covering two periods of intensive observation, i.e. 1941 to 1949, and 1961 to 1976. A synoptic table accompanied by detailed explanations was presented at the first symposium of European hydrobotanists at Illmitz in 1981.

I should like to point out two basic ideas derived from our long-term observations.

1. Evaluation of successive structural changes in vegetation units

In relatively species-poor aquatic communities, the structural change of the units, exposed to powerful anthropogenic factors, develops through the changes of individual structural elements, such as the decline or extinction of the dominant species followed by the decline of the whole stratum and/or synusia. Thus the entire associations do not alter, only after a total destruction they may disappear. In spite the fact that our observations extended over less than half a century, the successive series of vegetation relevés showed that two different types of destruction can take place:

a. Destruction of the community of which only one synusia can but may not regenerate. For example, in the association *Glycerietum aquaticae utricularietosum australis* the destruction by bottom scraping is usually followed by regeneration of a synusia of pleustophytes, which can later develop into the *Utricularietum australis.*

b. Destruction of a community of which one dominant species retreats and another regenerates, which results in the development of a new association. The declining *Potameto natantis–Nymphaeetum candidae* association can serve as an example. *Nymphaea candida* disappears, and over a certain period *Potamogeton natans* (*Potametalia*) persists as a basal cenose.

2. Evaluation of transformation of vegetation units

In this case, the changes are caused by anthropogenic factors, and they are manifest in the alteration of a relatively simple community into a new association that is structurally organized at the same or a slightly higher or lower level. Transformation de-

velops by 'transposition' or exchange of some species. We can illustrate this type of change on the structurally simple communities of the *Lemnion minoris, Utricularion vulgaris,* and *Hydrocharition morsus-ranae* alliances.

The development of the pleustophyte communities followed two pathways during the second period of our observations between 1961 and 1976.

a. Transformation of more complicated forms into simpler ones, e.g.,

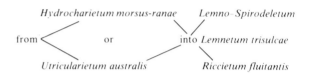

b. Development of complicated forms from simpler ones, e.g.

These cases have been taken from a hydrosere which can provide many other examples of this kind. These examples illustrate the necessity of careful observations of changes in vegetation units, and, also, the necessity of an adequate theoretical explanation of these changes. At this point, I should like to mention the concepts of 'decumbency' and 'incumbency' used by Soviet geobotanists.

Our long-term observations allow us to evaluate the decline and expansion of communities, a prominent chorological phenomenon. Generally, the decline of a plant association proceeds through two stages:

a. A stage, lasting about 10 to 20 yr, in which the surviving stands or communities are clustered, and, if illustrated on a phytocartogram, resemble a disjunctive area of distribution of a plant species.

b. A stage, lasting about 30 yr, manifested in a very scattered occurrence of the surviving stands or communities which resembles, if illustrated in a phytocartogram, a diffusive area of distribution.

Ultimately, the plant association may become extinct within the territory under study.

In spite of the fact that our detailed observations cover a comparatively small territory, we have verified a similar process of decline or expansion of a plant association over a larger region, such as the whole of southern Bohemia.

The decline of an association can be illustrated by the example of the *Potameto natantis–Nymphaeetum candidae*. In this case, the vanishing plant association is strongly affected by the gradual loss of adaptability of the dominant species to the environmental changes. High amounts of fertilizers and the winter drainage of the fishponds cause a rapid disappearance of *Nymphaea candida* and *Potamogeton acutifolius,* while *Potamogeton natans* declines in a much slower way.

The expansion of communities shows the following three stages:

a. The constituent species gradually attain a higher adaptability to the altered environment.

b. A species or several species invade the declining association, and gradually establish themselves as new dominant or co-dominant species.

c. A species or several species attain a high vitality following their improved adaptation, and the remains of the declining association are successively pushed back.

Progressive development can be demonstrated for three types of communities:

(A) The *Utricularietum australis,* arising through a kind of 'liberation' of the pleustophytes (*Utricularia australis* after the destruction of communities of the alliance *Phragmition.* The number of its stands increased from 17 to 28.

(B) *Potameto–Zanichellietum,* arising by a successive increase in adaptability of the populations of *Potamogeton pectinatus* and *Zanichellia palustris* following the application of nitrogen, phosphorus and potassium fertilizers to fishponds. This association spreads widely, replacing all declining vegetation types such as *Potameto natantis–Nymphaeetum candidae* or *Potametum lucentis.* Starting from 24 occurrences, the number of stands recorded in the period under study reached a total of 60. The occurrence of this association cannot be considered as a temporary phenomenon. The stabilization is affected by a richer species composition and higher net production.

(C) *Elodeetum canadensis* is a newly establishing association which continuously encounters favourable conditions for its further spreading in ponds. This association has four specialized developmental stages; namely, the initial, the invasion, the explosive and the degradation stages. It can pass through all four stages within less than 15 to 20 years.

A sudden fall in the water level gives the impulse to the transition from the initial stage to the invasion and explosive stages. The *Elodeetum* retreats and may be suppressed in years following combined summer and winter drainage. As a neophytic type, it shows a periodic development in the ponds where it is clearly a reversible type of plant community. Extinction of the association has been observed in 19.8% (probably only transient disappearance), stabilization in 27.3%, and progressive increase in 54.6% of the instances of its occurrence.

Accepted 19.1.1984.

Discussions

Paper by
D. Lausi & P. L. Nimis
Contribution to the History of Vegetation in NW Canada (Yukon Territory) on a Chorologial-Phytosociological Basis

E. HADAČ:

This paper is an excellent corroboration of the idea that plants usually do not migrate as individual species, but together with their whole community or better with the whole biocoenoses, step by step. I have shown this with the example of Spitsbergen plant communities on the Symposium on North Atlantic Biota twenty years ago, and later on with the vegetation of the Iraqi Kurdistan. There, surrounded by the indigenous and mostly endemic irano-turanian plant communities, a spring community occurs with *Juncus inflexus, J. gerardi, Mentha longifolia, Blysmus compressus* etc. together with Eurasian frogs and insects. The findings of the authors correspond well with these observations.

H. ELLENBERG:

Mr. Nimis showed us clearly that the vegetation types in Northern North America are closely related to pedological gradients on the one hand and to the areal types on the other hand. What proof can you give for your opinion, that the areal types are an expression of the history of the flora, i.e. of migration of plants?

Mr. Nimis showed slides giving a clear differentiation between steppe ridges and forest areas dominated either by *Populus tremuloides* or *Picea glauca*. Which factors are decisive for this mosaic of plant formations, especially for the steppe–forest limit?

P. L. NIMIS:

Reply to Prof. Ellenberg's remarks: From the methodological point of view, I think that an actualistic interpretation of chorological facts should always come first. In most cases, the introduction of the time factor implies the knowledge of ecological conditions in the past, and the hidden assumption that species did not change substantially their ecological requirements in time, so that the knowledge of the present relations between species ranges and ecological factors is basic also to historical phytogeography. History should be considered above all in those cases where chorological facts cannot be explained on the bases of actualistic hypotheses only (e.g. the absence of tree species common to both sides of the Bering Strait).

The main factor is water. In the Kluane area, *Picea*-woods are confined to the main strings of the hydrographic net. The slopes are covered by steppe.

Papers by
G. Jahn
Geographische Abwandlungen in Fichten- und Buchenwaldgesellschaften

and

J. Moravec
Chorological and Ecological Phenomena in the Differentiation and Distribution of the Fagion *Ass. in Western Czechoslovakia*

H. ELLENBERG:

Ein einheitliches Rezept, die regionale, höhenbedingte und edaphische Differenzierung einer Gruppe nahe verwandter Pflanzengesellschaften synsystematisch zu fassen, wird es kaum geben. Im Sinne von Braun-Blanquet sollten wir stets von der floristischen Differenzierung ausgehen. Bei den Fichtenwäldern ist diese stark vom Klima abhängig, weil die von Ihnen erzeugte Rohhumusschicht die Bodenunterschiede buchstäblich fast verdeckt. Bei den Buchenwäldern dagegen treten im Artengefüge zunächst die Bodenunterschiede hervor. Regionalklimatische Unterschiede können sich hier nur zeigen, wenn man Buchenwälder auf Böden gleicher Qualität (z.B. Kalkboden, Sandstein) betrachtet.

R. CARBIENER:

Was mich an den Ausführungen von Frau Jahn gefreut hat, ist die Feststellung, daß die klimatischen Bedingungen gegenüber den anderen Vorrang haben und die entsprechenden Einheiten eine hierarchisch höhere Stellung bekommen. Das, glaube ich, ist sehr wichtig, um Soziologie und Chorologie zusammenzubringen. Wir stellen in vielen Fällen fest, daß die klimatischen Bedingungen vorherrschen. In W-Französischen höheren Mittelgebirgen, z.B. im Massiv-Zentral und in den

Vogesen, haben einige Autoren vorgeschlagen, die bodensaueren Buchenwälder aufgrund der rein floristischen Merkmale mit den azidophilen Eichenwäldern im Rahmen der Klasse *Quercetea roboripetraeae* zu verbinden. Es ist jedoch eine Gliederung nur nach edaphischen Gesichtspunkten, die klimatisch ganz verschiedene Einheiten zusammenfaßt. Analoge Verhältnisse bestehen in Hochlagenbuchenwäldern, wo *Betulo-Adenostyletea*-Arten eine große Rolle spielen.

H. WAGNER:

Im Anschluß an die Bemerkungen von R. Carbiener: Der Vorrang klimatisch-chorologischer Abwandlung der Artenlisten gegenüber ökologischer Differenzierung legt die Forderung nahe, höhere Einheiten für vikariierende Gesellschaften zu bewahren und nicht für ökologische Abwandlungen, da letztere entsprechend der mehrdimensional unabhängig variierenden ökologischen Faktoren nicht in einfach linearer Hierarchie darstellbar sind. Gerade die mehrdimensionalen Verknüpfungen des *Fagion*-Verbandes – einerseits über das *Luzulo-Fagion* zu den *Quercetalia roboris* bzw. *Vaccinio-Piceetalia*, andererseits über 'Schluchtwälder' zu den *Betulo-Adenostyletea*, schließlich auch über das *Cephalanthero-Fagion* zu den *Erico-Pinetea* bzw. *Quercetalia pubescentis* – zeigen deutlich, daß wir bei der notwendigen Berücksichtigung nur einer Richtung und Trennung in 'unabhängig' nebeneinander stehenden Klassen wesentliche Zusammenhänge unterdrücken müssen. Auch die Vegetationssystematik hat primär der Erkenntnis der Zusammenhänge in der Natur und nicht einer Idee zu dienen.

Im Anschluß an Bemerkung von H. Ellenberg: Gerade in chorologischem Vergleich über weitere Räume kann die Floristik in einer bislang nicht genügend berücksichtigen Weise durch Heranziehung vikariierender Arten angewandt werden. Schließlich ist auch zu bedenken, daß die moderne Cytosystematik zur Auflösung vieler 'guter alter' in (meist nicht entsprechend in der Pflanzensoziologie beachtete) vikariierende Kleinarten geführt hat, die an sich bereits zu gewissen Neubewertungen von Charakterarten, bzw. Begleitern führen sollte.

P. L. NIMIS:

The discussions on the chorological versus ecological interpretation of syntaxonomy recalls the same questions that arose in plant and animal taxonomy chiefly after the introduction of numerical taxonomy in systematics, namely the controversy between 'phenetics' and 'cladistics', or, in other words, between 'omnispective general purpose classifications' and classifications oriented towards the elucidation of phylogenetic relationships. Numerical taxonomists tend to underline the fact that, from the operational point of view, these are quite different things. I share the same opinion. Synsystematics should be regarded as a 'general purpose classification', based on a purely floristic basis; as such, it can be used as a tool for drawing ecological or chorological interferences, when these are possible and in the limits of what is possible. That limits are present is obvious: in floristically homogeneous areas the ecological interpretation of syntaxonomy tends to prevail, in areas less homogeneous the chorological interpretation often marks ecological affinities. At the extreme, any meaning, either chorological or ecological, tends to disappear. An example: we were forced to include the thorny-cushions vegetation of mediterranean Italy in almost as many classes as associations, in spite of their ecological, structural and historical affinities. This because the long isolation on the mediterranean mountains led to imposing phenomena of endemism and vicarism at the species level. In such case, syntaxonomy has lost most of its value. There we are at the limits where neither ecology nor chorology enter the picture: all what we have, is a list of empty names.

H. DIERSCHKE:

Die beiden vorhergehenden Vorträge haben zwei Grundkonzepte bei der Berücksichtigung chorologischer Erscheinungen für die syntaxonomische Gliederung gezeigt. Frau Jahn hat mit Hilfe großer übersichten für weite Teile Europas die floristischen Abwandlungen vorgeführt. Sie hat sich sehr zurückhaltend zu einzelnen Assoziationen geäußert, wenn auch sicher einige gut erkennbar sind.

Herr Moravec hat in einem vergleichsweise recht kleinen Gebiet eine relativ große Zahl von Buchenwald-Assoziationen erwähnt, die sich zum Teil aber wohl recht ähnlich sind. Wenn wir dieses Prinzip kleiner Regional-Assoziationen im größeren Raum verfolgen würden, hätten wir eine verwirrende Vielfalt von Syntaxa. Kann man nicht z.B. Ihre zwei Linden-Buchenwälder als geographische Abweichung des Melico-Fagetum im Sinne von geographischen Rassen ansehen?

J. J. BARKMAN:

Ich glaube, es ist nicht richtig von 'chorologischen Faktoren' zu reden im Gegensatz zu ökologischen Faktoren. Ich bin darum auch nicht glücklich mit dem Namen des Themas für heute: 'Chorological influences in the formation and differentiation of natural, seminatural and anthropogenic plant communities'. Es gibt keine chorologische Faktoren oder Einflüße. Gegenüber ökologischen stehen historische Faktoren. Die chorologischen (geographischen) Unterschiede zwischen Pflanzengesellschaften sind entweder historischer Natur oder ökologischer Natur (Makroklimadifferenzen).

G. WIEGLEB:

Ich möchte auf das Schema von Moravec zurückkommen, das zwischen der 'Siebwirkung' der ökologischen Faktoren und der biotischen Interaktionen unterscheidet. Es gibt Gesellschaften, deren Zusammensetzung ausschließlich von den ökologischen Faktoren bestimmt wird. Man kann sie als 'außengesteuert' bezeichnen. Diese Gesellschaften lassen sich nur statistisch definieren. Sie zeigen meist eine stärkere Zerfaserung sowohl bezüglich edaphischer wie klimatischer Faktoren. Die autökologische Potenz der konstituierenden Arten stimmt eben mit den Gegebenheiten überein oder nicht, dann fällt die Art aus. In anderen Gesellschaften bestehen starke biotische Interaktionen. Ein Beispiel ist der von Ellenberg angesprochene Fichtenwald, der sich sein Milieu selbst schafft. Diesen Typ könnte man als 'binnengesteuert' bezeichnen, da auch funktionale Zusammenhänge bestehen. Es gibt alle Übergänge zu dem 'außengesteuerten' Typ. Diese Gesellschaften, die auch durch funktionale Zusammenhänge ausgezeichnet sind, zeigen meist eine gleichförmige Verbreitung über größere Areale.

Für die verschiedenen Typen der Gesellschaften müssen vielleicht auch verschiedene Methoden der Klassifikation angewendet werden. Bei den ausschließlich 'statistisch' definierten Wasserpflanzengesellschaften habe ich bereits darauf hingewiesen, daß ohne Rücksicht auf lebensformentypologische Kriterien keine hinreichende Klassifikation möglich ist.

S. PIGNATTI:

Ich finde es sehr anregend, daß dieses Problem der Beziehungen zwischen Chorologie und Ökologie einzelner Pflanzengesellschaften aufgetaucht ist. Ich glaube, daß es sich um eine ganz grundlegende Problematik handelt, die die Beziehungen zwischen Struktur und Funktion in den biologischen Systemen berührt. Ich werde morgen dieser Problematik meinen Vortrag widmen, dem ich jetzt nich vorgreifen möchte.

J. MORAVEC:

Antwort auf die Diskussionsbemerkung von G. Wiegleb: Ich habe das Beispiel mit zwei Sieben (ökologische Faktoren und Interaktionen zwischen den Pflanzen) nur deshalb wieder gestrichen, weil es die falsche Vorstellung hervorruft, als ob zuerst die Flora eines Gebietes durch Evolution und Migration entstanden ist und erst nachher die Agglomerationen und Phytozönosen entstehen mittels der Selektionsmechanismen des einen oder der zwei 'Siebe'.

Antwort auf die Bemerkung von H. Dierschke: Die Situation mit den breiten und enggefaßten Assoziationen ist ähnlich der Situation in der Sippentaxonomie, wo man Arten verschiedener Breite unterscheidet. Man arbeitet mit 'linneonen' (Großspezies) wie *Nardus stricta* – eine gute Art, die sogar allein die Gattung repräsentiert – man begegnet aber auch Kleinarten (Jordanonen), wie z.B. den Arten der Gattung *Alchemilla*. Das System sollte die Situation in der Natur widerspiegeln und nicht nach unserem Wunsch gebaut werden. Man kann zwar ein charakterartenloses bzw. kleinflächig vorkommendes Coenon einer bestimmten Assoziation als eine verarmte Rasse unterordnen, man stößt jedoch auf die Schwierigkeit, daß die Charakter- bzw. Differentialarten nich vorhanden sind (deshalb wurde dieses Coenon als eine selbständige Assoziation abgetrennt). Nur selten kann man beweisen, daß die Charakter- oder Differenzialarten ursprünglich vorhanden waren. Und auf welchen theoretischen Prinzipien darf man diejenigen Bestände zu einer Assoziation zusammenfassen, wenn die Assoziationscharakterarten und -differentialarten fehlen?

G. JAHN:

Zwei Bemerkungen zu Herrn Moravec: Die eine

berührt einen sehr wichtigen Punkt zu unserem Fragenkomplex. Sie erinnern sich, daß Herr Moravec im Norden Böhmens eine *Violo reichenbachianae-Fagetum* beschrieb, das er so benannte, weil dort auf dem Standort eines *Dentario enneaphyllidi-Fagetum* die namensgebende Charakterart *Dentaria enneaphyllos* fehlt. Eine Pflanzengesellschaft wird aber nach Braun-Blanquet nicht nur durch die Charakterarten, sondern durch die gesamte Artenkombination bestimmt. Nach der Artenkombination neigte Oberdorfer (schriftliche Mitteilung) dazu, diese Gesellschaften zum *Elymo-Fagetum* Kuhn 1937 zu stellen.

Hier möchte ich widersprechen: Wenn man chorologische Phänomene in Pflanzengesellschaften erkennen will, kann man nur Gesellschaften auf gleichem Substrat vergleichen, d.h. man muß, um die Auswirkungen des Klimas in der Abwandlung der Artenkombination zu erkennen, die übrigen Faktorenkomplexe, so auch den des Bodens, gleich halten. Man kann also nur vergleichen: Wälder auf reichen Silikatgesteinen oder Wälder auf Kalkböden, und diese dann von West nach Ost verfolgen. Man kann nicht Wälder auf Kalk im Westen – und das *Elymo-Fagetum* Kuhn 1937 ist ein *Fagetum* auf Karbonatgestein – mit Wäldern auf reichem Silikatgestein im Osten wie dem '*Violo-Fagetum*' vergleichen. Ich habe in meinem Referat versucht, Ihnen zu zeigen, daß die montanen Buchen-wälder auf reichem Silikatgestein von Westen nach Osten artenreicher werden. Dadurch nähern sie sich zwar in ihrer Artenkombination dem *Elymo-Fagetum* im Westen – enthalten auch in reichen Ausbildungen *Mercurialis perennis, Elymus europaeus, Lathyrus vernus* – sollten jedoch nach ihrer gesamten Artenkombination eher als Gebietsausbildung des *Dentario enneaphyllidi-Fagetum* angesehen werden.

Dann zu den Übergängen *Dentario enneaphyllidi-Fagetum* in den Sudeten, *Dentario glandulosae-Fagetum* in den Karpaten. Ich würde dieses Problem nicht für so gravierend halten. Denn da sich die Klimabedingungen nur allmählich ändern und die Bodenverhältnisse vergleichbar sind, kann man auch bei der Vegetation keine klare Trennlinie erwarten. Man sollte vermeiden, aus solchen Übergängen eigene Gesellschaften zu machen, allenfalls könnte man Gebietsausbildungen beschreiben.

Anwort auf die Diskussionsbemerkungen zu meinem Referat: Mit diesen Bemerkungen sind viele Fragen angeschnitten, die ich nur kurz und sicher unvollkommen beantworten kann.

Zur Herrn Carbiener: Ja, die Rolle des Klimas als in erster Linie für die Artenkombination von Waldgesellschaften maßgebender Standortsfaktor bestätigt sich immer wieder, sofern es sich nicht um extreme Böden handelt. Zu Ihren Fragen nach der Abgrenzung von Gesellschaften in den westfranzösischen höheren Mittelgebirgen: Das Problem Buchenwälder–Eichenwälder stellt sich auch bei uns besonders in Norddeutschland. Es ist aus zwei Gründen sehr schwer zu lösen: Erstens ist die Krautschicht sehr artenarm und wird nur von wenigen Arten gebildet, wie z.B. *Deschampsia flexuosa* und *Carex pilulifera* – in Eichenbeständen zusätzlich *Vaccinium myrtillus* – die aber auch nur in Bestandeslücken vorkommen, sonst besteht die Bodenvegetation in Buchenbeständen oft nur aus wenigen jungen Buchenpflanzen. Eine Abgrenzung nach der Bodenvegetation ist danach kaum möglich. Zweitens war der menschliche Einfluß auf die Wälder größer als anfänglich vermutet. Wegen ihrer Bedeutung für die Schweinemast und als Bauholz wurde die Eiche direkt, durch die Nieder- und Mittelwaldwirtschaft infolge ihrer besseren Ausschlagfähigkeit indirekt gefördert. Die echte Rolle der Eiche in der potentiellen natürlichen Vegetation ist daher sehr schwer einzuschätzen. Ein Ausweg aus diesem Dilemma und pflanzensoziologisch durchaus vertretbar wäre daher die Einordnung der beiden in ihrer Artenkombination so ähnlichen artenarmen Gesellschaften in das *Quercion roboripetraeae*.

Bei den Hochlagenbuchenwäldern ist die Buche das gesellschaftsaufbauende Element. Also sind es Fageten, die durch Arten aus den *Betulo-Adenostyletea* gegenüber anderen Buchenwäldern differenziert werden.

Zu Herrn Wagner: Sicher hat jedes System seine Schwächen, es dient aber der Verständigung, und deshalb sollte man mit Änderungen sehr vorsichtig sein.

Zu den Herren Barkman, Ellenberg und Nimis: Es ist ganz selbstverständlich, daß in der Pflanzensoziologie das floristische Prinzip die Priorität hat. Aber chorologische Phänomene sind floristische Probleme und wir müssen und können sie im System berücksichtigen, etwa nach dem Vorschlag von Matuszkiewicz: 'Ökologische Faktoren sind nicht Teils des Systems', aber ohne Kenntnis ihrer Synökologie sind Waldgesellschaften schwer interpretierbar, z.B. bei der Anwendung in der Forstwirtschaft. Das hängt mit der starken menschlichen Beeinflussung zusammen. Ich erinnere auch an Klötzli (1968), der selbst Verbände – *Fagion* und *Carpinion* – nur mit Hilfe der Synökologie trennen konnte. Wir Waldsoziologen arbeiten daher gut mit der Definition von Ellenberg: Eine Pflanzengesellschaft ist eine 'gesetzmäßig von der Umwelt abhängige konkurrenzbedingte Kombination von Pflanzenarten'.

Ob die Ökologie durchschlagen, also stärker als die Wirkung des Klimas sein würde, wenn man die Buchenwälder Europas so vergleichen würde wie die Fichtenwälder, das ist eine Frage, die noch nicht beantwortet werden kann. Bei einem Vergleich innerhalb der Unterverbände möchte ich es bezweifeln.

Ich hoffe, damit auf alle Fragen eingegangen zu sein. Mit Herrn Dierschke stimme ich in seiner Ansicht über sinnvolle Zusammenfassungen vikariierender Vegetationseinheiten überein.

Zum Schluss möchte ich noch einmal auf die Frage nach den Möglichkeiten der Berücksichtigung chorologischer Phänomene in Pflanzengesellschaften eine Antwort versuchen: Wie man aus Tabelle 9 ersieht, sind regionale Unterschiede vorwiegend auf der Ebene der Assoziation berücksichtigt, auch wenn, wie bie Neuhäusl (1977), ein großes Gebiet wie ganz Europa vergleichend bearbeitet wurde. Soó (1964) und Oberdorfer (1957) trennen schon auf Verbandsebene, was vielleicht der besseren Übersicht dient. Man könnte auch an eine Berücksichtigung auf Unterverbandsebene denken, aber dafür findet sich kein Beispiel, und man sollte den einmal eingeschrittenen Weg nicht verlassen.

Mein Vorschlag wäre, regionale, großräume überdeckende Verbände zu schaffen und in diese nach den Vorschlägen von A. und W. Matuszkiewicz (1973) die in Assoziationsgruppen zusammengefaßten vikariierenden Assoziationen einzuordnen. Diese könnten nach Bedarf weiter in geographische Gebietsausbildungen (Rassen, Lokalausbildungen) unterteilt werden, die aber keinen neuen Namen erhalten sollten. Auch Höhenabwandlungen sollten, soweit vertretbar, nur als Höhenform ohne neuen Namen bezeichnet werden. Damit wäre eine klare Gliederung erreicht, ohne eine neue Namens-

flut infolge chorologischer Phänomene hervorzurufen.

Späteren Zusatz: Herr Rameau hat in seinem Referat eine sehr schöne Gliederung in dem hier vorgeschlagenen Sinne gebracht.

Paper by
J. C. Rameau
L'intérêt chorologique de quelques groupements forestiers du Morvan

R. CARBIENER:

J'ai beaucoup aprécié, également, la claire et remarquable synthèse de M. Rameau. Cette synthèse contient des données sur les Hêtraies d'altitude de l'étage montagnard du Morvan entre 750 et 900 m. Or ces données illustrent une des questions posées ce matin, à savoir celle du choix des critères de classification syntaxonomique. Faut-il préférer une classification d'essence verticale à base edaphique à une classification horizontale à base climatique (étages de végétation), en montagne? En effet, les stations eutrophes de l'étage montagnard du Morvan portent des Hêtraies (colonne 1 du tableau III) que vous classez dans l'*Asperulo-Fagion*. Par contre, les stations oligotrophes correspondantes portent des Hêtraies acidophiles très pauvres en espèces (colonne 3 du tableau I) que le cortège d'espèces acidophiles) (peu discriminantes en fait) vous fait classer dans le *Quercion robori*. Or ces Hêtraies sont situées au dessus de la limite climatique des chênes. Il me semble peu opportun, en regard de l'information climatologique liée aux définitions respectives de l'étage du chêne et de l'étage du hêtre, de classer ces Hêtraies montagnardes dans une alliance de Chênaies (collinéo-planitiaires par définition!). Ceci d'autant plus que cette césure est basée sur des critères floristiques de très faible valeur discriminante. Dans le Massif Central (Monts Dore), ce même type de démarche – défendue il est vrai par Guinochet et aussie par Bartoli dans une étude des forêts des Préalpes françaises du massif de la Grande Chartreuse – aboutit à classer dans une alliance de Chênaies acidophiles, des Hêtraies des Monts Dore situées à la limite climatique générale des forêts (limite étage montagnard supérieur-subalpin) à 1 500–1 600 m d'altitude, 400–700 m au dessus de la limite climatique des Chênaies. Je ne veux pas revenir par la a la fausse querelle des faciès de dominance (Hêtraie sans hêtre, Chênaies-Charmaies du *Fagion*) mais poser la question du maintien d'une logique climatologique zonale qui doit rester selon nous, prééminente en syntaxonomie, y compris pour les choix nomenclaturaux. Les enclaves de Chênaies sessiliflores dans l'étage du Hêtre, sur les versants exposés au Sud, doivent selon cette conception se rattacher à l'étage collinéen, en Europe centrale et occidentale du moins (plus au Sud, les données changent).

J. C. RAMEAU:

La caractérisation sérieuse du groupement à *Fagus sylvatica* et *Deschampsia flexuosa* ne pourra se faire que lorsque nous aurons rassemblé des relevés floristiques provenant du Massif Central. Il serait plus correct actuellement de parler d'un groupement à *Fagus sylvatica* et *Deschampsia flexuosa*. En effet ce groupement très appauvri a sans doute des affinités avec certaines formes oligotrophes de *Luzulo niveae-Fagetum*, association des Pyrénées et du Massif Central.

J.-M. GEHU:

Il y a trop de différences floristiques, structurales, dynamiques .. entre les Chênaies-Hêtraies acidiphiles hyperocéaniques (Bretagne) et les Chênaies-Hêtraies subatlantiques (Allemagne du Nordouest) pour les réunir en une seule association. Il s'agit par contre surement d'un même groupe d'association territoriales vicariantes (*Fago-Querceta* ou *Ilici-Fagenion*) dans lequel le gradient de richesse floristique atlantique et la structure 'hemi-laurisylve' vont en s'atténuant d'ouest en est, avec quelqes 'ilots' à caractère de nouveau plus océanique au niveau des premiers reliefs montagnards (Morvan par example).

Par ailleurs, j'ai le plus souvent rencontré le *Peucedano-Quercetum* sur pseudogley et je me demande s'il ne possède pas avant tout une signification édaphique.

Un point de nomenclature, le nom d'alliance de *Rubio-Ruscion* proposé par Lapraz me parait impossible pour un ensemble forestier, d'autant plus qu'aucune de ces deux espèces n'est liée dans le Sud-ouest français à l'une ou l'autre des deux grandes séries sur substrat acidophile ou mésotrophe.

J.-C. RAMEAU:

Le *Peucedano-Quercetum* offre un très large éventail écologique à l'image des différentes races de *Fago-Quercetum* et ceci en fonction du sol:
- bien sûr: les sols à pseudogley sont fréquents avec *Quercus robur*
- mais on retrouve souvent des sols secs avec *Quercus petraea* dominant (forêt d'Orléans) ou avec *Fagus sylvatica* et *Quercus petraea* (forêts de Fontainebleau).

Race d'une grande association ou association territoriale: Je préfère la première conception – pour ces groupements les différentielles se retrouvent dans d'autres associations appartenant à une alliance différente. Sur le plan théorique je ne vois pas de difficultés à employer ce système: une grande association est représentée localement, régionalement par une race qui possède un éventail propre de sous-associations (cette race correspond à l'association territoriale).

G. JAHN:

Zu der Bemerkung von Herrn Carbiener über die sehr klaren und aufschlußreichen Ausführungen von Herrn Rameau: Das Problem der artenarmen Buchenwälder taucht immer wieder auf. Man hat früher die Rolle der Buche sowohl auf den armen Böden als auch in der Ebene und in den *Quercetalia robori-petraeae* gewaltig unterschätzt. Das war ein begreiflicher Fehler, der sich aus dem damaligen oft devastierten Waldzustand ergab. Die Buche geht aber im Gegensatz zu früheren Auffassungen sowohl auf sehr arme Böden als auch in die Ebene und in die *Quercetalia robori-petraeae*.

Tüxen hat diese Einsicht zum Teil berücksichtigt, als er das *Querceto petraeae-Betuletum* in *Fago-Quercetum* umbenannte. Man sollte aber weiter gehen, die Klasse *Quercetea robori-petraeae* fallen lassen und die Ordnung *Quercetalia robori-petraeae* in die Klasse *Querco-Fagetea* einordnen. Dann hätte man folgende Einteilung:
Klasse 1 *Querco-Fagetea*; Ordnung 11 *Quercetalia robori-petraeae*; 12 *Fagetalia*; Verband 111 *Quercion robori-petraeae*; 121 *Fagion*; 112 *Carici piluliferae-Fagion* (einschl. *Luzulo-Fagenion*); 122 *Carpinion* u.s.w. Dann könnten alle artenarmen Buchenwälder einschließlich des *Luzulo-Fagenion* – das nach seiner Artenkombination keine Daseinsberechtigung im *Fagion* hat – in das *Carici piluliferae-Fagion* eingeordnet werden, und wir hätten eine saubere Lösung. Ich sehe keine andere.

Paper by
J. J. Barkman
Geographical Variation in Associations of Juniper-Scrub in Europe

K. DIERßEN:

Im Zusammenhang mit den Ausführungen interessiert mich eine detailliertere Angabe zum Minimumareal homogener Aufnahmeflächen, da die mir bekannten Wacholderhaine eine zumeist sehr unregelmäßige Bestandesstruktur mit vermutlich kleinflächig stärker schwankendem Lichtklima aufweisen.

Die Untersuchung von Pilz-Gemeinschaften in den Wacholderbeständen stimuliert darüber hinaus zu der Frage, bei welchen Gesammtflächengrößen eine Absättigung mit den jeweils bezeichnenden Sippen erfolgt. Meines Wissens sind vergleichbare Untersuchungen über monophage Insektenpopulationen in Wacholderbeständen auf den Britischen Inseln durchgeführt worden.

G. SCHWERDTFEGER:

Liegt ein Artenverlust im Ost-Westgefälle im Hinblick auf die Nutzungsintensität vor dem *Juniperus*-Stadium vor?

Sind die Begleitpflanzen auf Anteile von Ruderal- und Pionierarten untersucht worden? Deren Anteil könnte ein Hinweis auf die Nutzungsintensität vor dem *Juniperus*-Stadium sein.

D. FIJALKOWSKI:

In SO-Polen kommen *Juniperus*-Büsche in mehreren Gesellschaften vor. Am häufigsten treten sie im *Carici-Inuletum, Brachypodio-Teucrietum* und *Ligustro-Prunetum* auf. Die Stetigkeit IV weisen sie in der Gesellschaft mit *Rosa* sp. und *Crataegus* sp. und im *Thalictro-Salvietum pratensis* auf, Stetigkeit III im *Festuco-Thymetum serpylli* und *Spergulo-Corynephoretum*. Seltener (Stetigkeit II) kommt *Juniperus communis* im *Carpino-Prunetum* und *Peucedano cervariae-Coryletum*, selten (Stetigkeit I) im *Festuco-Koelerietum glaucae, Cladonio-Pinetum, Dicrano-Pinetum, Peucedano-Pinetum, Pino-Quercetum* und *Potentillo albae-Quercetum* vor.

P. L. NIMIS:

Cryptogams are rarely used in synsystematics, above all in Southern Europe. Lichens, e.g., although abundant in xeric soils on limestone, are rarely listed in the tables. In general, they have a broather geographic distribution than higher plants. If lichens would be considered in the syntaxonomy of xeric grasslands on alkaline soils, for example, we would probably recognize a new class, bringing together the greatest part of the associations described for the Northern hemisphere, characterized by species of the *Toninion coeruleonigricantis*. Many examples of this type could be made. I would be interested in your opinion on this point.

H. ELLENBERG:

Juniperus ist ein Weideunkraut und wurde früher bekämpft. Wie alt sind die ältesten Wachholderhaine?

J. J. BARKMAN:

Die floristischen Unterschiede zwischen den Wacholderhainen im Westen (Niederlande, Bundesrepublik) und im Osten (Polen) hängen nicht mit der intensiveren Landwirtschaft im Westen zusammen, im Gegenteil: im Osten finden sich die Wacholderfluren oft auf verlassenen Äckern (Sozialbrache), im Westen auf Flugsanddünen, oft in Naturreservaten, also weniger von Menschen beeinflußt.

Die Probeflächen müssen für höhere Pflanzen etwa 100 m^2 groß sein. Wenn einzelne Wacholdergebüsche nicht so groß sind, werden mehrere zu eine Aufnahme zusammengenommen, unter Auslassung der Flächen zwischen den Gebüschen.

Juniperus communis ist eine Pionierholzart, die sich nach stärkerer Störung (Flugsand, Überbeweidung, verlassene Äcker) ansiedelt und selber von anderen Holzarten verdrängt wird: Eichen, Birken und Ebereschen in den Niederlanden, Buchen in Jütland, Fichten in Süd-Schweden, Zitterpappeln und Kiefern in Ostpolen. Ihre Gesellschaften können für längere Zeit nur durch Beweidung oder durch Abhauen der Folgebaumarten sichergestellt werden. Übrigens geht die Sukzession im Westen langsam vor sich: nach Messung der Jahresringe sind alle Wacholdergebüsche in den Niederlanden 60–80 Jahre alt, und die meisten sind noch nicht von anderen Holzarten verdrängt worden. In Polen geht die Sukzession auf Flugsand (Kiefer) und besonders auf Brachen (erst *Populus tremula*, dann *Pinus*) aber viel rascher vor sich, so daß nur ziemlich junge Wacholderhaine optimal ausgebildet sind. Dagegen erhalten sich die Wacholder sehr gut unter Kiefern (im Gegensatz zu den Niederlanden wo sie eingehen), wenn auch dann von einem *Carici-Juniperetum* nicht mehr die Rede ist.

Paper by
J. M. Géhu et J. Franck
Données synchorologiques sur la végétation littorale européenne

Without discussion

Paper by
J. M. Royer
Liens entre chorologie et différenciation de quelques associations du Mesobromion erecti *d'Europe occidentale et centrale*

H. WAGNER:

Gerade dieser Vortrag zeigt deutlich einen künftigen Weg für pflanzensoziologische Forschung. Viele Gesellschaften wurden ursprünglich nicht in ihrem Optimalbereich, sondern zufällig erstmals in einer verarmten Ausstrahlung beschrieben (z.B. *Lithospermo-Quercetum*). Ohne auch nur im geringsten die frühen Leistungen von Pflanzensoziologen in Zweifel zu ziehen, muß es gerade den kommenden Generationen ein Anliegen werden, nicht stehenzubleiben, sondern, auf dem heute viel reicheren Material aufbauend, neue Wege zu suchen, auch wenn dadurch die eine oder andere Einheit unhaltbar wird. Jedenfalls sollte uns stets die Ehrlichkeit der Suche nach Zusammenhängen wichtiger sein als ein Gesellschafsname.

J. MORAVEC:

I would like to support the finding of M. Royer concerning the occurrence of *Genista pilosa* on carbonate substrates. The same we can observe in the Slovak Carpathian where this species occurs in relic pine woods on limestone and dolomite rocks.

E. HÜBL:

Ähnlich wie in den Karpaten kommt auch in Österreich am Alpen-Ostrand *Genista pilosa* in Dolomit-Föhrenwäldern vor, außerdem in Kalk-Trockenrasen, aber auch in lichten bodenarmen Wäldern. Vielleicht hat *G. pilosa* in submediterran beeinflußten Gebieten eine breitere ökologische Amplitude als in nördlicheren, weil in den deutschen Florenwerken *G. pilosa* nur für saure Böden ange-

geben wird. Herr Schwaar machte mich in einem privaten Gespräch darauf aufmerksam, daß *G. pilosa* in der Eifel ebenfalls auf Kalk vorkomme, wo man nicht von einem submediterranen Einfluß sprechen könne; außerdem sei bei Oberdorfer *G. pilosa* auch für *Mesobrometen* angegeben.

J. M. ROYER:

Genista pilosa est très fréquent dans les associations françaises du *Mesobromion* et *Xerobromion* sur rendzine oolithique et sur sols marneux. On le trouve également dans les landes du *Calluno-Genistion*. Il existe peut être un problème taxonomique non résolu.

Il est probable que des études futures démontreront encore plus nettement les liens entre de *Mesobromion* du centre et de l'ouest de l'Europe et la végétation des pelouses du sud de la France. Cela est encore plus net pour le *Xerobromion* qui n'est bien développé que dans le sud et le centre de la France et qui n'est que relictuel en Suisse, en France ou dans le nord de la France.

Paper by
W. Pietsch
Chorologische Phänomene in Wasserpflanzengesellschaften Mitteleuropas

G. SCHWERDTFEGER:

Woher kommen die beobachteten Arten? Welche Transportmittel sind auf diesen Wegen möglich?

H. ELLENBERG:

Dadurch, daß Herr Pietsch ökologische Untersuchungen einbezog, wurde besonders deutlich, daß chorologische Differenzierungen von Pflanzengesellschaften zwei grundsätzlich verschiedene Ursachen haben können. Meistens handelt es sich um Abwandlungen der Standorts-Gegebenheiten, seien es klimatische, edaphische oder andere. In anderen Fällen sind jedoch Verbreitungsschranken entscheidend. Im Extrem können die Standortsbedingungen völlig gleich sein, aber die Artenkombinationen gänzlich verschieden (z.B. bei den tropischen Tieflands-Regenwäldern Afrikas, Südamerikas und Hawaiis).

Beide Ursachen haben ganz verschiedene syntaxonomische Konsequenzen und sollten auseinander gehalten werden.

U. ASMUS:

Ich möchte an Prof. Ellenberg anschließen und nochmals darauf verweisen, daß eine bestimmte Umweltsituation notwendig ist, um eine geographische Gebieterweiterung zu ermöglichen. Im fränkischen Weihergebiet, an der Südost-Grenze von *Pilularia globulifera*, insbesondere dort, wo neue Teiche angelegt werden und diese dann aus Gründen wie auch immer nicht genutzt werden, hat der Pillenfarn explosionsartige Entwicklungen für einige Zeit.

Eine ähnliche Situation ist im Bereich des Oberrheinischen Braunkohlegebietes festzustellen. In Restseen des Tagebaus und in Gräben und Plätzen in deren Bereich weist der Pillenfarn im Zentrum seines Verbreitungsgebietes eine ähnliche Verhaltensweise auf, wie sie von Herrn Pietsch dargestellt wurde.

O. WILMANNS:

Unzweifelhaft nehmen Ihren Daten zufolge der Elektrolyt-Gehalt und die K^+/Na^+-Relation nach Osten zu. Wollen Sie dies wirklich als Änderung der ökologischen Ansprüche von *Pilularia* und *Eleocharis multicaulis* mit dem Klima deuten? Ich möchte eher sagen: *Pilularia* benötigt (wie gewisse andere *Littorellion*-Arten) konkurrenzarme Situationen, dauernd hohes Wasserpotential, zeitweilige Überschwemmung, niedriges pH mit seinen Folgen, vor allem Stickstoff-Angebot in Form von Ammonium-Ionen; der sonstige Elektrolyt-Gehalt ist belanglos. Stimmt das mit Ihrer Auffassung überein?

K. DIERSSEN:

Die Tendenz zur Ausbildung azidophytischer *Eleocharis multicaulis*-Bestände scheint mir im nw-deutschen Raum in solchen Heideweihern ein natürlicher Vorgang, wo der Wasserkörper episodisch einen Stoffeintrag aus angrenzenden Feuchtheiden mit hoher Austauschazidität erfährt. Die Entwicklung solcher Flächen ist in der Regel verbunden mit dem Stufenweisen Ausfallen 'anspruchsvoller' Sippen wie *Apium inundatum, Baldellia ranunculoides, Hypericum elodes*. Mit Gülle oder Kunstdünger kontaminierte Gewässer verlanden dagegen über eutraphente *Phragmitetalia*-Gesellschaften, die ihrerseits die *Littorelletalia*-Bestände direkt verdrängen. Das stark intermittierende *Pilularietum globuliferae* findet in NW-Deutschland zur Zeit

kaum noch geeignete offene Wuchsorte, erweist sich daher im Regelfall als stark rückläufig.

G. WIEGLEB:

Mit einigen Schlußfolgerungen bezüglich des ersten Beispiels bin ich gar nicht einverstanden. Von einer 'auffälligen Verschiebung des diagnostischen Wertes gegenüber dem Arealzentrum' kann nicht die Rede sein. Auch was Herr Asmus gesagt hat, stützt meine Meinung, daß im Gegenteil die Arten sich über einen größeren Raum relativ gleichartig verhalten. Es handelt sich um störungszeiger, gewissermaßen Äquivalente zum *Agropyro-Rumicion*, unter nährstoffärmeren Bodenbedingungen, überwiegend an hydrogen-karbonatarmen Gewässern. Diese Art von Vergesellschaftungen sind weit verbreitet und wenig beschrieben. Im westlichen Europa siedelt sie meist in *Littorellion*-Gewässern, aber oberhalb im stärker amphibischen Bereich. In der Lausitz sind die echten *Littorellion*-Arten seltener, und deshalb findet man auch nicht die entsprechende Artenkombination. In komplexen '*Littorelletea*'-Tabellen sind vergleichbare Gesellschaften oft enthalten und weisen auch immer deutliche Anteile von *Agropyro-Rumicion*-Arten auf (*Agrostis stolonifera, A. canina* usw.).

Außerdem möchte ich noch vor der unkritischen Verwendung des Begriffes 'oligotroph' warnen. Was ist gemeint? Ca-arm, P-arm oder N-arm? Auch muß man unterscheiden zwischen dem Trophiegrad des Wassers und des Bodens, die durchaus unterschiedlich sein können. Die hier besprochenen Arten besiedeln z.B. Gewässer mit nährstoffarmen Böden, aber freien Nährstoffen im Wasser. *Myriophyllum alterniflorum* z.B. ist eine Art, die umgekehrte Verhältnisse anzeigt.

W. PIETSCH:

Antwort zu Prof. Schwerdtfeger: Eine Verbreitung auf dem Weg der Ornithochlorie erscheint durchaus möglich, ich will mich aber nicht festlegen. Außerdem spielt eine Verbreitung durch Verkehrsmittel im Zusammenhang mit bestimmten Transportgütern eine Rolle.

Zur Frage von Prof. Wilmanns: Ihre Auffassung stimme ich völlig zu. Diese Tatsache erklärt sich auch aus der steten Armut der Gewässer an Ammonium. Ein wesentlicher Einfluß geht auch von der spezifischen Beschaffenheit des Gewässersediments aus: Pioniersubstrate reich an Eisen und

Aluminium; außerdem von Schwankungen des Wasserspiegels und vor allem von dem pseudo-atlantischen Mikroklima der Standorte.

Zur Frage von Prof. Dierßen: Aufgrund extremer Standortsverhältnisse dürfte sich die Sukzession auch erst nach 50 bis 80 Jahren im Zusammenhang mit dem Metamorphoseprozeß der Wasserbeschaffenheit bemerkbar machen. In Moorgewässern tritt das ozeanische Florenelement als edaphische Dauergesellschaft auf.

Zur Frage von Dr. Wiegleb: Auf die Veränderung des diagnostischen Werts durch Beispiele in Tabellenform wurde verzichtet, da die ökologischen Faktoren vorrangig betrachtet wurden. In keinem Fall handelt es sich um Ausbildungen des *Agropyro-Rumicion*, eher dagegen im Flachwasserbereich um Durchdringungen zwischen *Littorelletea* und *Isoeto-Nano-juncetea*, wie sie von Braun-Blanquet et Vlieger und von Klika früher als Klasse *Isoeto-Littorelletea* einmal beschrieben wurden. Als oligotroph verstehe ich die Armut an Nährstoffen und Mineralstoffen. Auf die limnische Oligotrophie bin ich nicht eingegangen, d.h. Oligotrophie auf Grundlage des O_2-Gehaltes.

Papers by
E. Balátová-Tuláčková
Chorologische Phänomene in tschechoslowakischen Molinietalia-*Gesellschaften*

and

F. Krahulec
The Chorological Pattern of European Nardus-*rich Communities*

K. RYBNÍČEK:

Most of the *Nardus*-communities (and also the *Molinietalia* communities) are typical examples of a secondary substituting vegetation. Their species composition and variability has been formed under very different conditions and by plants which were present from different initial stages and which were able to assemble dominants. We have also to keep in mind the human influences during the evolution of a community. Seeing this we have to suppose that the species composition of *Nardus* stands, substituting, e.g. different spruce forest types or subalpine communities or some other vegetation can be very different. Therefore I would prefer to explain

the variability and, eventually, to base the differentiation of these secondary types of communities rather on historical than on chorological phenomena.

H. DIERSCHKE:

Molinietalia-Gesellschaften sind oft sehr feine standörtliche Zeiger bodenökologischer Gegebenheiten und ihrer Veränderungen. Der Vortrag hat sehr schön gezeigt, daß sie auch gute Zeiger klimatischer Verhältnisse sind. Ich muß aber an das, was ich gestern zum Vortrag von Herrn Moravec gesagt habe, noch einmal anschließen und fragen: muß man wirklich so viele Assoziationen haben? Genügen nich zumindest in manchen Fällen auch geographische Rassen, um eine zu starke syntaxonomische Zersplitterung zu vermeiden?

U. DEIL:

Die für die ČSSR sehr schön gelungene regionale Gliederung steht für die europäischen Molinieten noch aus. Die synsoziologische Gliederung sollte noch auf zwei Ebenen ausgeweitet werden: Zum einen auf die systematische Ebene, in diesem Fall etwa auf die Arten der *Cirsium tuberosum*-Gruppe, zum anderen auf die biosoziologische Ebene, nämlich auf die Phytophagenkomplexe der *Cirsium*-Arten, wo sich nach dem vorliegenden Material von Zwölfer eine räumliche Gliederung andeutet.

J. WILKOŃ-MICHALSKA:

Es kommen in Polen zwei Arten der Gattung *Trollius* vor, u.zw. *Trollius europaeus* im nördlichen Teil Polens und *Trollius transsilvanicus* in submontanen Gebieten (Gorce-Gebirge). Beide Arten sind mit *Molinietalia*-Assoziationen verbunden. Ist es in der Tschechoslowakei ähnlich oder nicht?

H. ELLENBERG:

Ich möchte Kollegen Rybniček widersprechen. Die vom Menschen mitbedingten Gesellschaften lassen sich nicht grundsätzlich schlechter gliedern als natürliche; im Gegenteil, sie sind oft artenreicher. An sehr armen, sauren Standorten sind sowohl natürliche als auch anthropo-zoogene Gesellschaften artenarm und geographisch wenig differenzierbar (z.B. *Luzulo-Fagion* und *Violo-Nardion*). An reicheren Standorten ist das Gegenteil der Fall (z.B. *Eu-Fagion* und *Calthion*).

E. BALÁTOVÁ:

Ich bin mit Herrn Prof. Ellenberg ganz einverstanden, daß das *Molinietum coeruleae* W. Koch 1926 s.l. eine natürliche, die Standortsverhältnisse gut widerspiegelnde Assoziation darstellt, sei es auf Moor- oder Mineralböden. Man kann es pflanzensoziologisch gut charakterisieren; die Kenntnis der Florenentwicklung ist nicht unbedingt nötig.

Mit Kleinassoziationen des *Molinietum coeruleae* habe ich mich in meinem Beitrag nich beschäftigt – sie sind bei uns noch nicht bearbeitet. Nach vorläufigen Untersuchungen ist aber anzunehmen, daß sie auch in der Tschechoslowakei enge phytogeographische Bindungen darstellen.

Was die Auffassung der Assoziationen des *Calthion*-Verbandes betrifft, so glaube ich nicht, daß es sich um geographische Rassen handelt. Die meisten von ihnen zeigen deutlich eine unterschiedliche Ökologie, was für die Anwendung der Erkenntnisse in der Grünlandwirtschaft von Bedeutung ist. Selbstverständlich, in einer Übersicht der Gesellschaften Europas können manche von ihnen als regionale Assoziationen betrachtet werden.

Mit *Trollius transsilvanicus* habe ich leider keine Erfahrungen. In der Tschechoslowakei habe ich in den *Molinietalia*-Wiesen nur *Trollius altissimus* Crantz, Kleinart von *Trollius europaeus* L. angetroffen. Seine Beschreibung ist in der Arbeit von Chrtek et Chrtková 1979 (Preslia 51: 97–106) zu finden.

R. NEUHÄUSL:

Zusätsliche Anmerkung: *Trollius transsilvanicus* Schur, Enum. Pl. Trans. 26, 1866 ist ein Synonymum von *Trollius altissimus* Crantz.

F. KRAHULEC:

Answer to Dr. Rybniček: I agree with Prof. Ellenberg about the anthropic influence on species richness. We may find e.g. in *Nardo-Agrostion* communities 40 species per relevé and some *Pilosella* species of hybridogene origin, also characteristic for these communities.

Paper by
L. Mucina & D. Brandes
Communities of Berteroa incana *in Europe: A Geographical Differentiation*

S. PIGNATTI:

May I ask you if classification and ordination have been carried out on the basis of relevés or synoptic tables? I think your picture of the ordination is quite convincing, but with an elaboration on the basis of small groups of relevés (5–10 relevés for each association) you may reach perhaps a better discrimination and ordination of the clusters.

L. MUCINA:

I have used the relevés for ordination and numerical classification treatments. With the last ordination the centroids of the local tables were plotted.

It might be that when using synoptic tables for ordination or cluster analysis you can get a clearer pattern, much better clear-cut clusters for example. Particularly when you use a ranking of characters to pick-up those characters of the most important information contents. However, I could not afford this kind of approach since those synoptic tables were rather different in terms of number of relevés included. Therefore, I was afraid not to bring noise into the analyses by adopting the tables with different total number of species.

Paper by
F. J. A. Daniëls
Floristic Relationship between Plant Communities of Corresponding Habitats in Southeast Greenland and Alpine Scandinavia

P. L. NIMIS:

I recently made a comparison between two transects, from bare limestone to tundra, respectively taken in Svalbard and in the Julien Alps. The results fully confirm your conclusion: similarity is highest between saxicolous communities, and progressively decreases towards more mesic environmental conditions. A difference was the higher number of species × unit of area in the Alps, as compared to Svalbard.

J. J. BARKMAN:

I am of course in favour of using similarity coefficients between syntaxa, but we must realise that the outcome may be affected by our species concept. In the case of a comparison between arctic dwarf shrub communities of Greenland and Scandinavia this is particularly evident: The affinity between these communities largely depends upon the question whether we consider *Salix callicarpaea* and *S. glauca* conspecific or not.

E. HADAČ:

The similarity of plant communities of extreme habitats is mainly caused by the prevailing of lichens or mosses, having circumpolar distribution. lichen communities like the *Umbelicarietum* in Greenland, Scandinavia and Central Europe belong to the same association, as lichen communities are very ancient and had enough time to spread on vast areas. The *Empetro-Vaccinietum* from C. Europe, Scandinavia and Iceland may belong to one broad association, as they have a common origin and were connected by migration, but similar communities, dominated by Greenlandic or Arcto-American species can hardly be united with the European ones as they have a different origin.

F. J. A. DANIELS:

It is interesting to hear that Dr. Nimis found the same phenomena during his studies. I have not studied myself the number of species in community stands in both the Alps and Scandinavia, but it is my impression that they don't differ strongly in this respect.

Of course, the similarity coefficients depend on the species concepts. I agree that we have to discuss on this symposium the pro and contra's of a division by differential species of an association according ecological or geographical point of views. However the decision is rather subjective. It still remains a matter of taste. The most important thing is, besides constructing a system, to get more insight in the ecology of the communities. I agree with Prof. Hadač that the uniformity in the floristics of communities of extreme habitats is strongly determined by the dominance of cryptogams, which show in general a broad ecological amplitude.

Papers by
K. Dierßen & B. Dierßen

Korrespondierende Caricion bicolori-atrofuscae-*Gesellschaften in W-Grönland, N-Europa und der zentraleuropäischen Gebirgen*

and

L. M. Fliervoet & M. J. A. Werger
Vegetation Structure and Microclima of Three Dutch Calthion palustris *communities*

S. PIGNATTI:

I consider your paper very interesting, because you made a careful analysis of structural factors in the plant community and found a correlation with climate by means of functional parameters. Therefore it seems that in this case a relation between structure and function in the ecosystem does exist.

Have you made a principal component analysis of your data? I think this may give a very significant picture of the results.

R. NEUHÄUSL:

Is it possible to generalize your results on Dutch *Calthion* communities as a whole? According to our experiences there are great differences in the biomass between different variants of one *Calthion* association.

The floristic composition of your stands from different territories with diverse climatic conditions suggests that some differences in soil properties control the vegetation structure also. The influence of soil conditions on the production of biomass cannot be undervaluated.

J. J. BARKMAN:

The differences in phenological development of vegetation structure in three *Calthion* stands is very interesting, but is it not rather speculative to ascribe them to climatic differences? After all, the coastal climate is rather different, but Nijmegen and Drenthe (both in the East) do not differ very much. On the other hand your *Calthion* stands do not belong to the same communities and therefore, edaphic and hydrological differences may be more important. Using a transplantation experiment, it would be possible to verify your hypothesis.

H. ELLENBERG:

Das von Herrn Barkman vorgeschlagene Experiment würde große Schwierigkeiten bereiten, weil *Calthion*-Gesellschaften oft sehr tief würzeln. Blöcke des Pflanzenbestandes mit ihrem gesamten Würzelboden könnte man nur mit größten Schwierigkeiten und Kosten verpflanzen.

L. FLIERVOET:

I agree with Prof. Pignatti that it is of great importance to look at the structural and functional aspects of vegetation communities. I used already PCA for analysing structural data but not especially for this small number of *Calthion* communities, I selected for the Prague symposium.

I cannot generalise from this data the strong difference in the total above ground biomass between associations and subassociations of the alliance *Calthion palustris*. Therefore I have to analyse more samples of different *Calthion* types, but especially striking is the variation in distribution of biomass both vertical and in growth forms between the studied *Calthion* types of the different phytogeographical districts.

Certainly, there should also be an influence of the amount of soil-nutrients on the vegetation structure, but in general the studied *Calthion* communities are located on similar peaty soils with a high groundwater level during the whole growing season. Therefore the variation in soil-nutrients will be rather small.

The comment of Prof. Barkman that macroclimate is not the only factor which influences the vegetation structure is correct, just as I mentioned in this lecture. Variation in structure of the studied *Calthion* vegetation is not even strong correlated for all mentioned variables with the climate of different districts. More striking is for example the relation between leaf size and the factor wind, and the distribution of biomass and the length of the growing season. I wonder to confirm the relation climate – structure on small distance. I agree with further experimental research, such as transplantation of the vegetation, but also less radical changes can be studied, e.g. measurements of leaf transpiration rates and leaf temperatures.

To Prof. Ellenberg: The impossibility of transplantations of *Calthion* stands in spite of deep rooting species applies only for a small number of species; more over there is almost an amount of 90% of root biomass in the upper 20 cm of the soil.

Paper by
H. Dierschke

Anthropogene Areal-Ausweitung mitteleuropäischer Gehölzpflanzen auf den Britischen Inseln und ihre Bedeutung für die Beurteilung der heutigen potentiell natürlichen Vegetation

E. HÜBL:

Die sekundäre Ausbreitung von *Carpinus betulus* in Irland mit dem Ausklingen gegen Westen geht konform mit dem gesamten Arealbild, wo *Carpinus* weiter nach Osten geht als die Buche, die sich in Irland trotz der Niederwaldwirtschaft stärker ausbreitet. Daß *Acer pseudoplatanus* nicht ursprünglich in Irland vorkam ist vielleicht doch nicht nur historisch erklärbar, weil *Acer pseudoplatanus* in Irland mehr ruderal verbreitet ist. Vielleicht wirkt sich die menschliche Eutrophierung günstig aus für die Ausbreitung mitteleuropäischer Gehölze. Physiognomisch erinnern die irischen Buchenwälder mit *Rhododendron ponticum* im Unterwuchs an die *Fagus orientalis*-Wälder im Pontus.

G. SCHWERDTFEGER:

Fehlende Bäume können nicht nur fehlen, weil sie dies Gebiet nicht erreicht haben, sondern auch aus anderen Gründen. Neben dem im Referat aufgeführten Punkte könnte auch das Schneiteln der Buche an ihrer alten Grenze der Grund sein. Eine starke Übernutzung verhindert die Verjüngung der Buche, so daß von den Wuchsorten in England keine Ausbreitung möglich war.

H. ELLENBERG:

Carpinus stellt relativ hohe Ansprüche an die Sommerwärme und fehlt wohl aus diesem Grund in Irland von Natur aus.

Das fehlen von *Fagus* in der Naturlandschaft Irlands und großer Teile Englands auf Spätfrostgefahr zurückzuführen (wie dies Watt tat), ist sicher abwegig. Denn *Fagus* ist gerade in SO-England natürlich vorhanden, dem relativ frostreichsten Teil dieses Landes. Und welcher Teil des temperierten Europas wäre weniger spätfrost-gefährdet als Irland?

R. NEUHÄUSL:

Ich möchte auf das Problem der spontanen Ausbreitung der florenfremden Holzarten mit Rücksicht auf ihre Eingliederung in die potentielle natürliche Vegetation aufmerksam machen. Die Konstruktion der potentiellen natürlichen Vegetation beruht im allgemeinen auf der Voraussetzung, daß nur die Arten der einheimischen Flora in Wettbewerb treten. Die angepflanzten *Robinia*-Bestände auf Sandstandorten wärmerer Gebiete der nemoralen Zone sind aber so konkurrenzkräftig, daß sie von standortsgemäßen einheimischen Holzarten ohne menschliche Hilfe kaum zurückgedrängt werden können. Trotzdem rechnen wir die 'Robinieten' nicht zu der potentiellen natürlichen Vegetation. Das ist zwar ein extremes Beispiel, aber im Prinzip handelt es sich um die gleiche Situation jenseits der natürlichen Arealgrenze von waldbildenden Holzarten. Und besonders einige vom Menschen erzeugte Biotope sind für die Einbürgerung von florenfremden Arten sehr günstig (z.B. das Eindringen von *Ailanthus glandulosa* in die Stadtbiotope).

G. JAHN:

Ich möchte eine Ergänzung zu Herrn Dierschkes Bemerkung machen, daß die Buche sich auf Rohhumus verjüngt. Ich habe bei meinen Untersuchungen in Nordwestdeutschland die Beobachtung gemacht, daß die Buche, je weiter sie in den atlantischen Klimabereich kommt, umso anspruchsloser wird. Sie findet sich auf sehr sauren Böden, wo sie Rohhumus bildet; sie kommt übrigens auch auf stark durch Stau- oder Grundwasser beeinflußten Böden vor, auf denen sie im weniger atlantischen mitteleuropäischen Bereich wegen der sommerlichen Trockenheit entweder gar nicht mehr gedeiht, oder der Konkurrenz anderer Arten, wie z.B. der Eiche, nicht mehr gewachsen ist. Daß die Buche sich im atlantischen Klimaeinflußbereich auf wechselfeuchten und sauren Böden und selbst auf Rohhumus verjüngt, ist vermutlich mit darauf zurückzuführen, daß es in diesem Klimagebiet nur selten eine ausgesprochene Sommertrockenheit gibt, daß also weder die oberen Bodenhorizonte der wechselfeuchten Böden noch die Rohhumusauflagen in der Vegetationszeit stark austrocknen. Die Voraussetzungen für die Naturverjüngung sind daher vergleichweise günstig. Wenn es trotzdem nur wenig Naturverjüngung gibt, so liegt das oft daran, daß das Wild die Rolle des früheren Weideviehs übernommen hat. Ein infolge übermäßiger Hege und/oder zu geringem Abschuß überhöhter Wildbestand, also ein anthropogener Einfluß, ist für die Naturverjüngung im subatlantischen Nord-westdeutschland eher ein Grund für das Fehlen der Naturverjüngung als saurer Boden und Rohhumus.

J. TÜXEN:

Wir haben gehört, daß es auf den Britischen Inseln keine ursprünglichen Wälder mehr gibt. Ich

kann bestätigen, daß auch die berühmtesten Erlenwälder Schottlands vor wenigen Jahrhunderten gepflanzt worden sind. Daneben gibt es in Schottland durchaus natürliche und ursprüngliche *Pinus*-Wälder, die als *Pinus-Quercus*-Wälder dem *Quercion robori-petraeae*, als reine *Pinus*-Wälder aber dem *Piceion septentrionalis* angehören.

A. SCAMONI:

Über die natürliche Ausbreitung der Buche (Verbreitung durch Vögel, Eichhörnchen, vielleicht auch anthropogen) ist wenig bekannt, namentlich zeitmäßig.

H. DIERSCHKE:

Zunächst zu den Fragen über die Ausbreitung von *Fagus sylvatica*: Die meisten der in der englischen Literatur zu findenden Argumente, die gegen eine Ausbreitung vorgebracht werden, erscheinen mir ebenso wenig einleuchtend wie Herrn Ellenberg. Die Beobachtungen von Frau Jahn zeigen, daß eine Rohhumusdecke nicht verjüngungsfeindlich sein muß. Sicher ist bei den herrschenden Klimabedingungen eine ausreichende Feuchtigkeit in der Auflage zu erwarten.

Herr Scamoni hat darauf hingewiesen, daß schon eine extensive Wald-Weidewirtschaft eine Verjüngung der Buche verhindern kann. In der englischen Literatur wird zum Teil auch angenommen, daß die frühe Besiedlung und Offenstellung der Landschaft für *Fagus* ein großes Hindernis dargestellt hat. Gleiches gilt für die langzeitige Niederwald-Wirtschaft.

Wie ich bereits kurz ausgeführt habe, ist die Buche möglicherweise selbst dort, wo man sie heute in Südengland als natürlich ansieht, erst durch Besiedlung aufgelassener Kulturflächen sekundär zur Herrschaft gelangt. In diese Richtung zielte in gewisser Weise die Bemerkung von Herrn Neuhäusl. Weitere Förderung hat *Fagus* dann später durch direkte Anpflanzung bis weit in den Norden der Britischen Inseln erfahren. Natürlich kann man, wie es Herr Barkman sagte, aus Anpflanzungen nicht direkt auf das natürliche Verhalten einer Baumart schließen. Ich kann mir aber keinen ökologisch einleuchtenden Grund vorstellen, der zumindest auf basenreichen, nicht zu feuchten Standorten das Wachstum der Buche entscheidend behindern sollte. Sicher ist vieles hypothetisch, genauso wie auch die angeführte Vegetationskarte von Ozenda et al.

Was *Acer pseudoplatanus* betrifft, mag der Baum vielleicht durch Eutrophierung der Kulturlandschaft gefördert worden sein. Es gab aber auch vor Eingriffen des Menschen genügend Standorte, wo der Ahorn hätte wachsen können.

Daß die klimatischen Bedingungen für die Hainbuche nicht günstig sind, da sie zu etwas wärmeren und kontinentaleren Gebieten hin tendiert, mag ein wichtiger Grund für ihre geringe Ausbreitung sein. Sie wurde aber wohl auch weniger angepflanzt als die stattlich wachsende, dekorative Buche.

Zur Bemerkung von Herrn Tüxen über *Pinus*-Urwälder: Ich habe sie auch vor einigen Jahren auf einer IVV-Exkursion gesehen. In Irland ist jedoch ein durchgehend natürliches Vorkommen nach der letzten Eiszeit bis heute umstritten.

Paper by
J. Duty
Die Fagus-*Sippen Europas und ihre geographisch-soziologische Korrelation zur Gesellschaftsverbreitung der Assoziationen des* Fagion *s. latiss.*

A. SCAMONI:
Meiner Erfahrungen nach muß *Fagus intermedia* als eine der *Fagus silvatica* untergeordnete Sippe aufgefaßt werden.

Paper by
N. Boşcaiu und F. Täuber
Die zönologischen Verhältnisse der dazischen und dazisch-balkanischen Arten aus dem rumänischen Karpatenraum

S. PIGNATTI:
Ich habe sehr geschätzt, daß Sie endemische Syntaxa aus den Karpaten aufgrund von endemischen Charakterarten gekennzeichnet haben. In mehreren rezenten Arbeiten über die Vegetation der drei südeuropäischen Halbinseln werden nämlich endemische Syntaxa mit Artenkombinationen von nicht endemischen Arten belegt; dieses Verfahren betrachte ich als logischen Fehler. Ich möchte Sie fragen, ob der Eindruck, den ich habe, stimmt, daß die Differenzierung der dazischen Provinz gegenüber der mitteleuropäischen relativ gering ist?

H. ELLENBERG:
Bei der chorologischen und soziologischen Betrachtung der Wälder auf dem Balkan sollte beachtet werden, daß während der Eiszeiten die Wälder, insbesondere Buchenwälder, nirgends in ihrer heutigen Form überleben konnten. Die Buchen z.B. mußten völlig neu einwandern. Die Pflanzengesellschaften sind dort also nicht viel älter als in Mitteleuropa. Neuere vegetationsgeschichtliche Untersuchungen von verschiedener Seite, auch von Beug in Göttingen, haben dies erwiesen.

F. TÄUBER:
Auch für die Buchenwälder aus dem Südostkarpatenraum wird kein voreiszeitliches Alter angenommen, sondern ihre postglaziale Einwanderung aus ihrem Refugial-Areal vorausgesetzt. Aber trotzdem haben zahlreiche nemorale südostkarpatische Endemiten in diesen Wäldern ihr zönotisches Optimum wiedergefunden.

Die Differenzierung der dazischen Provinz gegenüber der mitteleuropäischen stützt sich eher auf ihre endemischen Taxa, die zwar öfters auch als vikariierende endemische Mikrospezies erscheinen, aber gute Gesellschaften aufbauen.

Paper by
J. Looman
*Distribution of Plant Species and Vegetation Types in Relation to Climate**

J. J. BARKMAN:
You showed us the distribution of different species of one genus in the prairies of the Middle West (Canada) and tried to correlate these patterns with monthly rainfall in various months. In The Netherlands I found it extremely useful to use the precipitation/saturation deficit quotient as a measure for water economy (input/output) and to average it for either the whole growing season or the period when drought is most critical.

Paper by
H. D. Krausch
Ozeanische Florenelemente in aquatischen Pflanzengesellschaften der DDR

S. HEJNÝ:
Es wurde gesagt, daß eigentliche ozeanische Arten aus der Subgattung *Batrachium* in der DDR fehlen. *Ranunculus hederaceus* gehört aber zu der

* Published in Vegetatio 54/1: 17–25, 1983.

Subgattung *Ranunculus* (morphologisch sehr ähnlich der Art *Ranunculus sceleratus*).

Juncus bulbosus hat zwei ökologische Nischen: a) auf verletzten periodisch überfluteten Böden (*Juncion bulbosi*), b) in tieferen saueren, naturnahen oder anthropogenen Gewässern.

J. J. BARKMAN:

In den Niederlanden hat die Azidität der oligotrophen Heidetümpel nicht so stark durch saure Regen zugenommen wie z.B. in den Seen Südschwedens. Wahrscheinlich sind sie dank der dicken organischen Bodenschicht besser gepuffert. Nur im stark betroffenen Süden (Provinz Brabant, eingeschlossen vom Ruhrgebiet und von den Industriegebieten Antwerpens und Rotterdams) ist der pH-Wert fast um eine Einheit gesunken, im Norden (Drenthe) nur um etwa 0,2–0,5 Einheiten. Rezente Untersuchungen haben gezeigt, daß die Planktonflora hier seit 1927 wenig abgenommen hat, im Gegensatz zu Brabant.

G. WIEGLEB:

Das Problem der Versauerung ist komplex und wird vielfach falsch eingeschätzt. Die echten *Littorellion*-Arten (Isoetiden) sterben bei starker Versauerung aus. Sinkt der pH-Wert dauernd unter 4,5, bleibt nur *Juncus bulbosus*, der sich stark ausbreitet. Bei einer Untersuchung in Holland (Roelofs, Nijmegen) hat sich gezeigt, daß viel mehr *Littorellion*-Standorte durch Versauerung vernichtet worden sind als durch Eutrophierung. Ich selbst sah ein Gewässer (das letzte *Isoetes*-Vorkommen in den Niederlanden), dessen floristischer Reichtum nur dadurch aufrecht erhalten wird, daß sich dort auf der einen Seite eine Badeanstalt befindet. Auf dieser Seite ist der pH noch über 4,5 und die Isoetiden wachsen gut. Auf der entgegengesetzten Seite des Gewässers ist der pH bereits unter 4,5 gesunken, und sowohl *Juncus bulbosus* als auch *Sphagnum cuspidatum* breiten sich aus.

W. PIETSCH:

Die Diskussionen bestätigen die Notwendigkeit, aufgrund des spezifischen Verhaltens des azidophilen ozeanischen Florenelements in Mitteleuropa eine Aufteilung der Klasse der *Littorelletea* in zwei Ordnungen vorzunehmen. Die *Eu-Littorelletalia* umfassen die eigentlichen *Littorella*-reichen Isoetiden-Gesellschaften des ozeanischen Europas. Die *Juncetalia bulbosi* umfassen die im Wasser flutende Vegetation des azidophilen Florenelements. Die Gesellschaften dieser Gruppe dürften sich in Zukunft in Ausbreitung befinden.

Paper by
G. Karrer
Soziologie und Chorologie von kontrastierenden Pflanzengesellschaften im südlichen Wienerwald (Niederösterreich)

Without discussion

Paper by
J. E. Bjørndalen
Some Synchorological Aspects of Basiphilous Pine Forests in Fennoscandia

H. DIERSCHKE:

I have a question on the behaviour of *Trifolio-Geranietea* species in your pine forests: I assume that in the dry and warm climate of Southern Europe they are more woodland species, e.g. of the *Quercetalia pubescenti-petraeae*, in Middle Europe they behave as species of marginal communities and coming to the north or northwest they are to be found more and more in open areas like the grassland of the *Festuco-Brometea*. My question concerns the behaviour of these species in Norway. Where is their optimum, in the woodland, at woodland edges or in open grassland?

K. ZUKRIGL:

There are apparently different factors which make pine dominate on those rather rich soils, where also *Picea* or other tree species could exist. In many cases it is of course dryness but not in all of them. What about human influence?

Besides, also in the Alps we can remark that more or less thermophilous species out of the *Quercetalia pubescentis*, the *Trifolio-Geranietea* or from grasslands move into the pine forests.

J. E. BJØRNDALEN:

Reply to Prof. Dierschke: Good developed *Trifolio-Geranietea* communities occur only in southern Fennoscandia, e.g. the Oslo region, Central Sweden, the Stockholm region, Gotland and Öland. These communities belong to different associations, both xerophilous *Geranium sanguineum* communities and more mesic *Trifolium medium-Inula salicina* communities. The differentiation between *Geranion sanguinei* and *Trifolion medii* is very difficult in Scandinavia. *Trifolio-Geranietea* species form marginal communities which occupy mostly smaller areas, bordering different forest types, meadows, road sides, etc. However, if some of the basiphilous pine forests (especially *Convallario-Pinetum* and *Seslerio-Pinetum*) are regarded as *Trifolio-Geranietea* communities, the forest-rim class *Trifolio-Geranietea* has a kind of optimum in the basiphilous pine forests.

Reply to Prof. Zukrigl: Drought seems to be one of the most important ecological factors for the basiphilous pine forests in Fennoscandia. The soil is usually shallow, with a rendzina-like profile. Drought occurs mostly in early summer, but may persist during the whole summer in the more continental areas. Even the moister types seem to be seasonal hygrophilous and are adapted to endure longer dry periods. The soil depth will in most cases prevent the spruce from domination. Human influence is evident in many areas, and strongly cultural types occur. Many of the stands are regularly grazed. On the other hand, the basiphilous pine forests occur also in localities where the human influence is minimal.

Paper by
A. Miyawaki & Y. Sasaki
Die Abwandlung der immergrünen Laubwaldgesellschaften auf den Japanischen Inseln von Süden nach Norden

H. ELLENBERG:

Welches sind die Ursachen dafür, daß der Sekundärwald des immergrünen Lorbeerwaldes aus laubwechselnden Arten besteht? Offenbar gilt dies nur für die Grenzbereiche des Lorbeerwaldes in mittleren Japan, wo dieser Wald ohnehin klimatisch unter Schwierigkeiten lebt. Im südlichen Japan ist der Sekundärwald doch wohl ebenfalls immergrün?

O. WILMANNS:

Die Frage der Überlegenheit der Sommergrünen bei häufigem Schlag dürfte sich aus dem Verteilungsschlüssel der Assimilate erklären: der höhere Aufwand beim Bau des Hartlaubes 'lohnt' sich nur, wenn das Blatt mehrere (2–3) Jahre hindurch aktiv

sein kann. Gegen Norden nimmt die Assimilationsdauer – wie die Koinzidenz mit dem Kälte-Index zeigte – ab, die Überlegenheit der Sommergrünen steigt also. Dies gilt umso mehr, als die Assimilationsintensität des Hartblattes mit dem Alter sinkt.

E. HÜBL:

Es könnte sein, daß die Sommergrünen mehr Reservestoffen im Holz gespeichert haben als die immergrünen, so daß sie ausschlagskräftiger sind. Im Randbereich der immergrünen Vegetation handelt es sich anscheinend um eine ähnliche Erscheinung, die wir auch in den Alpen beobachten können, daß auf ungünstigen Standorten die Vegetation der höheren Stufe tiefer hinab geht.

Sind die natürlichen Pionierholzarten im nördlichen Randgebiet der immergrünen Zone Japans sommergrüne Arten?

A. MIYAWAKI:

Zuerst danke ich sehr für mehrere Anregungen. Auf das, was Herr Professor Ellenberg und Frau Professor Wilmanns gefragt und erklärt haben, möchte ich folgendes antworten: Das Phänomen, daß die Immergrünen im *Camellietea*-Gebiet zurückgehen und die Sommergrünen sekundär eindringen, ist eine Konkurrenzfrage; die Vitalität nimmt durch wiederholten Schlag im nördlichen Grenzgebiet der Lorbeerwälder ab bzw. zu. Die immergrünen Blätter leben normalerweise 2–3 Jahre.

Zu Herrn Professor Hübl: Es ist eine allgemeine Erscheinung, daß die Pioniere nach Kahlschlag und Brand laubwerfende Arten sind; wir haben dies nicht nur in Japan, sondern auch z.B. in Borneo und Thailand beobachtet.

Paper by
G. Wiegleb & W. Herr
Die Verbreitung von Ranunculus *Subgenus* Batrachium-*dominierten Gesellschaften in Mitteleuropa*

W. PIETSCH:

Woraus erklären Sie sich das unterschiedliche Verhalten von *Ranunculus*-Arten in Fließ- und Stillgewässern?

S. HEJNÝ:

Die Lebensform vom *Batrachium*-Arten in Stillgewässern ist amphibisch, es handelt sich also nicht um eine echte Wasserpflanze. Manche Arten aus dem Subgenus *Batrachium* weisen einen synantropen Charakter auf, und ihr Anteil steigt mit der Stufe der Eutrophierung (*Ranunculus peltatus, R. aquaticus, R. circinatus, R. fluitans*).

G. WIEGLEB:

Bezüglich des Verhaltens von *Ranunculus circinatus* und *R. trichophyllus* kann ich keine Auskunft geben. Hierfür wären ausgeweitete autökologische Untersuchungen notwendig. Die Beobachtungen von Herrn Hejný bezüglich der Lebensformen kann ich nur bestätigen. Ich habe in den Niederlanden *R. aquatilis* beobachten können. Die Art keimt noch im gleichen Jahr, im Spätsommer/-Herbst, jedoch nur wenn der Wasserstand niedrig ist. Ist dieser hoch (z.B. wegen Rheinhochwasser), so keimen die Pflanzen nicht und im nächsten Jahr gibt es wenig *Ranunculus*.

Ranunculus fluitans scheint doch Gewässer einer gewissen Größe, wobei zonale Breite, Tiefe wie auch Abfluß eine Rolle spielen, zu besiedeln. Ich halte ihn für etwas 'aquatischer' als die übrigen Arten, da Landformen selten zu sein scheinen.

Vikariierende Taxa bilden vitale Bestände in verschiedenen Einzugsgebieten. Das Problem ist ähnlich wie in dem Beispiel von Pignatti mit den Mittelmeerinseln. Einzugsgebiete sind auch 'Inseln' im biogeographischen Sinne. Durch anthropogene Einflüsse werden sie oft verbunden (Gewässer), wodurch sich nach längerer Isolation wieder Kreuzungsmöglichkeiten für bestimmte Arten ergeben.

Paper by
S. Hejný

Expansion and Retreat of Aquatic Macrophyte Communities in South Bohemian Fishponds during 35 Years (1941–1976)

W. PIETSCH:

Zunächst möchte ich Ihnen dafür danken, daß Sie durch Ihre Arbeiten mir den Anstoß gaben, die vielen Fischteiche in der Lausitzer Niederung zu bearbeiten. Diese werden alle 5 bis 8 Jahre von Schlammablagerungen und *Phragmitetea*-Beständen geräumt. Was passiert mit der Vegetationsentwicklung in Ihren Teichen bei einer periodischen Räumung?

S. HEJNÝ:

Nach der Räumung der Litoralgesellschaften in Fischteichen entsteht eine ganze Reihe von Sukzessionsstadien, besonders des oligo-mesotrophen Bereiches (*Eleocharitetum acicularis, Littorelletum, Ranunculo flammulae-Juncetum bulbosi*, später in mesotrophen Bereich übergehend, usw.), mit den Regenerationsphasen der Riedgesellschaften der *Phragmiti-Magnocaricetea*-Klasse. Dieser Regenerationsprozeß von Pflanzengesellschaften weist fast im jeden Teich spezifische Merkmale auf.

List of participants

Algeria

Moravec, J., Dr., Centre Universitaire, B.P. 89 Sidi-Bel-Abbes.

Wojterski, T., Prof. Dr., Departement de Botanique, Institut National Agronomique, El Harrach-Alger.

Australia

Barson, M., School of Botany, University of Melbourne, Parkville Victoria 3052, Melbourne.

Austria

Hübl, E., Prof. Dr., Botanisches Institut der Universität für Bodenkultur, Gregor-Mendelstr. 33, 1180 Wien.

Frau Hübl.

Karrer, G., Institut für Botanik der Universität Wien, Rennweg 14, 1130 Wien.

Niklfeld, H., Univ. Doz. Dr., Institut für Botanik der Universität Wien, Rennweg 14, 1130 Wien.

Wagner, H., Prof. Dr., Akademiestr. 15, 5020 Salzburg.

Zukrigl, K., Prof. Dipl. Ing. Dr., Botanisches Institut der Universität für Bodenkultur, Gregor-Mendelstr. 33, 1180 Wien.

Canada

Looman, J., Dr., Agriculture Canada Research Station Swift Current, Sask., P.O. Box 1030, S9H 3X2.

Czechoslovakia

Antoš, T., Ing., Botanická zahrada UPJŠ, Mánesova, Košice.

Balátová, E., Dr., Botanický ústav ČSAV, Stará 18, Brno.

Banásová, V., Dr., Ústav experimentálnej biológie a ekológie SAV, Sienkiewiczova 1, 885 34 Bratislava.

Blažková, D., Dr., Botanický ústav ČSAV, Průhonice.

Eliáš, P., Dr., Ústav experimentálnej biológie a ekológie SAV, Sienkiewiczova 1, 885 34 Bratislava.

Hadač, E., Prof. Dr., Ústav krajinné ekologie ČSAV, Malá plynárni 2, Praha-Holešovice.

Hejný, S., Prof. Dr., Botanický ústav ČSAV, Průhonice.

Herben, T., Dr., Botanický ústav ČSAV, Průhonice.

Holubová, E., Ing., Botanický ústav ČSAV, Průhonice.

Hroudová, Z., Dr., Botanický ústav ČSAV, Průhonice.

Husáková, J., Dr., Botanický ústav ČSAV, Průhonice.

Husák, Š., Dr., Botanický ústav ČSAV, Průhonice.

Husová, M., Dr., Botanický ústav ČSAV, Průhonice.

Jehlik, V., Dr., Botanický ústav ČSAV, Průhonice.

Jenik, J., Doc. Ing. Botanický ústav ČSAV, Dukelská 145, Třeboň.

Kolbek, J., Dr., Botanický ústav ČSAV, Průhonice.

Kopecký, K., Ing., Botanický ústav ČSAV, Průhonice.

Krahulec, Fr., Dr. Botanický ústav ČSAV, Průhonice.

Krippel, E., Dr., Geografický ústav SAV, Obrancov mieru 49, Bratislava.

Krippelová, T., Dr., Ústav experimentálnej biológie a ekológie SAV, Obrancov mieru 3, Bratislava.

Kropáč, Z., Dr., Botanický ústav ČSAV, Průhonice.

Kubiková, J., Dr., Žateckých 14, Praha 4.

Kühn, Fr., Doc. Dr., Vysoká škola zemědělská, Zemědelská 1, Brno.

Maglocký, Š., Dr., Ústav experimentálnej biológie a ekológie SAV, Obrancov mieru 3, Bratislava.

Mucina, L., Dr., Ústav experimentálnej biológie a ekológie SAV, Sienkiewiczova 1, Bratislava.

Neuhäusl, R., Dr., Botanický ústav ČSAV, Průhonice.

Neuhäuslová, Z., Dr., Botanický ústav ČSAV, Průhonice.

Oťahelová, H., Dr., Ústav experimentálnej biológie a ekológie SAV, Obrancov mieru 3, Bratislava.

Rambousková, H., Dr., Ústav krajinné ekologie ČSAV, Malá plynárni 2, Praha-Holešovice.

Rybníček, K., Dr., Botanický ústav ČSAV, Stará 18, Brno.

Rybníčková, E., Dr., Botanický ústav ČSAV, Stará 18, Brno.

Slaviková, J., Přírodovědecká fakulta KU, Benátská 2, Praha 2.

Švarcová, J., Botanický ústav ČSAV, Průhonice.

Vojtuň, A., Ing., Botanická zahrada ÚPJŠ, Mánesova ul., Košice.

Volf, F., Prof. Ing., Vysoká škola zemědělská, Praha-Suchdol.

Zaliberová, M., Dr., Ústav experimentálnej biológie a ekológie SAV, Obrancov mieru 3, Bratislava.

Zlinská, J., Dr., Ústav experimentálnej biológie a ekológie SAV, Obrancov mieru 3, Bratislava.

Zelená, V., Dr., Botanický ústav ČSAV, Stará 18, Brno.

Federal Republic of Germany

Asmus, U., Dr., Lehrstuhl für Landschaftsökologie, Reiffmuseum, Schinkelstr. 1, 5100 Aachen.

Braun, W., Dr., Lessingstr. 19, 8047 Karlsfeld.

Bohn, U., Dr., Bundesforschungsanstalt für Naturschutz und Landschaftsökologie, Konstantinstr. 110, 5300 Bonn 2.

Deil, U., Universität Bayreuth, Lehrstuhl für Biogeographie, Postfach 3008, 8580 Bayreuth.

Frau Deil.

Dierschke, H., Prof. Dr., Lehrstuhl für Geobotanik, Systematisch-Geobot. Institut, Untere Karspüle 2, 3400 Göttingen.

Dierssen, K., Prof. Dr., Biologiezentrum der Universität, Olshausenstrasse 40–60, 23 Kiel 1.

Frau Dierssen, B., Dr., Biologiezentrum der Universität, Olshausenstrasse 40–60, 23 Kiel 1.

Ellenberg, H., Prof. Dr., Hasenwinkel 22, 3400 Göttingen.

Frau Ellenberg, Ch., Hasenwinkel 22, 3400 Göttingen.

Griese, F., Institut für Waldbau II der Universität Göttingen, Büsgenweg 1, 3400 Göttingen.

Jahn, G., Prof. Dr., Institut für Waldbau, Universität Göttingen, Büsgenweg 1, 3400 Göttingen-Weende.

Lieth, H., Prof. Dr., FB 5 – Biologie, Universität, Postfach 4469, 4500 Osnabrück.

Rudolph, M., Am Brachfelde 4, 3400 Göttingen.

Schönfelder, P., Prof. Dr., Universität Regensburg, Institut f. Botanik, Fachbereich Biologie, Botanik II, Universitätsstr. 31, Postf. 397, 8400 Regensburg 2.

Schuhwerk, F., Dr., Holzgartenstrasse 58, D 8400 Regensburg.

Frau Schuhwerk.

Schwaar, J., Dr., Niedersächs. Landesamt f. Bodenforschung, Friedrich Missler Str. 46/50, 2800 Bremen.

Schwerdtfeger, G., Prof. Dr., Am Tannenmoor 34, 3113 Suderburg 1.

Frau Schwerdtfeger.

Tüxen, J., Prof. Dr., Forrelweg 41, 3004 Isernhagen 1.

Wiegleb, G., Dr., FB 7 der Universität Oldenburg, Ammerländer Heerstr. 67–99, 2900 Oldenburg.

Wilmanns, O., Prof. Dr., Biologisches Institut II der Universität, Schänzlestr. 1, 7800 Freiburg/Br.

Winterhoff, W., Prof. Dr., Keplerstrasse 14, 6902 Sandhausen.

Frau Winterhoff.

France

Carbiener, R., Prof. Dr., Faculté de Pharmacie, Laboratoire de Botanique, 74, route du Rhin – BP 10, 670 48 Strasbourg Cedex.

Madame Carbiener.

Géhu, J.-M., Prof. Dr., Hendries, F 59270 Bailleul.

Madame Géhu, J., Prof. Dr., Hendries, F 59270 Bailleul.

Rameau, J. C., Labor. de Phytosociologie de Besançon, 40 Rue Kennedy, 52 000 Chaumont.

Royer, J.-M., Dr., 42 BIS Ru Mareschal, 52 000 Chaumont.

German Democratic Republic

Casper, J., Dr., ZIMET, Abt. Limnologie, Beutenbergstr. 11, 69 Jena.

Duty, J., Dr. Ing., bei der Tweel 11, 25 Rostock 1.

Gutte, P., Sektion Biowissenschaften der KMW, Talstr. 33, 7010 Leipzig.

Knapp, H.-D., Dr., Lange Str. 56, 2060 Waren/Müritz.

Krausch, H.-D., Dr. hab., Wilhelm-Pieck-Strasse 32, 1500 Potsdam.

Müller-Stoll, W., Prof. Dr. hab., Am Drachenberg 1, 1500 Potsdam.

Passarge, H., Dr., Schneiderstr. 13, 13 Eberswalde 1.

Pietsch, W., Dr., Am Tälchen 16, 8027 Dresden.

Frau Pietsch, K., Am Tälchen 16, 8027 Dresden.

Scamoni, A., Prof. Dr., Schwappachweg 2, 1300 Eberswalde-Finow 1.

Schlüter, H., Dr. hab., Schillbachstr. 39, 69 Jena.

Frau Schlüter, J., G. Fischer-Verlag, 69 Jena.

Hungary

Kárpáti, I., Prof. Dr., Agr. Universität, H-8360, Keszthely.

Frau Kárpáti, V., Dr., Agr. Universität, H-8360, Keszthely.

Kovács, M., Dr. Botanisches Institut der Ungarischen Akademie der Wissenschaften, H-2163 Vácrátót.

Ireland

Moore, J. J., Prof., Department of Botany, University College, Belfield, Dublin 4.

Italy

Lausi, D., Prof. Dr., Istituto Botanico, Università di Trieste, V. Valerio 30, I 34100 Trieste.

Nimis, P. L., Dr., Istituto Botanico, Università di Trieste, V. Valerio 30, I 34100 Trieste.

Pedrotti, F., Prof. Dr., Istituto di Botanica, Via Pontoni 5, 62032 Camerino /MC/.

Frau Pedrotti, C., Prof. Dr., Istituto di Botanica, Via Pontoni 5, 62032 Camerino /MC/.

Pignatti, S., Prof. Dr., Dipartimento di Biologia Vegetale, Largo Christina di Svezia 24, 00165 Roma.

Frau Pignatti, Erica, Prof. Dr., Istituto ed Orto Botanico, Cos. Università, I 34100 Trieste.

Japan

Miyawaki, A., Prof. Dr., Yokohama National University, Dpt. of Vegetation Science, Inst. of Environmental Science and Technology, Yokohama 240.

The Netherlands

Arnolds, E. J. M., Dr., Biologisch Station, Kampsweg 27, 9418 PD Wijster.

Barkman, J. J., Prof. Dr., Biologisch Station, Kampsweg 27, 9418 PD Wijster.

Cleef, A. M., Dr., Institute of Systematic Botany, Heidelberglaan 2, Postbus BO-102, 3508 TC Utrecht.

Daniels, F. J. A., Dr., Botanical Laboratory, Lange Nieuwstraat 106, 3512 PN Utrecht.

Fliervoet, L. M., Drs., Dept. of Geobotany, Fac. of Science, University of Nijmegen, Nijmegen.

Greven, H. C., Drs., Ministry of Cultural Affairs, Recreation and Social Welfare Directors for Nature and Landscape Conservation, Postbus 5406, Rijswijk.

Schouten, M. G. G., Drs., Dep. of Geobotany, University of Nijmegen, 6525 ED Toernooiveld.

van Opstal, A. J. F. M., Drs., Natuurwetenschappelijke Commissie van de Natuurbeschermingsraad, Maliebaan 12, 3581 CN Utrecht.

van Wijngaarden, W., I.T.C. Enschede.

Norway

Baadsvik, K., Dr., Department of Botany, University of Trondheim, 7055 Dragvoll.

Madame Baadsvik.

Bjørndalen, J. E., Sogn of Fjordane Regional Coll., P.O. Box 39, 5801 Sogndal.

Sather, B., oand. real., DKNVS, the Museum, Botanical Dep., N-7000 Trondheim.

Vevle, O., Telemark distriktshøgskole, N-3800 BØ i Telemark.

Poland

Fabiszewski, J., Doc., Instytut Biologii Roślin, ul. Cybulskiego 32, 50–205 Wrocław.

Fijałkowski, D., Prof. Dr. hab., Akademicka 19, 20–033 Lublin.

Szwed, W., Mgr. Zakład Ekologii Roślin i Ochrony Środowiska Uniw. im. A. Mickiewicza, Al. Stalingradzka 14, 61–713 Poznań.

Wikoń-Michalska, J., Dr. Inst. Biologii Zakład Taksonomii, Ekologii Roślin Ochr. Pr. UMK, Ul. Gagarina 11, 87–100 Toruń.

Wojterska, M., Mgr., Zakład Ekologii Roślin i Ochrony Środowiska, Al. Stalingradzka 14, 61–713 Poznań.

Rumania

Boşcaiu, N., Prof. Dr., Acad. RSR Subcomisia Monumentelor Naturii, Str. Republicii No. 48, 3400 Cluj-Napoca.

Täuber, F., Dr., Acad. RSR, Subcomisia Monumentelor Naturii, Str. Republicii No. 48, 3400 Cluj-Napoca.

Sweden

Spada, F., Institute of Ecological Botany, University of Uppsala, Box 559, 75122 Uppsala.

Zhang, L., Institute of Ecological Botany, University of Uppsala, Box 559, 75122 Uppsala.
Present address: Dept. of Biology, East China Normal University, Shanghai, 200062 China.

List of lectures presented at the 26th International Symposium on Chorological Phenomena in Plant Communities

D. Lausi & P. L. Nimis: Contribution to the History of Vegetation in North-West Canada (Yukon Territory) on a Chorological-Phytosociological Basis*

G. Jahn: Geographische Abwandlung von Buchen- und Fichtenwaldgesellschaften**

J. Moravec: Chorological and Ecological Phenomena in the Differentiation and Distribution of the *Fagion* Ass. in Western Czechoslovakia*

J.-C. Rameau: L'intéret chorologique de quelques groupements forestiers du Morvan*

J. J. Barkman: Geographical Variation in Associations of Juniper Scrub in Europe*

A. M. Cleef: Chorological Phenomena in Tropical Andean *Sphagnum* Bogs (with Special Reference to the Columbian Páramos)

J.-M. Géhu & J. Franck: Données synchorologiques sur la végétation littorale européenne

J.-M. Royer: Liens entre chorologie et différenciation de quelques associations du *Mesobromion erecti* d'Europe occidentale et centrale*

W. Pietsch: Chorologische Phänomene in Wasserpflanzengesellschaften Mitteleuropas*

E. Balátová-Tuláčková: Chorologische Phänomene in tschechoslowakischen *Molinietalia*-Gesellschaften**

F. Krahulec: The Chorological Pattern of European *Nardus*-rich Communities*

U. Asmus: Stadtvegetation – ein chorologisches Phänomen!

L. Mucina & D. Brandes: Communities of *Berteroa incana* in Europe: A Geographical Differentiation*

J. Tüxen: Schottische Salzwiesen als vikariierende Assoziationen der europäischen Küsten

H. Passarge: Syntaxonomische Wertung chorologischer Phänomen* (published only)

F. J. A. Daniëls: Floristic Relationship between plant Communities of Corresponding Habitats in Southeast Greenland and Alpine Scandinavia*

E. Hübl: Wie weit entspricht die Vegetation der Alpen derjenigen höheren Breiten?

K. Dierßen & B. Dierßen: Korrespondierende *Caricion bicolori-atrofuscae*-Gesellschaften in W-Grönland, N-Europa und den zentraleuropäischen Gebirgen**

L. M. Fliervoet & M. J. A. Werger: Vegetation Structure and Microclimate of three Dutch *Calthion palustris* Communities under Different Climatic Conditions*

D. Fijałkowski: Übergangsgesellschaften zwischen *Magnocaricion elatae* und *Alnetea glutinosae*

G. Schwerdtfeger: Die Erfassung der räumlichen Verteilung der Böden (Bodenassoziationen) in Beziehung zur pflanzensoziologischen Systematik

S. Pignatti: the Use of Chorological Groups for the Problem Structure versus Function in Mediterranean Ecosystems

J. J. Moore: Changes in the Diagnostic Value of Species in the Irish Plant Communities

H. Dierschke: Anthropogene Areal-Ausweitung mitteleuropäischer Gehölzpflanzen auf den Britischen Inseln und ihre Bedeutung für die Beurteilung der heutigen potentiell natürlichen Vegetation*

W. Hilbig & H.-D. Knapp: Pflanzengeographische Elemente des Vegetationsmosaiks am Südrand des Chentej-Gebirges (Mongolische Volksrepublik)

J. Duty: Die *Fagus*-Sippen Europas und die geographisch-soziologische Korrelation zur Gesellschaftsverbreitung der Assoziationen des *Fagion* s. latiss.*

N. Bosçaiu & F. Täuber: Die zönologischen Verhältnisse der dazischen und dazisch-balkanischen Arten aus dem rumänischen Karpatenraum*

J. Looman: Distribution of Plant Species and Vegetation Types in Relation to Climate

H.-D. Krausch: Ozeanische Florenelemente in aquatischen Makrophytengesellschaften der DDR*

G. Wiegleb & W. Herr: Die Verbreitung von *Ranunculus* Subgenus *Batrachium*-dominierten Gesellschaften in Mitteleuropa**

J. Kubíková: The Incorporation and Significance of Some Thermophilous Species in Various Phytocoenoses throughout their Distribution Area

A. Miyawaki & Y. Sasaki: Die Abwandlung der immergrünen Laubwaldgesellschaften auf den Japanischen Inseln von Süden nach Norden**

J. Wilkón-Michalska: Die Veränderungen der Halophytenareale in Polen

J. E. Bjørndalen: Some Synchorological Aspects of Basiphilous Pine Forests in Fennoscandia*

G. Karrer: Soziologie und Chorologie von kontrastierenden Pflanzengesellschaften im südlichen Wienerwald (Nieder-österreich)**

S. Hejný: Expansion and Retreat of Aquatic Macrophyte Communities in South Bohemian Fishponds during 35 Years (1941–1976)*

K. Kopecký: Die syntaxonomischen und synchorologischen Aspekte des sog. Apophytisierungsprozesses am Beispiel de Saumgesellschaften mit *Chaerophyllum aromaticum* L.

F. Kühn: Die Süd-nord-, West-ost- und Höhenveränderungen in der Unkrautvegetation (*Secalietea*) in Mähren

Author index

Balátová-Tuláčkova, E., 111
Barkman, J. J., 67
Bjørndalen, J. E., 211
Boscaiu, N., 185
Brandes, D., 125
Daniëls, F. J. A., 145
Dierschke, H., 171
Dierssen, B., 151
Dierssen, K., 151
Duty, J., 177
Fliervoet, L. M., 159

Franck, J., 73
Géhu, J.-M., 73
Hejný, S., 243
Herr, W., 235
Jahn, G., 21
Karrer, G., 199
Krahulec, F., 119
Krausch, H.-D., 193
Lausi, D., 9
Miyawaki, A., 225
Moravec, J., 39

Mucina, L., 125
Neuhäusl, R., 3
Nimis, P. L., 9
Passarge, H., 137
Pietsch, W., 97
Rameau, J. C., 47
Royer, J. M., 85
Sasaki, Y., 225
Täuber, F., 185
Werger, M. J. A., 159
Wiegleb, G., 235

* Published in the Symposium Proceedings
** English translation published